Therapeutic Angiogenesis

Yukihito Higashi • Toyoaki Murohara
Editors

Therapeutic Angiogenesis

Springer

Editors
Yukihito Higashi
Research Institute for Radiation Biology
and Medicine
Hiroshima University
Hiroshima
Japan

Toyoaki Murohara
Department of Cardiology
Nagoya University
Aichi
Japan

ISBN 978-981-10-2743-7 ISBN 978-981-10-2744-4 (eBook)
DOI 10.1007/978-981-10-2744-4

Library of Congress Control Number: 2017947050

Printed on acid-free paper

This Springer imprint is published by Springer Nature
The registered company is Springer Nature Singapore Pte Ltd.
The registered company address is: 152 Beach Road, #21-01/04 Gateway East, Singapore 189721, Singapore

Contents

1 **Introduction Section: Overview of Therapeutic Angiogenesis** 1
Yukihito Higashi and Toyoaki Murohara

Part I Cell Therapy

2 **Autologous Bone Marrow Mononuclear Cell Implantation in Extremities with Critical Limb Ischemia** 5
Kenji Yanishi and Satoaki Matoba

3 **Peripheral Blood Mononuclear Cells for Limb Ischemia** 25
Masayoshi Suda, Ippei Shimizu, Yohko Yoshida, and Tohru Minamino

4 **Endothelial Progenitor Cells for Ischemic Diseases** 45
Takayuki Asahara and Haruchika Masuda

5 **Therapeutic Angiogenesis with Adipose Tissue-Derived Regenerative Cells** 67
Toyoaki Murohara and Kazuhisa Kondo

6 **Cell Therapy for Ischemic Heart Disease** 81
Hiroshi Kurazumi, Tohru Hosoyama, and Kimikazu Hamano

7 **Angiogenic Therapy for Ischemic Cardiomyopathy with Cell Sheet Technology** 99
Shigeru Miyagawa and Yoshiki Sawa

Part II Gene Therapy

8 **The Role of VEGF in the Extremities** 111
Brendan A.S. McIntyre, Takayuki Asahara, and Cantas Alev

9 **HGF Gene Therapy for Therapeutic Angiogenesis in Peripheral Artery Disease** 133
Jun Muratsu, Fumihiro Sanada, Yoshiaki Taniyama, Yuka Ikeda-Iwabu, Rei Otsu, Kana Shibata, May Kanako Brulé, Hiromi Rakugi, and Ryuichi Morishita

10 **Fibroblast Growth Factor in Extremities** 145
Michiko Tanaka and Yoshikazu Yonemitsu

Part III Other Therapy

11 **Low-Intensity Pulsed Ultrasound** 161
Akimichi Iwamoto, Takayuki Hidaka, Yasuki Kihara, Hiroshi Kubo, and Yukihito Higashi

12 **Low-Energy Extracorporeal Shock Wave Therapy** 177
Kenta Ito, Tomohiko Shindo, and Hiroaki Shimokawa

13 **Granulocyte Colony-Stimulating Factor** 191
Yasuyuki Fujita and Atsuhiko Kawamoto

14 **Waon Therapy: Effect of Thermal Stimuli on Angiogenesis** 217
Masaaki Miyata, Mitsuru Ohishi, and Chuwa Tei

15 **Exercise** 229
Tatsuya Maruhashi, Yasuki Kihara, and Yukihito Higashi

16 **Nanoparticle-Mediated Endothelial Cell-Selective Drug Delivery System** 247
Kaku Nakano, Jun-ichiro Koga, and Kensuke Egashira

Chapter 1
Introduction Section: Overview of Therapeutic Angiogenesis

Yukihito Higashi and Toyoaki Murohara

The book discusses recent findings and current perspectives in therapeutic angiogenesis. Generally, surgical bypass and percutaneous transluminal angioplasty alone or in combination with pharmacological therapy are options for revascularization and improvement in limb ischemic symptoms in patients with peripheral arterial disease. Unfortunately, patients with peripheral arterial disease with no other treatment option are subjected to amputation.

Recently, clinical studies have shown that novel therapies, including implantation of autologous bone marrow mononuclear cells, peripheral mononuclear cells, endothelial progenitor cells, mesenchymal stem cells, and adipose-derived stem cells; transfer of genes encoding for angiogenic growth factors, such as vascular endothelial growth factor, hepatocyte growth factor, and fibroblast growth factor; and other therapies (e.g., irradiation of pulsed ultrasound and shock wave, injection of granulocyte colony-stimulating factor, Waon therapy, aerobic exercise, new drug delivery system, and use of tissue engineering) are effective for improvement of clinical symptoms in patients with critical limb ischemia who previously had no treatment option other than amputation. This concept is called "therapeutic angiogenesis." In 2002, it was reported for the first time that implantation of autologous bone marrow mononuclear cells increases collateral vessel formation and improves ischemic symptoms in patients with peripheral arterial disease who have no other treatment option.

Almost 15 years have passed since the first cell therapy for angiogenesis in humans. Results of studies have shown that therapeutic angiogenesis in humans is

Y. Higashi, M.D., Ph.D. (✉)
Research Institute for Radiation Biology and Medicine, Hiroshima University,
Kasumi 1-2-3, Minami-ku, Hiroshima 734-8553, Hiroshima, Japan
e-mail: yhigashi@hiroshima-u.ac.jp

T. Murohara, M.D., Ph.D.
Department of Cardiology, Nagoya University Graduate School of Medicine,
Tsurumai-cho 65, Showa-ku, Nagoya 466-8550, Aichi, Japan

Y. Higashi, T. Murohara (eds.), *Therapeutic Angiogenesis*,
DOI 10.1007/978-981-10-2744-4_1

being established. Despite significant advances in therapeutic angiogenesis since the first clinical studies at the start of the twenty-first century, it has been largely ignored in the literature. These therapies are discussed, and data collected in clinical studies over the past decade are presented in this book. The book has 3 parts and 16 chapters including an overview of therapeutic angiogenesis, cell therapy, gene therapy, and other intriguing therapies. In this book, we focus on recent findings of therapeutic angiogenesis and present clinical perspectives of therapeutic angiogenesis. We believe that this comprehensive book fills that gap, making it a valuable resource for both researchers and practitioners alike. The door is now open to expand therapeutic angiogenesis beyond the area of experimental study.

We thank all authors, who are specialists in this field, and the editorial staff of Springer Japan, especially Ms. Chihiro Haraguchi and Ms. Kanako Honma, for their excellent assistance.

Part I
Cell Therapy

Chapter 2
Autologous Bone Marrow Mononuclear Cell Implantation in Extremities with Critical Limb Ischemia

Kenji Yanishi and Satoaki Matoba

Abstract Critical limb ischemia (CLI) is a terminal stage of peripheral artery disease (PAD), and the number of patients diagnosed with this condition is on the rise. CLI exerts a major impact on the patient's quality of life and has poor prognosis. The number of CLI cases showing resistance to existing treatments (medical therapy, percutaneous transluminal angioplasty, and surgical bypass) is also increasing. Many of these cases lead to exacerbation of the ischemia and subsequent lower-limb amputation. To treat these patients, cell-based angiogenesis is becoming an attractive new strategy. Cell-based therapy has also been used to treat ischemic myocardium. Bone marrow cells possess the ability to differentiate into various tissue types and can thus regenerate the myocardium by inducing angiogenesis and myogenesis, as shown by recent accumulating evidence reporting improved cardiac function and myocardial perfusion in animals and humans. Therapeutic angiogenesis with autologous bone marrow mononuclear cells (BMMNCs) is a term that has become widely used in the last decade. The safety and efficacy of therapeutic angiogenesis using autologous BMMNCs have been supported by clinical studies and meta-analyses.

Keywords Therapeutic angiogenesis • Bone marrow mononuclear cells • Critical limb ischemia

2.1 Introduction

Critical limb ischemia (CLI) is defined as rest pain or tissue necrosis with ulceration or gangrene due to peripheral artery disease (PAD). CLI is estimated to increase at a rate of 200–500 cases per million persons per year [1]. Treatments

K. Yanishi, M.D. (✉) • S. Matoba, M.D., Ph.D.
Department of Cardiovascular Medicine, Kyoto Prefectural University School of Medicine, Kawaramachi-Hirokoji, Kamigyo-ku, Kyoto 602-8566, Japan
e-mail: yanishi@koto.kpu-m.ac.jp

Y. Higashi, T. Murohara (eds.), *Therapeutic Angiogenesis*,
DOI 10.1007/978-981-10-2744-4_2

for PAD include removal of risk factors, exercise therapy, drug therapy using antiplatelet agents, vasodilators, percutaneous transluminal angioplasty, and surgical bypass. However, in a case where PAD is accompanied by diffuse coarctation or a lesion extending beyond a below-knee artery, the rate of restenosis is high, and the disease is less likely to be indicated for surgical treatment or for intravascular treatment. Cases showing resistance to existing treatments are on the rise. Many of these cases result in lower-limb amputations due to ulcers or infection caused by the evolution of the disease. To treat these ischemic tissues that lack sufficient arteries to allow for direct intervention, cell-based angiogenesis has become a new strategy that can be used safely and effectively for revascularization of tissues. This cell-based "therapeutic angiogenesis" originates from several new studies on neovascularization. Until the end of the 1990s, the differentiation of mesodermal cells into angioblasts and their subsequent endothelial differentiation were believed to occur exclusively during embryonic development [2]. However, this dogma was invalidated when human adult peripheral blood mononuclear cells (PBMNCs) were demonstrated to differentiate into endothelial lineages [3]. These cells, named "endothelial progenitor cells (EPCs)," showed expression of endothelial markers and were incorporated into ischemic sites [4, 5]. Furthermore, Kamihata and Quin demonstrated that bone marrow mononuclear cells (BMMNCs) also contain EPCs in their CD34-positive ($CD34^+$) cell fraction as well as various proangiogenic factors, such as basic fibroblast growth factor (bFGF), vascular endothelial growth factor (VEGF), and angiopoietin 1 (Ang-1) in their CD34-negative ($CD34^-$) cell fraction and that transplantation of BMMNCs into an ischemic site enhances angiogenesis via synergistic effects of EPCs and angiogenic factors [6, 7]. BMMNCs-based therapy has been applied clinically and has developed into a useful therapeutic option against human critical limb ischemia [8]. Moreover, cell-based therapy using bone marrow cells has also been applied for the treatment of ischemic myocardium. Bone marrow cells are able to differentiate into various tissue types and can regenerate the myocardium by inducing angiogenesis and myogenesis, as shown by accumulating recent evidence reporting improved cardiac function and myocardial perfusion in animals and humans [9–15].

Recently, autologous bone marrow cells have been used to promote angiogenesis resulting in the clearance of lower-limb ulcers, pain relief, and the avoidance of lower-limb amputation. As a consequence of the treatment, the patients' quality of life is improved considerably.

2.2 Revascularization Therapy for Critical Limb Ischemia

The beginning of revascularization therapy is an attempt to promote angiogenesis from tissue surrounding a location of ischemia as well as collateral blood circulation in order to secure blood flow in the ischemic tissue and relieve the tissue damage and necrosis caused by severe PAD. In addition, revascularization therapy aims to reduce accompanying rest pain, lower-limb ulcers, and gangrene due to ischemia (in PAD and Buerger's disease). These treatments are called therapeutic angiogenesis and are important treatment strategies for critical limb ischemia. Revascularization therapy, developed in Japan, has its roots in the discovery that adult peripheral blood contains vascular EPCs from the bone marrow and that these cells contribute to angiogenesis after birth [3]. Likewise, pluripotent stem cells in the bone marrow have also been shown to undergo differentiation into myocardial cells and endothelial cells [16]. Therefore, cardiovascular medicine in general can benefit from regenerating medicine that employs bone marrow cells. In particular, since the effectiveness of angiogenesis using BMMNCs-based transplantation for lower-limb ischemia due to PAD and Buerger's disease was published, the use of therapeutic angiogenesis has spread all over the world, and many medical facilities have reported relief of lower-limb ischemia and functional recovery in treated patients [8, 17, 18].

2.3 Mechanisms of Angiogenesis Through Cell Transplantation

Generally, there are three types of stem cells: embryonic stem cells (ES), inducible pluripotent stem (iPS) cells, and adult stem cells. Due to the ethical controversy and the need for strict examination of feeder-free and xenogenic-free culture methods, ES cells are currently not ready for clinical use. iPS cells are still being examined intensively with the aim of finding clinical applications for them. In contrast, clinical applications have been found for adult stem cells. Based on numerous investigations in animal models, BMMNCs and PBMNCs are now being widely used as sources of adult stem cells to treat cardiovascular diseases.

There is increasing evidence that postnatal neovascularization occurs not only by migration and proliferation of resident endothelial cells but also by the in situ differentiation of EPCs. Asahara et al. showed that peripheral blood contains EPCs which can develop into fully differentiated and functional endothelial cells [3]. It has been demonstrated that EPCs are involved in postnatal neovascularization in many animal models of bone marrow transplantation. It has also been shown that EPCs are mobilized into the circulation and participate in the formation of new vascular structures in several physiologic and pathophysiological conditions [19–22]. On the other hand, Tateishi-Yuyama et al. have demonstrated that BMMNCs have a paracrine effect on resident endothelial cells by the secretion of angiogenic growth factors and cytokines [6]. Vascular endothelial growth factor (VEGF), angiopoietin-1 (Ang-1), placenta growth factor (PlGF), fibroblast growth factor 2 (FGF-2), and other numerous growth factors are at present being extensively investigated. The roles of BMMNCs as a target or as a source of cytokines and growth factors are not fully understood and need to be further elucidated for the improvement of BMMNC-based therapy.

In these studies, the contribution of EPCs to neovascularization has focused on therapeutic angiogenesis. Indeed, previous studies using a hindlimb ischemia model reported that implantation of human BMMNCs increases neovessel formation at the capillary level [23, 24]. Furthermore, recent data have demonstrated that bone marrow cells contribute to the repair of severely damaged endothelial cells in high-turnover areas [25].

2.4 Indications for Autologous BMMNCs-Based Therapeutic Angiogenesis

Currently, patients aged 20–80 years old whose lower-limb ischemia has not improved even after surgical or medical treatment (patients with PAD or Buerger's disease or collagen disease) are eligible to receive advanced medical treatment. However, in special cases where infection control is impossible, the disease is not indicated for cell transplantation. Since worsening of a location of ischemia occurs rapidly, it is necessary to find an indicated treatment promptly through discussions with the departments of dermatology, plastic surgery, orthopedic surgery, and radiology as well as with the departments of cardiovascular internal medicine and vascular surgery. Due to concerns about adverse effects caused by angiogenesis throughout the body, indications for treatment exclude complications of untreated diabetic proliferative retinopathy and malignancies. Despite these concerns, there has been no reported increase in cancers due to bone marrow cell transplantation [17, 18]. Recently, the effects of therapeutic angiogenesis through cell transplantation for the treatment of a peripheral vascular lesion due to collagenosis were reported [26]. In view of this, a future expansion of the indications for therapeutic angiogenesis is expected.

2.5 Details of Autologous BMMNCs Transplantation

For the collection of bone marrow cells, approximately 600 ml of autologous bone marrow aspirate is obtained from the patient under general anesthesia. After collection, BMMNCs are promptly separated. Approximately two billion BMMNCs are injected separately at 120 positions above the muscles of an ischemic limb. The number of cells obtained and their breakdown (e.g., the number of $CD34^+$ cells that gives an indication of vascular EPCs) may vary between individuals due to factors such as age and clinical conditions (Fig. 2.1).

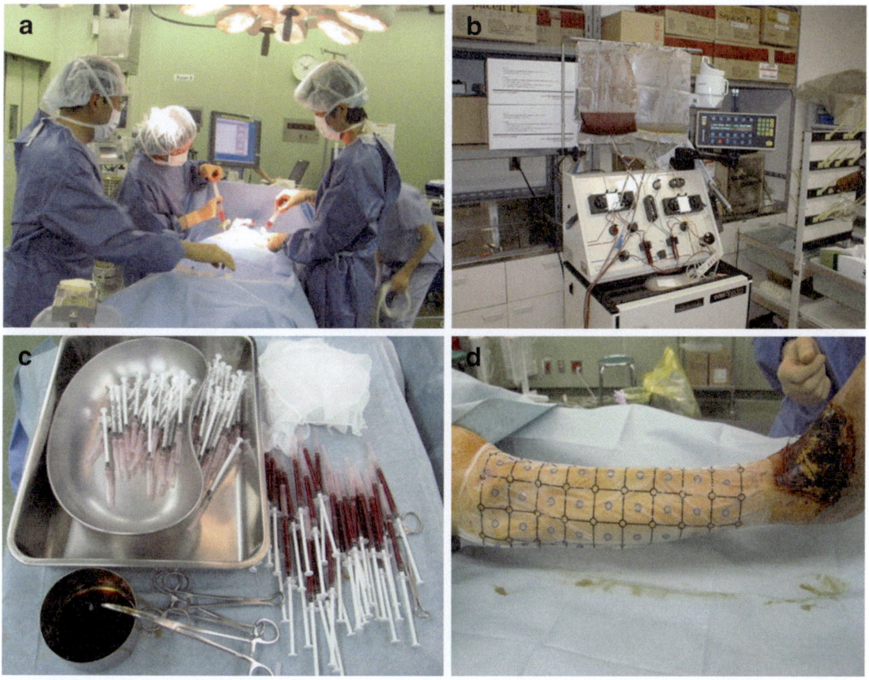

Fig. 2.1 Therapeutic angiogenesis using bone marrow mononuclear cells (BMMNCs) implantation. (**a**) Extraction of bone marrow aspirate from an ilium. Approximately 600 ml of bone marrow aspirate is collected from a patient's ilium under general anesthesia. (**b**) Separation and concentration of BMMNCs. A blood cell separation machine separates the BMMNCs promptly after collection. (**c**) Loading of BMMNCs into 27G syringes. (**d**) Injection of BMMNCs in the muscles of an ischemic limb. Approximately two billion BMMNCs are injected

2.6 Clinical Trials of Therapeutic Angiogenesis Using Autologous BMMNCs

2.6.1 First Clinical Trial of Therapeutic Angiogenesis Using Autologous BMMNCs

The duration of the effects of therapeutic angiogenesis treatment in patients has been validated at follow-up periods during clinical trials that range from 3 weeks up to 4 years (Table 2.1). The first report of an initial randomized, clinical pilot study for angiogenic cell therapy using intramuscular injections of autologous BMMNCs into 45 PAD patients with critically ischemic legs (Japan Trial for Therapeutic Angiogenesis Using Cell Transplantation (J-TACT study)) was published by our group in 2002 [8]. In this study, 29 patients (group A) with

Table 2.1 Clinical trials (including >ten patients) of therapeutic angiogenesis with autologous bone marrow mononuclear cells (BMMNCs) for critical limb ischemia

Year	Source	Pts	No. of pts	Total no. of cells [$\times 10^9$] (SD)[a]	No. of CD34+ cells [$\times 10^7$] (SD)[a]	Outcome	Follow-up period
2002	Tateishi-Yuyama et al. [8]	CLI	Total			Significant improvement in ABI, TcO2, VAS, walking time, and angiography	4 wks.
		(unilateral)	45				24 wks.
		(bilateral)	25	1.6 (0.6)	3.9 (2.2)		
			20	1.5 (0.6)	3.5 (1.3)	Significant improvement in ABI, TcO2, VAS, and walking time	
2004	Miyamoto et al. [31]	CLI	12	4.0 (0.3)	1.9 (0.3)	Significant improvement in ABI, VAS, walking time, Tc-TF perfusion, and scintigraphy	4 wks.
2006	Bartsch et al. [32]	PAD	10	0.1 (–)	–	Significant improvement in ABI and walking distance	8 wks.
2006	Durdu et al. [33]	TAO	28	1.7 (0.9)	5.3 (3.6)	Significant improvement in VAS, walking time, ulcer size, angiography, and QOL improvement in ABI	3 mos. 6 mos.
2006	Barc et al. [34]	CLI	15	–	–	Marginal improvement in ulcer size No improvement in ABI, TcO2, or angiography	(1 mos.) 3 mos. (6 mos.)
2006	Gu et al. [35]	CLI	35	5.6	–	Significant improvement in ulcer size, VAS, ABI, and TcO2	2 mos.

Table 2.1 (continued)

Year	Source	Pts	No. of pts	Total no. of cells [×10^9] (SD)[a]	No. of CD34+ cells [×10^7] (SD)[a]	Outcome	Follow-up period
2007	Bartsch et al. [36]	CLI	13	0.1 (0.03)	–	Significant improvement in ABI, walking distance, and capillary-venous SaO2	2 mos., 4 mos. 13 mos.
2007	Oda et al. [37]	CLI	21	3.0 (1.4)	6.7 (7.6)	Significant improvement in ABI and TcO2	4 wks.
2007	Saito et al. [38]	CLI (TAO)	14	4.2 (4.0)	–	Significant improvement in VAS, ulcer size, and angiography	4 wks., 24 wks.
2008	Gan et al. [39]	CLI	15	0.2–0.9	0.9–11.0	Significant improvement in ABI and walking distance	12 wks.
2008	Matoba et al. [27]	CLI	Total			Significant improvement in VAS, walking time, and ulcer size	3 yrs.
		PAD	115				
		TAO	74	3.1 (2.8)	2.6 (2.3)		
			41	4.1 (3.6)	4.6 (4.0)		
						No improvement in ABI and TcO2	
2011	Idei et al. [40]	CLI	Total			Significant improvement in VAS	4 yrs.
		PAD	51	1.8 (0.5)	3.5 (1.4)		
		TAO	25	–	–		
			26	–	–		
						No improvement in ABI and TcO2	

CLI critical limb ischemia, *PAD* peripheral artery disease, *wks.* weeks, *mos.* months, *yrs.* years, *ABI* ankle brachial index, *VAS* visual analog scale

[a]Values shown are expressed as mean ± SD

unilateral limb ischemia were recruited, and 25 of them were treated with BMMNCs into the gastrocnemius of the ischemic limb (ankle brachial index [ABI] < 0.6). Saline was injected as a control treatment into the opposite, less ischemic leg (ABI > 0.6). In addition, 22 patients with bilateral leg ischemia were also recruited (group B), and BMMNCs (as an active treatment) or PBMNCs (as a control treatment) were randomly injected into their ischemic legs. Overall, the implantation of autologous BMMNCs was safe and effective. The results of angiography showed a marked increase in the number of visible collateral vessels (Fig. 2.2). Significant increases in pain-free walking time, rest pain, and tissue oxygen pressure were observed 6 months after treatment, while injection of PBMCs (as a control) exerted less significant effects (Fig. 2.3). Because BMMNCs preparations contain EPCs and can release various angiogenic factors, we suggested that the incorporation of EPCs into newly formed vessels as well as the angiogenesis/arteriogenesis induced by angiogenic factors secreted from the injected cells contributed to an increase in blood flow and that this novel cell therapy constitutes a promising therapeutic strategy for treating PAD patients with critically ischemic legs.

2.6.2 Long-Term Clinical Outcomes of Autologous BMMNCs-Based Therapeutic Angiogenesis

In the world's first multicenter study (J-TACT study), autologous bone marrow cell transplantation was conducted on 45 patients whose lower-limb ischemia (Fontaine classification stage III to IV) had not improved even after surgical or standard medical treatment. As a result of the transplantation, 18 out of 20 patients showed complete relief of lower-limb pain. Overall, walking distance on a treadmill until the patient felt pain increased approximately 2.6-fold. Angiography also showed a significant increase in collateral blood circulation in 27 out of 45 cases. After these results were obtained, the number of participating institutions was increased to 11, and the number of participating patients was increased to 115, of which 74 were diagnosed with PAD and 41 with Buerger's disease. Effectiveness and safety of the treatment over 3 years were also confirmed. Figure 2.4 shows overall survival and lower-limb salvage rates 3 years after therapeutic angiogenesis. The 3-year overall survival rates were 80% (95% CI (68–91)) for patients with PAD and 100% for patients with Buerger's disease. During the 3 years, 11 of the 74 PAD patients died, whereas none of the 41 patients with Buerger's disease died (Fig. 2.4a). These results show an outstanding progress in terms of positive effects exerted on patients who had no other treatment option available to induce revascularization and would otherwise had to undergo lower-limb amputation. There were 19 time-dependent serious adverse events (excluding limb amputation) in the PAD patient group but only one case in the Buerger's disease patient group (Fig. 2.4b).

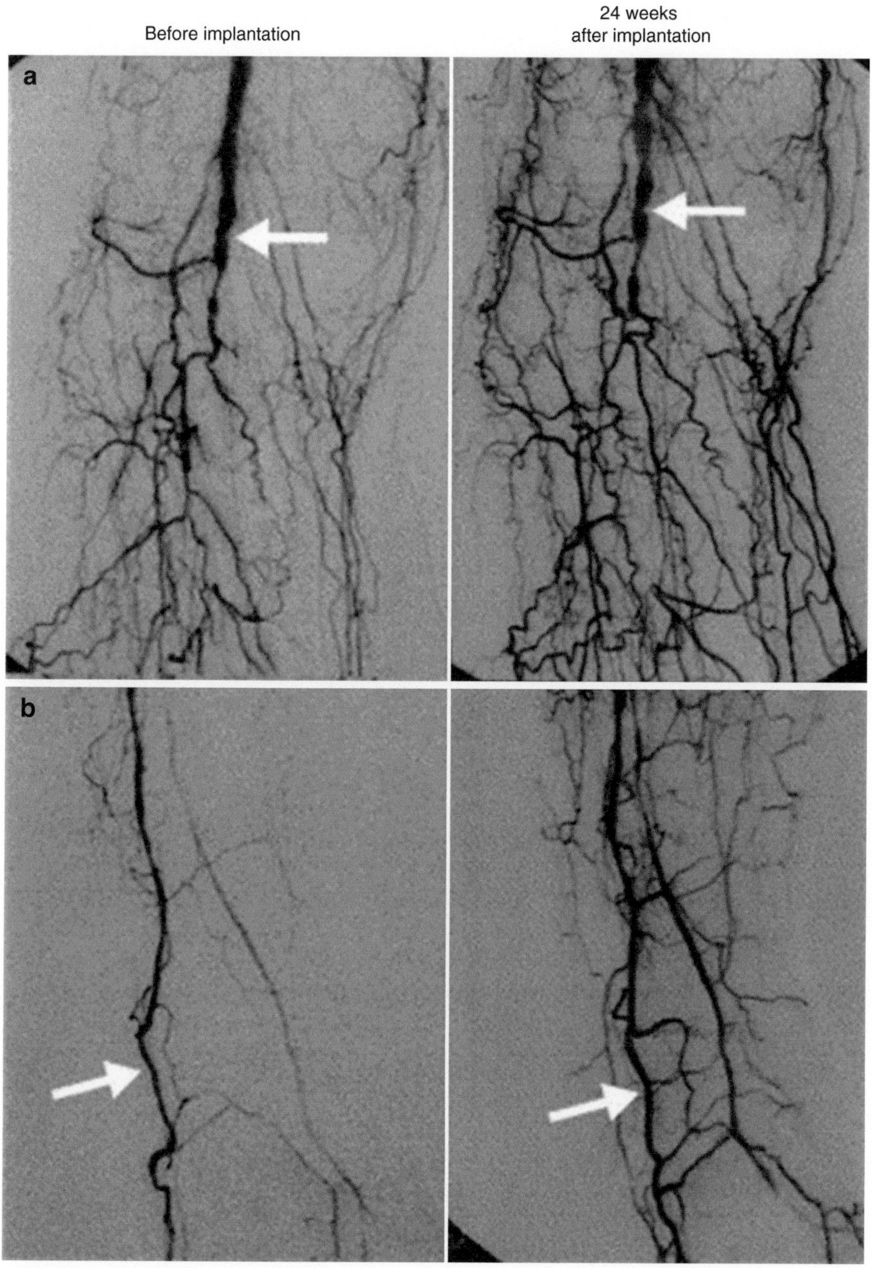

Fig. 2.2 Angiographic analysis of collateral vessel formation. Collateral branches were markedly increased in the (**a**) knee and upper tibia and (**b**) lower tibia, ankle, and foot 24 weeks after bone marrow cell implantation (*right column*) compared to before implantation (*left column*). Contrast densities in suprafemoral, posterior tibial, and dorsal pedal arteries (*arrows*) are similar before and after implantation. Reprinted with permission from The Lancet 2002

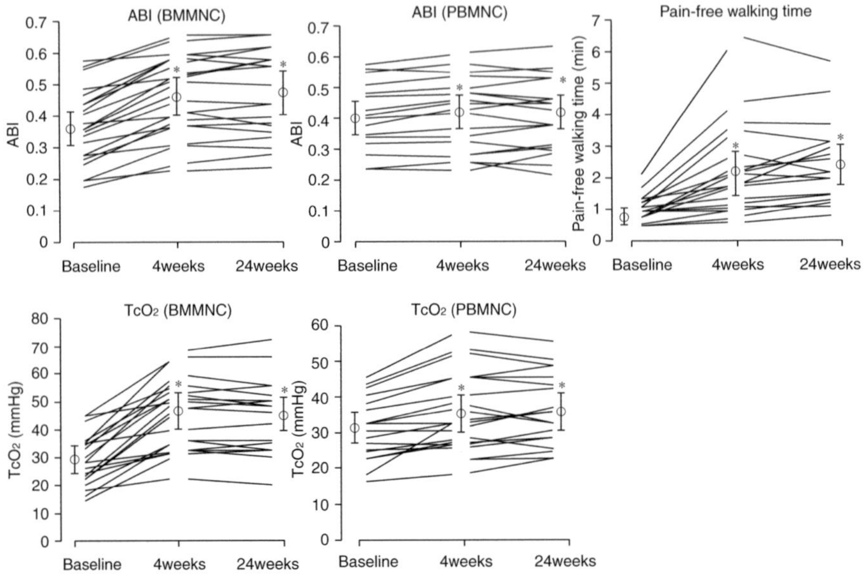

Fig. 2.3 Improvement of parameters after the implantation of bone marrow mononuclear cells (BMMNCs). Data are expressed as mean ± SD. Reprinted with permission from The Lancet 2002

Particularly, patients showed relief of pain, increase in walking distance, and ulcer shrinkage, among other positive effects. Angiogenesis effects varied depending on the conditions of the autologous bone marrow cells and the clinical condition of their lower-limb vessels. Examination of 115 cases in this 3-year period demonstrated that several factors such as a history of past lower-limb bypass, severity of ischemia, dialysis treatment, etc. affect prognosis and lower-limb preservation. After the BMMNCs transplantation, examination was conducted on the ratio of upper-limb blood pressure to lower-limb blood pressure (measured using the ABI value), ulcer size (cm^2), rest pain (measured using the visual analog scale [VAS]), and claudication distance until the onset of lower-limb pain. As a result, no significant difference was found in ABI values of patients with PAD or Buerger's disease during these 3 years. However, rest pain, ulcer size, and claudication distance were significantly improved, and this positive effect continued [27] (Fig. 2.5). Furthermore, a multicenter clinical study is scheduled in the future, which will reveal additional facts.

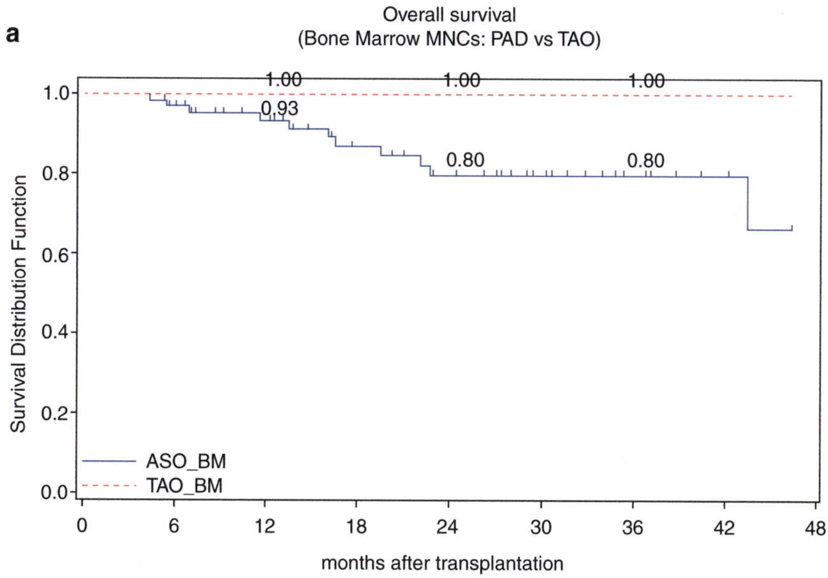

No. at Risk									
ASO_BM	74	60	51	37	31	20	12	7	4
TAO_BM	41	39	36	27	24	18	15	10	6

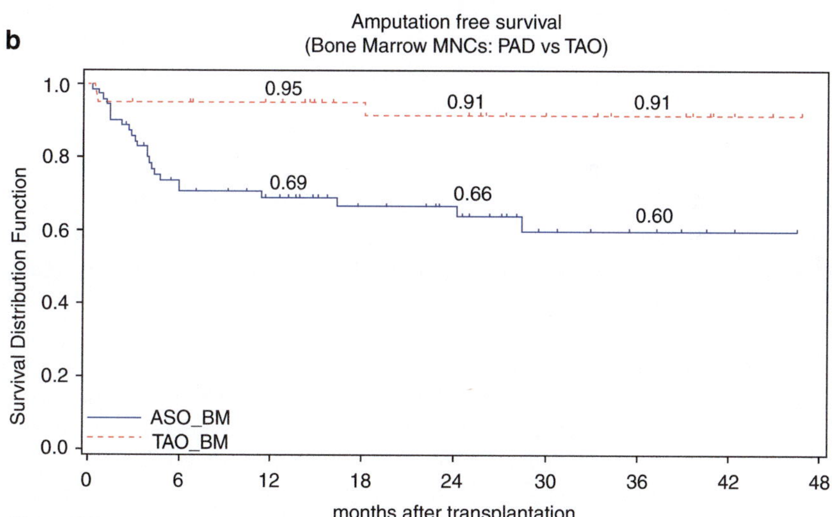

No. at Risk									
ASO_BM	74	44	38	28	24	13	9	6	4
TAO_BM	41	38	35	26	23	17	14	10	6

Fig. 2.4 Three-year overall survival rates and amputation-free survival rates after the implantation of bone marrow mononuclear cells (BMMNCs) in patients with atherosclerotic peripheral artery diseases (PAD) or Buerger's disease. (**a**) Three-year overall survival rates were 80% (95% CI (68–91)) in patients with atherosclerotic PAD (11 out of 74 patients died) and 100% (no deaths) in 41 patients with Buerger's disease. (**b**) Three-year amputation-free rates were 60% (95% CI (46–74)) in PAD patients and 91% (95% CI (82–100)) in patients with Buerger's disease

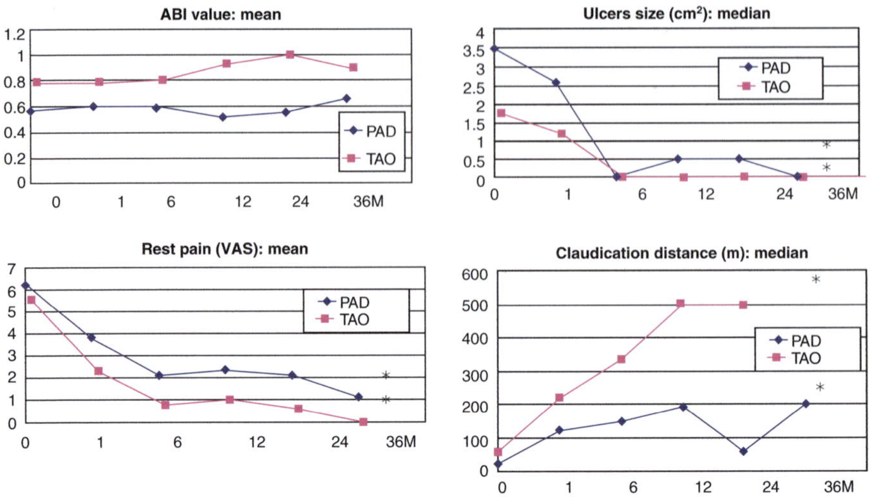

Fig. 2.5 Time-dependent changes in ABI, rest pain (VAS), ulcer size, and claudication distance after the implantation of bone marrow mononuclear cells (BMMNCs)

2.7 Persistence of the Effects Induced by Therapeutic Angiogenesis Treatment

When the therapeutic effects occur early after cell transplantation, it is possible to observe a resolution of lower-limb pain, bleeding due to increased local blood flow, and granulation growth within 1–2 weeks. When these effects are obtained, in many cases, it is also possible to observe resolution of the symptoms and ulcer shrinkage within 2 months at the latest. The healing process of an ulcer varies by case and depends on its area, depth, and site. Therefore, it is important, as part of infection prevention, to closely communicate with the departments of dermatology and plastic surgery in order to manage the granulation around the ulcer or to conduct a skin transplantation that will aid in the healing process. A small ulcer and a wound due to toe amputation heal relatively fast. However, it is important to continue to conduct rigorous disinfection and meticulous checks, paying careful attention to signs of infection until the epidermis covers the ulcer. Our examination showed that many of the cases showing a blood flow increase 6 months after surgery had no recurrence for over 3 years due also in part to subsequent exercise therapy and the improvement on the patient's living habits. Figure 2.6 shows the case of a patient with an ulcer on a heel caused by PAD and complicated by diabetes. As a result of therapeutic angiogenesis with BMMNCs, clearance of the ulcer and resolution of pain were observed which led to an increase in the patient's daily life activity. Moreover, the patient was still free of symptoms 4 years after treatment. Figure 2.7 shows a case of Buerger's disease. An initial bulky ulcer was cleared by the combined use of BMMNC implantation and skin transplantation. Even both of these big ulcer-healing cases needed a

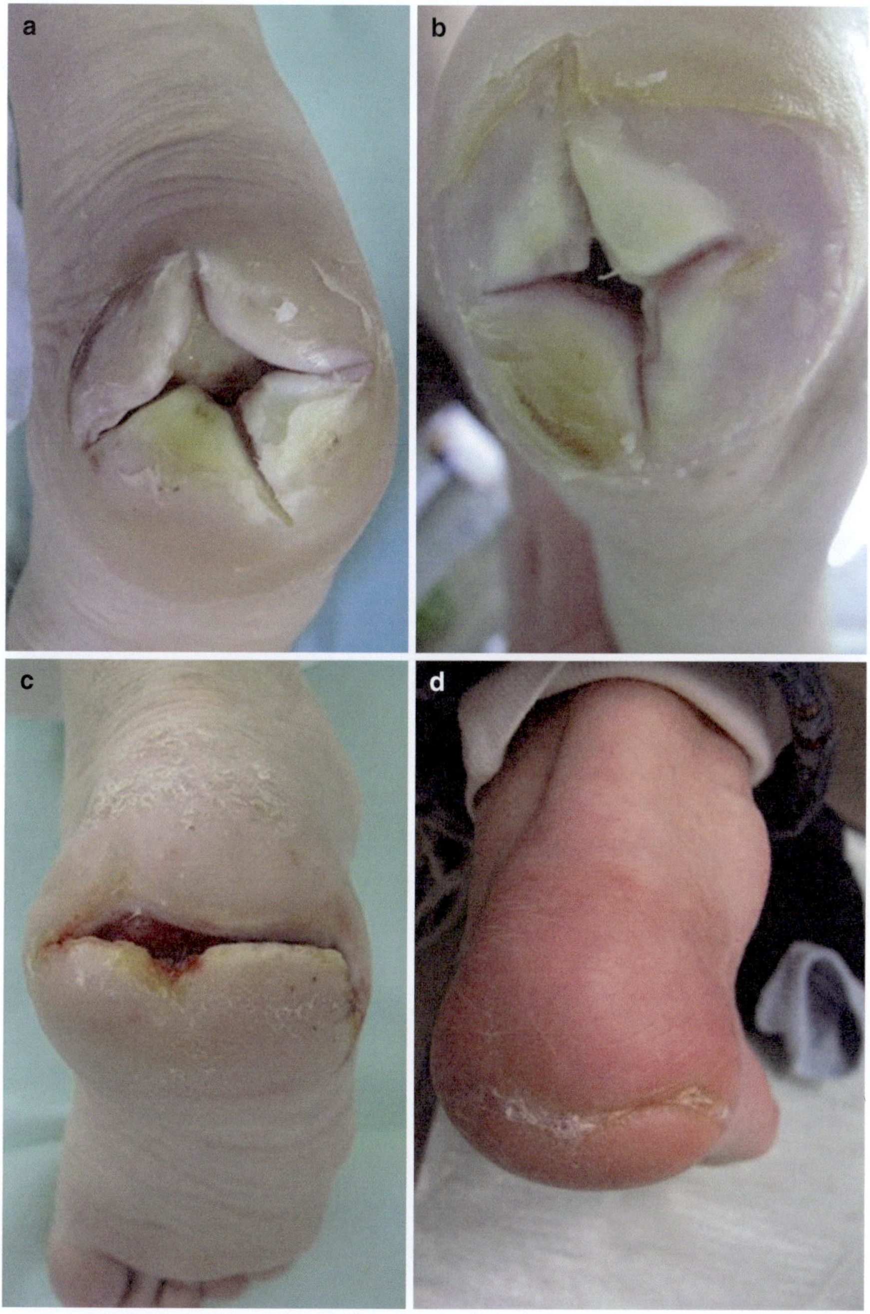

Fig. 2.6 Effects of bone marrow mononuclear cell (BMMNCs) implantation in the case of a 67-year-old male with diabetes mellitus and atherosclerotic peripheral artery diseases (PAD). Although a deep ulcer was observed in the heel of the patient at hospitalization, clearance and healing of the ulcer were observed after the implantation of BMMNCs. (**a**) Deep ulcer present on the heel of the patient at the time of hospitalization, before BMMNCs implantation, and at 20 days (**b**), 4 months (**c**), or 11 months (**d**) after the implantation of BMMNCs

treatment period for 11 months (Fig. 2.7). Nevertheless, the treatments were conducted on an outpatient visit basis thus allowing the patients to keep working as usual during the treatment period. It should be noted that the most important point during lower-limb angiogenesis treatment is to prevent deterioration of the patient's general condition due to local infection or pain. If sepsis occurs as a complication of a local infection or if the deterioration of ischemic cardiac disease is anticipated due to the worsening of lower-limb pain, amputation should be chosen without regard to lower-limb preservation.

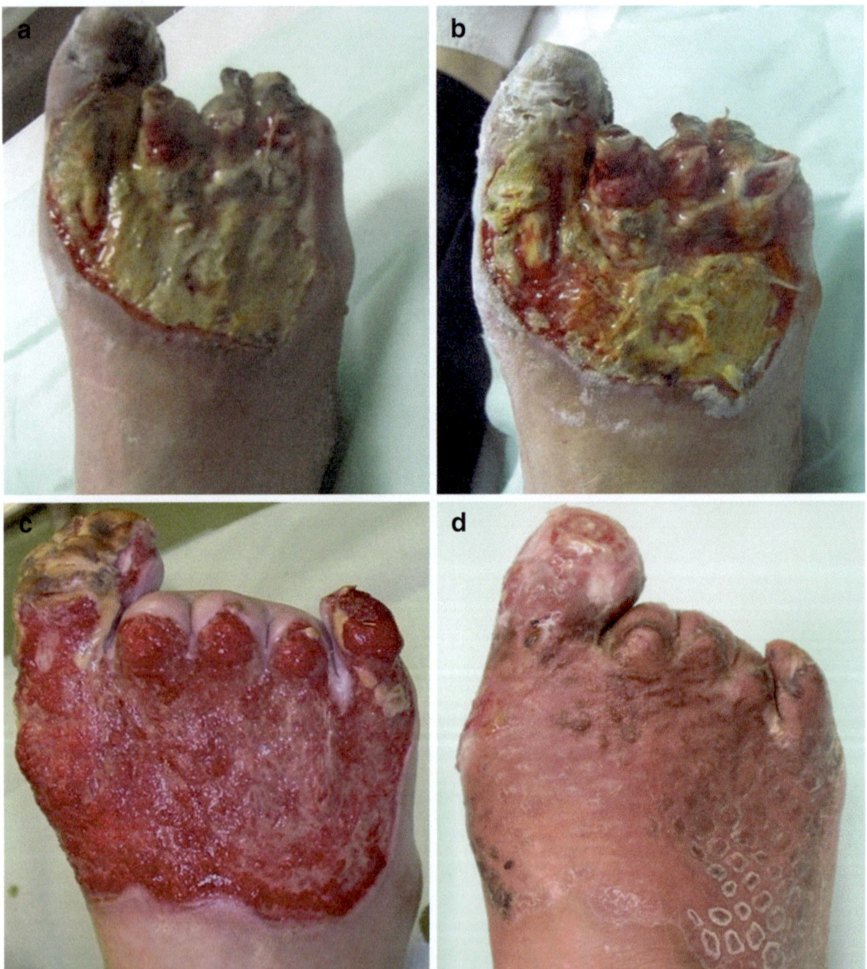

Fig. 2.7 Effects of bone marrow mononuclear cells (BMMNCs) implantation in the case of a 48-year-old male with Buerger's disease. Although a large ulcer was observed in the dorsum part, clearance of the ulcer was observed after the implantation of BMMNCs and a skin transplant. (**a**) At the time of hospitalization, before BMMNCs implantation, and at 7 days (**b**), 7 months (**c**), or 11 months (**d**) after the implantation of BMMNCs

2.7.1 *Drug Therapies Used in Combination with Revascularization Therapy*

The first choice for critical limb ischemia is vigorous revascularization. However, it is not always possible to conduct it. On the other hand, many cases can be improved by drug therapy. In particular, some severe cases can be resolved through drip infusion of prostaglandin El preparation during hospitalization. Therefore, positive use of it is desirable as long as there is no contraindication. Generally, the most common drugs in Japan against PAD and Buerger's disease are antiplatelet agents. They are important in terms of reduction of systemic cardiovascular events, particularly with regard to PAD, which accompanies many complications. Recommended antiplatelet agents include cilostazol, ticlopidine hydrochloride, sarpogrelate hydrochloride, and ethyl icosapentate. However, for cases in which an epidural anesthetic or the like is applied for relief of pain due to lower-limb cell transplantation, it is necessary to cautiously discontinue the use of the antiplatelet agent to avoid the formation of a hematoma as a complication after spinal anesthesia. In special cases where it is not possible to discontinue the use of the antiplatelet agent or to administer a spinal anesthetic, cell transplantation is conducted under general anesthesia after cell collection. Nonsteroidal anti-inflammatory drugs (NSAIDs), which are commonly used for pain control, mechanically inhibit angiogenesis. Hence, when possible, it is preferable to discontinue the use of NSAIDs or to reduce the dosage amount during revascularization therapy. Furthermore, a switch to opioid analgesic may be considered in such cases. In the cases where it is not possible to discontinue the use of NSAIDs, their continuous administration is necessary. In PAD patients, the administration of beta-blockers as antihypertensive drugs for managing high blood pressure was not recommended because they could worsen a patient's claudication symptoms. However, in a randomized controlled clinical trial, this effect was not observed. Hence, it is possible to safely administer beta-blockers to patients with claudication. In addition, when vascular complications are considered, a beta-blocker should be chosen in view of its cardioprotective effect [28].

2.8 Exercise Therapy for Ischemic Lower Limbs

Originally, exercise therapy was contraindicated in patients with critical limb ischemia that was accompanied by rest pain or an ulcer. It was postulated that worsening of ischemia and pain causes a cycle of vasoconstriction and the further worsening of ischemia. Nevertheless, a vicious cycle of muscle weakness, worsening of lifestyle diseases, and further progression of the lesion may arise as a result of this contraindication. Contrary to this opinion, positive exercise therapy during therapeutic angiogenesis treatment is viewed as a start and is currently considered to be important for the regeneration of skeletal muscles and vessels. In this regard, it must be noted that local contact between a brace and a shoe during exercise may worsen incision ischemia. Therefore, it is necessary to improve braces, choose the best exercise therapy method, and carefully monitor the exercise therapy.

2.9 Points to Note for Cases with Frequent Cardiovascular Complications

Reportedly, 25% of patients with critical limb ischemia suffering from peripheral arterial obstruction die 1 year later [29]. Because of this, past reports in regeneration medicine focused on life prognosis and cardiovascular complications [17]. Particularly, differences in the effects due to the presence or absence of dialysis were reported by a multicenter study. These complications should be considered in the actual treatment [27, 30].

2.10 Indicators of Peripheral Blood Flow Improvement

It usually takes approximately 6 months for an increase of collateral pathways to notably appear in an angiographic image after therapeutic angiogenesis treatment in PAD patients with critical limb ischemia. CT angiographic analysis is a useful follow-up method after therapeutic angiogenesis. However, this method is invasive and thus is not always used as a recurrent test. For this reason, indicators suitable for use as follow-up tests are still under discussion. Physiological tests such as ABI, transcutaneous oxygen pressure (tcPO2), and skin perfusion pressure (SPP) are used as indicators in lower-limb ischemia to determine whether lower-limb amputation is indicated and to determine the sites for amputation. However, these indicators do not always correlate with therapeutic effects although they are useful for severe cases. A recent meta-analysis showed that cell transplantation has positive effects on ABI, tcPO2, rest pain, walking distance until the onset of lower-limb pain, ulcer healing, and lower-limb amputation [20]. Nevertheless, it is important to comprehensively evaluate individual cases by taking into account also cyanosis, skin temperature, and the severity of lower-limb pain without regard to one specific indicator. Besides, engraftment of a transplanted skin section, speed of granulation regeneration around an ulcer, etc. provide also useful information with regard to dermatology.

2.11 Issues in Therapeutic Angiogenesis

The safety and feasibility of therapeutic angiogenesis using autologous bone marrow cells were partially confirmed by past clinical studies and meta-analyses [17, 18]. Still, many cases are not improved even with therapeutic angiogenesis as treatment. Past examinations suggest that the main reason why therapeutic angiogenesis induces only a minimal effect on PAD when compared to Buerger's disease is that PAD causes more advanced lower-limb ischemia due to arteriosclerosis in addition to vascular complications. In order to develop and employ a less invasive and more reliable treatment as the world's standard of care for critical limb ischemia, additional clinical studies that focus on aspects such as combination therapies involving cell and cytokine treatments, its timing, and its long-term effect as well as basic studies on angiogenesis mechanisms are needed. In Japan, this therapy has been approved as advanced medical treatment by the Ministry of Health, Labour and Welfare and can be combined with covered treatment by health insurance. In the future, it is expected that experimental facilities will increase and long-term strict management will be ensured.

2.12 Conclusion

This review provides an overview on the evolvement of therapeutic lower-limb angiogenesis which began with basic research studies and is now being tested and applied in the clinic for the actual treatment of patients. Recent studies have demonstrated promising results for cell therapy using BMMNCs for the treatment of patients with CLI resistant to surgery. In these studies, intramuscular implantation of BMMNCs led to the extension of the amputation-free interval and an improvement in ischemic pain, ulcer size, and pain-free walking distance. Since the beginning of the twenty-first century, progress in regeneration medicine has accelerated. The number of applicable patients is expected to increase at a pace exceeding that in the future. The most urgent issue in the treatment of critical limb ischemia is conducting the treatment in the most simple and effective way. In this regard, regeneration medicine is expected to fulfill an important mission. Hence, researchers should continue to perform basic research studies, clinical studies, and multicenter studies in order to advance the current knowledge in this field.

References

1. Dormandy J, et al. Clinical patterns and predictors. Semin Vasc Surg. 1999;121:154–61.
2. Cleaver O, et al. Endothelial signaling during development. Nat Med. 2003;9:661–8.
3. Asahara T, et al. Isolation of putative progenitor endothelial cells for angiogenesis. Science. 1997;275:964–7.
4. Asahara T, et al. VEGF contributes to postnatal neovascularization by mobilizing bone marrow endothelial progenitor cells. EMBO J. 1999;18:3964–72.
5. Asahara T, et al. Endothelial progenitor cells for postnatal vasculogenesis. Am J Physiol Cell Physiol. 2004;287:C572–9.
6. Kamihata H, et al. Implantation of bone marrow mononuclear cells into ischemic myocardium enhances collateral perfusion and regional function via side supply of angioblasts, angiogenic ligands, and cytokines. Circulation. 2001;104:1046–52.
7. Qin SL, et al. In vitro assessment of the effect of interleukin-1beta on angiogenic potential of bone marrow cells. Circ J. 2006;70(9):1195.

8. Tateishi-Yuyama E, et al. Therapeutic Angiogenesis using Cell Transplantation (TACT) Study Investigators. Therapeutic angiogenesis for patients with limb ischemia by autologous transplantation of bone-marrow cells: a pilot study and a randomised controlled trial. Lancet. 2002;360:427–35.
9. Orlic D, et al. Bone marrow cells regenerate infarcted myocardium. Nature. 2001;410:701–5.
10. Orlic D, et al. Mobilized bone marrow cells repair the infarcted heart, improving function and survival. Proc Natl Acad Sci U S A. 2001;98:10344–9.
11. Stamm C, et al. Autologous bone-marrow stem-cell transplantation for myocardial regeneration. Lancet. 2003;361:45–6.
12. Tse HF, et al. Angiogenesis in ischaemic myocardium by intramyocardial autologous bone marrow mononuclear cell implantation. Lancet. 2003;361:47–9.
13. Strauer BE, et al. Repair of infarcted myocardium by autologous intracoronary mononuclear bone marrow cell transplantation in humans. Circulation. 2002;106:1913–8.
14. Assmus B, et al. Transplantation of progenitor cells and regeneration enhancement in acute myocardial infarction (TOPCARE-AMI). Circulation. 2002;106:3009–17.
15. Tatsumi T, et al. Intracoronary transplantation of non-expanded peripheral blood mononuclear cells promotes improvement of cardiac function in patients with acute myocardial infarction. Circ J. 2007;71:1199–203.
16. Jiang Y, et al. Pluripotency of mesenchymal stem cells from adult marrow. Nature. 2002;418:41–9.
17. Matoba S, et al. Therapeutic angiogenesis for peripheral artery diseases by autologous bone marrow cell transplantation. Curr Pharm Design. 2009;15:2769–77.
18. Fadini CP, et al. Autologous stem cell therapy for peripheral arterial disease meta-analysis and systematic review of the literature. Atherosclerosis. 2010;209:10–7.
19. Asahara T, et al. Bone marrow origin of endothelial progenitor cells responsible for postnatal vasculogenesis in physiological and pathological neovascularization. Circ Res. 1999;85:221–8.
20. Rookmaaker MB, et al. Bone-marrow cells contribute to glomerular endothelial repair in experimental glomerulonephritis. Am J Pathol. 2003;163:553–62.
21. Takahashi T, et al. Ischemia- and cytokine-induced mobilization of bone marrow endothelial progenitor cells for neovascularization. Nat Med. 1999;5:434–8.
22. Jeon O, et al. Additive effect of endothelial progenitor cell mobilization and bone marrow mononuclear cell transplantation on angiogenesis in mouse ischemic limbs. J Biomed Sci. 2007;14:323–30.
23. Iba O, et al. Angiogenesis by implantation of peripheral blood mononuclear cells and platelets into ischemic limbs. Circulation. 2002;106:2019–25.
24. Shintani S, et al. Augmentation of postnatal neovascularization with autologous bone marrow transplantation. Circulation. 2001;103:897–903.
25. Foteinos G, et al. Rapid endothelial turnover in atherosclerosis-prone areas coincides with stem cell repair in apolipoprotein E-deficient mice. Circulation. 2008;117:1856–63.
26. Takahashi M, et al. Therapeutic neovascularization by the implantation of autologous mononuclear cells in patients with connective tissue diseases. Curr Pharm Des. 2009;15:2778–83.
27. Matoba S, et al. Long term clinical outcome after intramuscular implantation of bone marrow mononuclear cells (TACT trial) in patients with chronic limb ischemia. Am Heart J. 2008;156:1010–8.
28. Singer DR, et al. Management of hypertension in peripheral arterial disease: does the choice of drugs matter? Eur J Vasc Endovasc Surg. 2008;35:701–8.
29. Norgren L, et al. Inter-society consensus for the management of peripheral arterial disease (TASC II). Eur J Vasc Endovasc Surg. 2007;33(suppl l):Sl–S75.
30. Horie T, et al. Long term clinical outcomes for patients with lower limb ischemia implanted with G-CSF mobilized autologous peripheral blood mononuclear cells. Atherosclerosis. 2010;208:461–6.
31. Miyamoto M, et al. Therapeutic angiogenesis by autologous bone marrow cell implantation for refractory chronic peripheral arterial disease using assessment of neovascularization by 99mTc-tetrofosmin (TF) perfusion scintigraphy. Cell Transplant. 2004;13:429–37.

32. Bartsch T, et al. Transplantation of autologous adult bone marrow stem cells in patients with severe peripheral arterial occlusion disease. Med Klin (Munich). 2006;101(suppl 1):195–7.
33. Durdu S, et al. Autologous bone-marrow mononuclear cell implantation for patients with Rutherford grade II–III thromboangiitis obliterans. J Vasc Surg. 2006;44:732–9.
34. Barc P, et al. Bone-marrow cells in therapy of critical limb ischemia of lower extremities-own experience. Acta Angiol. 2006;12:155–66.
35. Gu Y, et al. A clinical study on implantation of autologous bone marrow mononuclear cells after bone marrow stimulation for treatment of lower limb ischemia. Chin J Reparative Reconstr Surg. 2006;20:1017–20.
36. Bartsch T, et al. Transplantation of autologous mononuclear bone marrow stem cells in patients with peripheral arterial disease (the TAM-PAD study). Clin Res Cardiol. 2007;96:891–9.
37. Oda M, et al. Prognostic factors of critical limb ischemia after autologous bone marrow implantation. J Cardiol. 2007;50:235–42.
38. Saito Y, et al. Effect of autologous bone-marrow cell transplantation on ischemic ulcer in patients with Buerger's disease. Circ J. 2007;71:1187–92.
39. Gan Y, et al. Autologous bone marrow stem cell transplantation for severe chronic lower limb ischemia in 15 cases. J Clin Rehab Tissue Eng Res. 2008;12:1541–4.
40. Idei N, et al. Autologous bone-marrow mononuclear cell implantation reduces long-term major amputation risk in patients with critical limb ischemia a comparison of atherosclerotic peripheral arterial disease and Buerger disease. Circ Cardiovasc Interv. 2011;4:15–25.

Chapter 3
Peripheral Blood Mononuclear Cells for Limb Ischemia

Masayoshi Suda†, Ippei Shimizu†, Yohko Yoshida, and Tohru Minamino

Abstract There is accumulating evidence that the peripheral blood is a source of pro-angiogenic mononuclear cells (MNCs). These cells were initially described as endothelial progenitor cells (EPCs) expressing CD34 and vascular endothelial growth factor (VEGF) receptor 2. Pro-angiogenic MNCs are now known to represent a mixed population of cells, including hematopoietic stem cells, mesenchymal stem cells, and EPCs. The therapeutic potential of bone marrow or peripheral blood MNCs has been tested in patients with severe peripheral artery disease, showing that implanted cells promote the production of pro-angiogenic molecules in an autocrine and/or paracrine fashion and contribute to the recovery of blood flow in ischemic limbs. The molecular mechanisms underlying therapeutic angiogenesis are yet to be defined, but a number of studies have indicated that injection of either bone marrow or peripheral blood MNCs improves the clinical outcome in patients with severe limb ischemia and importantly achieves this effect with minor adverse events. In addition to controlling classical risk factors such as diabetes and/or hypertension, recent studies have suggested several combination therapies that can contribute to improving the therapeutic effects of these pro-angiogenic cells. Therapeutic angiogenesis promoted by the injection of autologous peripheral blood MNCs is an essential and practical treatment for patients who have no other options.

Keywords Therapeutic angiogenesis • Peripheral blood mononuclear cells • Endothelial progenitor cells • Cell transplantation • Limb ischemia

---

†Masayoshi Suda and Ippei Shimizu contributed equally to this work.

M. Suda • T. Minamino, M.D., Ph.D. (✉)
Department of Cardiovascular Biology and Medicine, Niigata University Graduate School of Medical and Dental Sciences, 1-757 Asahimachidori, Chuo-ku, Niigata 951-8510, Japan
e-mail: t_minamino@yahoo.co.jp

I. Shimizu • Y. Yoshida
Department of Cardiovascular Biology and Medicine, Niigata University Graduate School of Medical and Dental Sciences, 1-757 Asahimachidori, Chuo-ku, Niigata 951-8510, Japan

Division of Molecular Aging and Cell Biology, Niigata University Graduate School of Medical and Dental Sciences, Niigata, Japan

Y. Higashi, T. Murohara (eds.), *Therapeutic Angiogenesis*,
DOI 10.1007/978-981-10-2744-4_3

3.1 Introduction

The prevalence of peripheral artery disease (PAD) continues to increase. The clinical outcome of critical limb ischemia remains poor, especially in the setting of limb amputation, and it is an urgent task to develop new therapies for this condition. Induction of angiogenic proteins such as vascular endothelial growth factor (VEGF) or fibroblast growth factor (FGF) contributed significantly to restoring the perfusion of ischemic limbs in preclinical studies performed in rodents. However, clinical studies designed to assess the augmentation of a single angiogenic factor have so far shown only a mild effect or no effect [1–8]. In contrast, there is accumulating evidence from both preclinical and clinical studies that indicates that cell therapy for critical limb ischemia is a promising option to improve perfusion of ischemic limbs, leading to better clinical outcomes [9–13]. Mononuclear cells extracted from the bone marrow or peripheral blood are the chief source of the cells utilized for these therapies. Implanted cells contribute to promoting the formation of collateral arteries by activating the secretion of pro-angiogenic factors from myocytes and/or implanted mononuclear cells, and it has also been demonstrated that these cells have the potential to differentiate into vascular endothelial cells [14–16]. It has been reported that injection of peripheral blood mononuclear cells (PB-MNCs) is as effective as use of bone marrow mononuclear cells (BM-MNCs) [10, 17]. Considering the accessibility of these highly pro-angiogenic cells, therapy with PB-MNCs is a less invasive option that should be considered for patients with severe limb ischemia. In this section, the role of PB-MNCs in the treatment of limb ischemia is outlined.

3.2 Pro-angiogenic Properties of PB-MNCs

Since the discovery of endothelial progenitor cells (EPCs) in the peripheral circulation in 1997, the peripheral blood continues to be highlighted as a source of pro-angiogenic cells that can potentially be used to treat peripheral artery disease [18]. This landmark study by Asahara and Isner demonstrated that cultured CD34 ($CD34^+$) and VEGF receptor 2 positive circulating mononuclear cells (MNCs) can differentiate into endothelium-like cells. $CD34^+$ cells only account for 1% of all PB-MNCs, but make up around 60% of the attached cells in vitro. These cells were also shown to be incorporated into new vessels in both mouse and rabbit models of ischemia. Accordingly, these cell populations were initially designated as EPCs. Studies have indicated that circulating EPCs are increased by injection of granulocyte colony-stimulating factor (G-CSF) or granulocyte-macrophage colony-stimulating factor (GM-CSF) [19, 20], and extensive preclinical and clinical investigations have assessed the therapeutic potential of pro-angiogenic cells originating from the bone marrow or peripheral blood mobilized by these stimulatory factors. Hur et al. described two distinct types of endothelial progenitor cells among human PB-MNCs. They extracted PB-MNCs by centrifugation on Histopaque gradients and cultured the extracted cells. Initially, the seeded cells were elongated and spindle-shaped, similar to the EPCs reported by Asahara et al. in 1997 (Hur et al. described these cells as "early EPCs"). Early EPCs increased for 2 weeks and then stopped replicating and disappeared by 4 weeks after plating. Hur et al. found that another population of cells with a different morphology and growth pattern appeared at 2–4 weeks after plating. These cells showed a cobblestone appearance similar to HUVECs, and Hur et al. described them as "late EPCs." Late EPCs showed higher expression of VE-cadherin and VEGF receptor 2 than early EPCs, while expression of VEGF receptor 1, endothelial nitric oxide synthase, and von Willebrand factor was similar for both EPC types. Early EPCs secreted more angiogenic molecules such as VEGF and interleukin-8 than late EPCs, while late EPCs were better at forming capillary tubes in vitro. Interestingly, both early and late EPCs showed comparable vasculogenic activity in vivo [21]. Bone morphogenetic protein (BMP) 2 and BMP 4 are selectively expressed by late EPCs, and the BMP pathway has been found to augment the expansion, proliferation, and migration of late EPCs in vitro and to promote their angiogenic potential in vivo [22].

There is accumulating evidence that $CD34^+$ EPCs are not the only cellular components contributing to vasculogenesis, and PB-MNCs are currently accepted to include a broad spectrum of cells with pro-angiogenic potential, such as hematopoietic cells, myeloid cells, side population cells, and circulating mature endothelial cells [5, 13]. PB-MNCs contain many populations of cells, including EPCs, and no single marker or combination of markers has yet been found for a pure EPC population. Pro-angiogenic PB-MNCs can be extracted from peripheral blood through several cell processing methods, such as Ficoll density gradient centrifugation or by use of a cell separator, and can then be delivered to patients by intramuscular or intra-arterial injection. Studies have suggested that delivery of isolated PB-MNCs to the ischemic limb, generally in combination with G-CSF mobilization, contributes to better clinical outcomes in patients with critical limb ischemia, and they are summarized in the following sections.

3.3 Clinical Studies of Unselected PB-MNCs for Limb Ischemia

Preclinical studies performed in rodents suggested that implantation of bone marrow mononuclear cells contributes to promoting collateral vessel formation in ischemic limbs. The therapeutic potential of pro-angiogenic MNCs was first demonstrated in patients by the Therapeutic Angiogenesis using Cell Transplantation (TACT) study [9]. In patients with limb ischemia, autologous implantation of BM-MNCs improved rest pain, the pain-free walking time, and transcutaneous oxygen pressure ($TcPO_2$). These effects were achieved without major adverse events. The TACT study compared BM-MNCs with PB-MNCs and demonstrated that the effects of cell therapy were superior in patients treated with BM-MNCs, although patients injected with PB-MNCs also showed some benefit [9]. Subsequent studies have indicated that PB-MNCs are as effective as BM-MNCs for promoting the recovery of blood flow in patients with severe limb ischemia. In one study, PB-MNCs were collected with an AS104 cell separator after mobilization with G-CSF, and 4×10^{10} cells (including 2×10^8 $CD34^+$ cells) were injected intramuscularly in patients with limb ischemia. One patient with atherosclerosis and five patients with Buerger's disease (Fontaine stage 3–4) underwent this therapy and showed significant improvement of the ankle-brachial index (ABI) at 4 weeks, as well as improvement of the pain scale and walking distance at 24 weeks without serious adverse events [23]. The therapeutic potential of intramuscular injection of PB-MNCs into the ischemic limb was also tested in other studies and was shown to significantly improve the ABI, pain scale, laser Doppler perfusion, and limb salvage without severe adverse effects [11, 12, 24, 25]. Lenk et al. investigated the effects of intra-arterial infusion of PB-MNCs in PAD patients. PB-MNCs were extracted by Ficoll density gradient centrifugation after recruitment with G-CSF, and $3.9 \pm 2.4 \times 10^7$ cells were injected into the ischemic limbs of patients in Rutherford classes 4–5, leading to significant improvement of the ABI, $TcPO_2$, and walking distance [26]. These results clearly indicate that unselected PB-MNCs are pro-angiogenic and contribute to better clinical outcomes in patients with critical limb ischemia.

3.4 Clinical Studies Comparing BM-MNCs with PB-MNCs for Limb Ischemia

Some studies have focused on comparing the efficiency of cell therapy with BM-MNCs or PB-MNCs. In the abovementioned TACT study, implantation of autologous BM-MNCs achieved better clinical outcomes in patients with limb ischemia compared with implantation of PB-MNCs [9]. However, a randomized comparison of G-CSF-mobilized PB-MNCs versus BM-MNCs showed that PB-MNCs significantly improved the ABI, skin temperature, and rest pain in patients with peripheral artery disease. The authors conducted a comparative analysis and concluded that use of PB-MNCs for the treatment of PAD is more practical than employing BM-MNCs [10]. Kamata et al. compared these two treatments in connective tissue disease patients with severe ischemia of the fingers and/or toes and observed significant improvement in both groups [17]. These results suggest that PB-MNC-based therapy is equivalent in efficacy with BM-MNC-based therapy for PAD. Considering that harvesting PB-MNCs is less invasive compared with harvesting BM-MNCs, PB-MNC-based cell therapy seems to be a more practical option for patients with severe limb ischemia.

3.5 Clinical Studies of PB-MNCs for Limb Ischemia

Because PB-MNCs contain various types of cells, some studies have focused on the therapeutic potential of specific cellular components. In a single-blind dose-escalation study, G-CSF-mobilized $CD34^+$ MNCs were extracted by magnetic separation and injected intramuscularly in patients with PAD or Buerger's disease, leading to significant improvement of the efficacy score (based on changes of the toe-brachial pressure index, pain scale, and total walking distance). Cells were injected at three doses (1×10^5, 5×10^5, and 1×10^6/kg), but no significant dose-response relationship was observed [27]. Losordo et al. investigated the pro-angiogenic potential of $CD34^+$ MNCs in a randomized controlled study involving intramuscular administration of autologous $CD34^+$ cells (1×10^5 or 1×10^6 cells/kg) to patients with critical limb ischemia. Compared with the control group, the cell-treated groups showed a lower incidence of amputation ($p = 0.125$ at 6 months and $p = 0.054$ at 12 months). Among a total of 28 patients, 8 subjects (66.7%) in the control group, 3 subjects (42.9%) in the low-dose group, and 2 subjects (22.2%) in the high-dose group underwent amputation by 6 months after injection, so implantation of $CD34^+$ cells led to a better clinical outcome with no major adverse events [28]. In another study, autologous $CD133^+$ cells were used to treat patients with critical limb ischemia, including seven patients with arteriosclerosis obliterans, one patient with thromboangiitis obliterans (Buerger's disease), and one patient with a thromboembolic disorder. After injection of G-CSF for mobilization, peripheral blood $CD133^+$ cells were harvested, and $2.5–5 \times 10^6$ cells were injected intramuscularly. One year later, seven of the nine patients were free from leg amputation, and the pain-free treadmill walking time and exercise capacity both showed improvement [29].

Thus, there is considerable evidence that PB-MNC-based cell therapy is safe and efficient, contributing to better clinical outcomes in patients with severe limb ischemia (Table 3.1). Cell therapy for critical limb ischemia is intended to increase blood flow in the affected limb by promoting the formation of collateral vessels, and the mechanisms involved are discussed next.

Table 3.1 Clinical studies for patients with ischemic limbs who underwent PB-MNC therapies

Trials (references)	*n*	Types of cells	Dose of cells	Delivery method	Indications or severity classes	Follow-up	Major adverse events	Factors for evaluation	Results
Tateishi-Yuyama et al. (TACT trial) [7]	20	Unselected BM-MNCs vs. PB-MNCs	1.5×10^9 BM-MNCs (including 3.5×10^7 CD34+ cells) or 1.5×10^9 PB-MNCs	IM	Fontaine 3–4 (=Rutherford 4–6)	24 weeks	None	ABI	Better in BM-MNCs
								$TcPO_2$ pressure	
								Rest pain	
								New collaterals	
								Pain-free walking distance	
Ishida et al. [8]	6	G-CSF-mobilized unselected PB-MNCs	$1.9–7.15 \times 10^{10}$ PB-MNCs (including $0.97 \times 2.8 \times 10^8$ $CD34^+$ cells)	IM	Fontaine 3–4 (=Rutherford 4–6)	24 weeks	None	ABI	Improved
								$TcPO_2$ pressure	
								Pain scale	
								Ulcer healing	
								Walking distance	
Lara-Hernandez et al. [9]	28	G-CSF-mobilized unselected PB-MNCs	Not listed (high $CD34^+$ and $CD133^+$ rates)	IM	CLI stage III or IV (possibly meaning Fontaine 3–4 (=Rutherford 4–6))	14 months	None	ABI	Improved
								Pain scale	
								Ulcer healing	
								Limb salvage	
								Flow-dependent vasodilation	
Huang et al. [10]	28	G-CSF-mobilized unselected PB-MNCs	3×10^9 PB-MNCs (including 0.4% $CD34^+$ cells)	IM	Fontaine 3–4 (=Rutherford 4–6)	3 months	None	ABI	Improved
								Ulcer healing	
								Walking distance	
								Laser Doppler blood perfusion	
								Angiographic scores	
								Limb salvage	

Ozturk et al. [11]	40	G-CSF-mobilized unselected PB-MNCs	PB-MNCs including CD34 + cells were injected. Dose was not clearly described	IM	Fontaine 3–4 (=Rutherford 4–6)	3 months	None	ABI	Improved
								$TcPO_2$ pressure	
								Pain scale	
								Ulcer healing	
								6 min walk distance	
Mohammadzadeh et al. [12]	21	G-CSF-mobilized unselected PB-MNCs	0.9–1.2 × 10^9 PB-MNCs	IM	CLI	3 months	None	ABI	Improved
								Pain scale	
								Ulcer healing	
								Amputation rates	
Lenk et al. [13]	7	G-CSF-mobilized unselected PB-MNCs	Mean 3.9 × 10^7 CPCs (including $CD34^+$(32%), $CD133^+$(23%) cells)	IA	Rutherford 4–5	12 weeks	None	ABI	Improved
								$TcPO_2$ pressure	
								Pain scale	
								Pain-free walking distance	
Huang et al. [14]	150	Unselected BM-MNCs vs. G-CSF-mobilized PB-MNCs	7.2 × 10^9 PB-MNCs (including 2.3 × 10^8 $CD34^+$ cells and 1.2 × 10^8 $CD133^+$ cells) vs. 5.8 × 10^8 BM-MNCs (including 2.7 × 10^7 $CD34^+$ cells and 0.8 × 10^7 $CD133^+$ cells)	IM	Lower limb arterio-sclerosis obliterans	12 weeks	None	ABI	Better in PB-MNCs
								Skin temperature	
								Rest pain	
								$TcPO_2$ pressure	Equally improved in both groups
								Pain-free walking distance	
								Ulcer healing	
								Amputation rates	

(continued)

Table 3.1 (continued)

Trials (references)	*n*	Types of cells	Dose of cells	Delivery method	Indications or severity classes	Follow-up	Major adverse events	Factors for evaluation	Results
Kamata et al. [15]	6	Unselected BM-MNCs vs. PB-MNCs	$0.5–1.1 \times 10^9$ BM-MNCs (including $2.9–5.3 \times 10^6$ $CD34^+$ cells) or $1.0–4.8 \times 10^8$ PB-MNCs (including $0.1–3.7 \times 10^7$ $CD34^+$ cells)	IM	All patients had digital ulcers and finger necrosis due to connective tissue disorders	1 month	None	Pain scale	Equally improved in both groups
								$TcPO_2$ pressure	
								Skin temperature	
Kawamoto et al. (EPOCH-CLI) [16]	17	G-CSF-mobilized PB-$CD34^+$ cells	Dose escalation, 1×10^5, 5×10^5, 1×10^6 cells/kg (body weight)	IM	Rutherford 4–6	12 weeks	None	Toe-brachial index	Improved in cell therapy groups
								Pain scale	
								Total walking distance	
								Ulcer healing	
								$TcPO_2$ pressure	
Losordo et al. (ACT34-CLI) [17]	28	G-CSF-mobilized PB-$CD34^+$ cells	Dose escalation, 1×10^5, 1×10^6 cells/kg (body weight)	IM	Rutherford 4–5	12 months	None	Amputation rates	Showed a trend toward dose-dependent improvement
Burt et al. [18]	9	G-CSF-mobilized PB-$CD133^+$ cells	$2.5–5.0 \times 10^6$ $CD133^+$ cells per injection. Total dose delivered is not clearly described	IM	CLI	12 months	None	Pain-free walking time	Showed a trend toward improvement
								6 min walk distance	Not improved
								ABI	
								Rest pain	Improved
								Limb salvage	

$TcPO_2$ transcutaneous oxygen pressure

3.6 Mechanisms of PB-MNC-Induced Neovascularization

Transplantation of autologous expanded PB-MNCs has been reported to promote neovascularization in several animal models of hind limb ischemia [14, 15, 30]. Partly because PB-MNCs are a mixed population of cells that have not been completely characterized, the mechanisms through which injection of these cells improves tissue blood flow are yet to be fully defined. Differentiation of injected cells to form new capillaries is reported to be rare, which suggests that angiogenic factors secreted by the injected cells or tissue resident cells contribute to neovascularization in the ischemic limb [14]. It has been reported that intramuscular injection of PB-MNCs promotes the production of angiogenic factors, such as VEGF and interleukin-1 beta (IL-1β), by skeletal muscle. Injection of IL-1β-deficient PB-MNCs into wild-type mice increased the expression of IL-1β in ischemic limbs. In contrast, injection of wild-type PB-MNCs into IL-1β-deficient mice failed to upregulate IL-1β expression in the ischemic limb, and neovascularization was impaired. Furthermore, neovascularization was significantly improved by co-injection of wild-type PB-MNCs and IL-1β, suggesting that the implanted mononuclear cells did not secrete cytokines at levels sufficient to promote neovascularization. Importantly, implantation of PB-MNCs led to an increase in the number of regenerating myocytes expressing various pro-angiogenic factors. Therefore, this study indicates that injection of PB-MNCs promotes the production of angiogenic factors by muscle cells, especially regenerating myocytes, and this mechanism is crucial for neovascularization in ischemic limbs [15].

Recently, human peripheral blood CD31-positive cells were reported to have sufficient angiogenic and vasculogenic properties to promote the recovery of blood flow in a murine model of ischemic vascular disease [14]. CD31 is a ubiquitous endothelial cell marker, and studies have suggested that it is also expressed on the surface of cells from various hematopoietic lineages, including hematopoietic stem cells and embryonic stem cells in mice [31–33]. Kim et al. reported that peripheral blood $CD31^+$ ($PB\text{-}CD31^+$) cells extracted by a magnetic-activated cell sorter expressed endothelial and hematopoietic stem/progenitor markers. Approximately 40% of $PB\text{-}CD31^+$ cells expressed CD14 (a monocyte/macrophage marker), and over 99% of $PB\text{-}CD31^+$ cells expressed CD45 (a well-known pan-hematopoietic marker). These cells were also positive for endothelial cell markers, such as von Willebrand factor, CD105, CD141, and CD144, but predominantly showed high expression of stem cell or progenitor markers such as CD34, CD133, CD117, and VEGF receptor 2. Taken together, these results suggested that $PB\text{-}CD31^+$ cells are not circulating endothelial cells. $PB\text{-}CD31^+$ cells displayed significantly higher expression of pro-angiogenic molecules, such as VEGF-A, hepatocyte growth factor (HGF), FGF-2, and angiopoietin-1 (Ang-1), compared with $PB\text{-}CD31^-$ MNCs. Cultured $PB\text{-}CD31^+$ cells formed vascular tubelike structures and exhibited an endothelial cell phenotype. In addition, $PB\text{-}CD31^+$ cell transplantation markedly improved blood flow and reduced limb loss in a murine model of hind limb ischemia, in association with high levels of pro-angiogenic cytokines and an increased capillary density. After treatment, FACS analysis revealed that 2.2% of the endothelial cells in ischemic skeletal muscle were derived from implanted PB-CD31$^+$ cells. These results indicate that PB-CD31$^+$ cells mainly promote neovascularization via transdifferentiation-independent mechanisms, and the findings of Kim et al. suggest the promising therapeutic potential of PB-CD31$^+$ cells for treatment of PAD [14].

3.7 Modifiers of Cell Therapy

3.7.1 Influence of Systemic Conditions

It has been reported that smoking, metabolic stress, and hypertension have a negative impact on the health of endogenous EPCs. The relative risk of amputation is 40-fold higher in patients with diabetes [34]. Cell therapy with both BM-MNCs and PB-MNCs can improve ischemia in diabetic patients [35], but metabolic stress is well known to impair the ability of progenitor cells to promote neovascularization. It has been reported that the fasting glucose level negatively correlates with the results of PB-MNC transplantation in patients with critical limb ischemia [36]. The diabetic state leads to a defective response of EPCs and peripheral tissues to hypoxia. Mobilization of cells from the bone marrow and recruitment to ischemic tissues are significantly reduced in diabetic mice compared with controls, and injection of healthy EPCs into these mice does not normalize the viability of ischemic tissue [37]. A high glucose level is reported to reduce the proliferative capacity and angiogenic activity of early and late EPCs [38]. Cellular senescence is well accepted to have a central role in the progression of age-related disorders [39–43], and EPC senescence has been identified in hypertensive rats and patients with hypertension [44]. Smoking is reported to be associated with decrease of EPCs, and it also impairs EPC differentiation and function [45]. It is well known that systemic metabolic dysfunction is associated with high levels of reactive oxygen species, which are thought to contribute to the progression of diabetes. Antioxidant therapy combined with BM-MNC achieved synergistic improvement of neovascularization in mice with diabetes and dyslipidemia [46–48]. These reports clearly indicate that in addition to cessation of smoking, treatment of systemic metabolic disorders and hypertension is important for therapeutic angiogenesis in patients with limb ischemia.

Exercise is the only therapy with a class I recommendation for patients with stable PAD. Interestingly, Schirmer et al. recently reported that monocyte-derived inducible nitric oxide synthase (iNOS) promotes exercise-induced growth of collateral vessels [49]. In a murine treadmill exercise model with femoral artery ligation, iNOS expression by PB-MNC was increased, while depletion of monocytes/macrophages by liposomal clodronate suppressed peri-collateral macrophage accumulation and attenuated upregulation of iNOS, leading to the suppression of exercise-induced restoration of perfusion and collateral development. Based on extensive studies using iNOS-depleted mice or bone marrow transplantation models, Schirmer et al. concluded that iNOS derived from circulating mononuclear cells is an important mediator of exercise-induced collateral vessel development [49].

3.7.2 *Mobilizers*

It has been demonstrated that circulating EPCs are dose-dependently increased by injection of G-CSF or GM-CSF [19, 20]. $CD34^+$ cells mobilized into the peripheral blood by G-CSF were expanded in vitro and injected into the hearts of mice with myocardial infarction, leading to the formation of vascular structures in the myocardium and improvement of cardiac function [50]. Arai et al. reported that subcutaneous injection of recombinant human G-CSF improves the clinical signs and symptoms of patients with severe PAD, including the ABI and transcutaneous oxygen tension, and they found that G-CSF treatment was as effective as intramuscular injection of BM-MNCs [51]. In addition, injection of GM-CSF significantly increased circulating $CD34^+$ cells and contributed to the improvement of endothelial dysfunction and exercise capacity in PAD patients [52]. However, a double-blind, randomized, placebo-controlled study showed that GM-CSF treatment did not improve the ABI in 40 PAD patients with moderate or severe intermittent claudication [53]. Also, a phase 2 double-blind, placebo-controlled study of 159 PAD patients with claudication showed that GM-CSF treatment did not improve the treadmill walking performance at 3 months after injection [54]. Thus, further studies are needed to better define the therapeutic potential of G-CSF and GM-CSF.

CXC chemokine receptor 4 (CXCR4) is involved in retention of stem cells/progenitor cells in the bone marrow. A single injection of a CXCR4 antagonist (AMD3100) was shown to augment mobilization of bone marrow-derived EPCs into the peripheral blood and also contributed to enhancement of neovascularization and functional recovery after myocardial infarction. This CXCR4 antagonist increased the expression of VEGF and metalloproteinase 9 (MMP9) by both PB-MNCs and BM-MNCs, and its beneficial effects were dependent on MMP9 expression in the bone marrow [55]. Accordingly, it is highly possible that CXCR4 antagonist therapy would also be effective for promoting neovascularization in patients with limb ischemia. Various angiogenic molecules, including $VEGF_{165}$, angiopoietin-1, fibroblast growth factors, hypoxia-inducible factor, and stromal cell-derived factors, have also been reported to recruit EPCs into the peripheral blood [56–58]. The precise mechanisms by which these mobilizers contribute to promotion of the homing, transdifferentiation, incorporation, and survival of injected progenitor cells have yet to be defined. However, recent studies have suggested that combinations of several agents need to be considered to enhance the efficacy of cell therapy.

3.7.3 *Genetic Engineering*

Genetic engineering of injected cells could also enhance the efficiency of cell therapy. Transfection of a plasmid encoding VEGF and insulin-like growth factor I (IGF-I) into BM-MNCs reduced apoptosis and promoted cell survival, leading to greater improvement of left ventricular function in a rat model of myocardial infarction [59]. Bone marrow-derived mesenchymal stem cells (MSCs) transfected with the human HGF gene promoted angiogenesis in a rat model of hind limb ischemia [60]. In addition, treatment with MSCs transduced with adenoviral vectors encoding fibroblast growth factor-2 (FGF-2) and platelet-derived growth factor-BB (PDGF-BB) significantly improved capillary and arteriole density in a rabbit hind limb ischemia model compared with control MSCs [61]. Furthermore, intramyocardial VEGF gene transfer with injection of G-CSF and stem cell factor synergistically promoted the mobilization of endothelial progenitor cells into ischemic myocardium after myocardial infarction [62]. Moreover, the introduction of Ppargc1a into skeletal muscle via an adenoviral vector improved blood flow in a murine hind limb ischemia model of PAD by promoting the release of secreted phosphoprotein 1 (SPP1) and recruitment of macrophages secreting monocyte chemoattractant protein-1 (MCP-1) that contributed to the recruitment of endothelial cells, pericytes, and smooth muscle cells [63].

3.7.4 Hypoxic Preconditioning

Several studies have demonstrated that hypoxic preconditioning promotes an angiogenic profile of cells utilized for cell therapy. A hypoxic preconditioning-induced increase of glycogen storage improved MSC survival and promoted perfusion in murine ischemic limbs. Hypoxic preconditioning induces glycogenesis in MSCs, and when these cells are subsequently exposed to ischemia, glycogenolysis releases glycogen to maintain the energy supply and promote cell survival [64]. According to another report, hypoxic preconditioning of PB-MNCs enhanced resistance to oxidative stress, increased cell viability, and promoted the production of VEGF, while autologous transplantation of preconditioned PB-MNCs efficiently promoted neovascularization and improved limb blood flow. The mechanism of such beneficial effects is yet to be defined, but this could be a therapeutic strategy worth considering to increase the angiogenic profile of implanted PB-MNCs [65].

3.7.5 Ultrasound

Ultrasound is known to have a therapeutic effect on limb ischemia by promoting tissue perfusion. Low-frequency (<100 MHz) ultrasound promotes peripheral arterial dilatation in rodents and humans [66, 67]. Exposure to ultrasound acutely increases tissue perfusion, and this effect is partially mediated by oscillatory pressure fluctuations within the blood vessels. The presence of microbubbles within the vessels is known to amplify this effect by increasing shear stress and activating eNOS [68]. Song et al. reported that rupture of capillaries elicited by microbubble destruction with 1 MHz ultrasound contributed to the stimulation of arteriolar remodeling in skeletal muscle [69]. Enomoto et al. examined the therapeutic potential of microbubble destruction with ultrasound and found that it could augment neovascularization after bone marrow cell transplantation in a rat model of hind limb ischemia [70]. In another study, plasmid DNA encoding $VEGF_{165}$ was charge-coupled to cationic lipid microbubbles, and ultrasound-mediated destruction of intravenous microbubbles carrying the $VEGF_{165}$ plasmid was found to be as effective for delivery as intramuscular injection of this plasmid [71]. Thus, PB-MNC implantation combined with ultrasound could be an additional strategy to promote the recovery of blood flow in patients with limb ischemia.

3.7.6 *Cell Scaffolds and Biomaterials*

Tracking studies performed in rodents indicate that the majority of transplanted cells do not remain in the ischemic region for more than 24 h after injection. It was reported that organs with a high content of transplanted progenitor cells were the liver and spleen at 24 h after injection of EPCs into the infarcted heart [72, 73]. Clinical studies have demonstrated that intracoronary infusion of unselected BM-MNCs and $CD34^+$-enriched progenitor cells results in the retention of only 2% and 14–39% of these cells, respectively, within the ischemic myocardium at 1 h after injection [74, 75]. Thus, development of bioactive materials is essential to improve the efficiency of cell therapy. A self-assembling, bioactive peptide nanofiber matrix presenting an integrin-binding domain of fibronectin was shown to enhance limb perfusion and reduce ischemic limb necrosis/amputation in a murine model of limb ischemia with injection of BM-MNCs [76]. An injectable elastin-like polypeptide capable of carrying pro-angiogenic genes was developed to enhance gene therapy for limb ischemia and was employed to deliver a plasmid encoding eNOS, leading to upregulation of major pro-angiogenic growth factors such as VEGF, PDGF-B, and FGF-1, whereas induction of IL-10 using the same system led to reduced infiltration of inflammatory cells [77]. These findings suggest that combining gene therapy and cell scaffolds could contribute to a better outcome of cell-based therapy for ischemia.

3.8 Conclusion and Future Directions

The peripheral blood is a rich source of pro-angiogenic MNCs. Early studies suggested that BM-MNCs had a higher angiogenic potential than PB-MNCs, but subsequently evidence was obtained that PB-MNC-based cell therapy is equally effective for the treatment of severe PAD. Injection of PB-MNCs into ischemic limbs promotes the formation of collateral vessels and increases blood flow, leading to significant improvement of clinical outcomes for patients with severe limb ischemia. Importantly, these beneficial effects are achieved without major adverse events related to cell therapy. The mechanisms by which PB-MNCs promote collateral vessel formation have not been clearly defined. PB-MNCs show relatively low potential for transdifferentiation into mature endothelial cells, and studies have suggested that transplanted PB-MNCs mainly augment neovascularization via transdifferentiation-independent mechanisms. PB-MNCs secrete growth factors and cytokines with paracrine effects that promote neovascularization and also promote regeneration of myocytes that produce angiogenic molecules. PB-MNCs contain a broad spectrum of cells with pro-angiogenic potential mediated by both cell-autonomous and non-cell-autonomous mechanisms. Further efforts are needed to identify the cell populations chiefly involved in promoting neovascularization in ischemic limbs. PB-MNCs can be harvested with relatively little invasion and show equal efficacy for improving clinical outcomes compared with BM-MNCs, making PB-MNC-based therapy a more practical option for the treatment of severe limb ischemia. Various factors and/or techniques have been reported to enhance the efficiency of cell therapy, and combined treatment is an interesting field that needs to be explored. Large-scale randomized clinical trials are required to better define the potential of PB-MNCs for treating critical limb ischemia.

Acknowledgments This work was supported by a Grant-in-Aid for Scientific Research, a Grant-in-Aid for Scientific Research on Innovative Areas (Stem Cell Aging and Disease), and a Grant-in-Aid for Exploratory Research from the Ministry of Education, Culture, Sports, Science and Technology (MEXT) of Japan and grants from the Takeda Medical Research Foundation, the Japan Diabetes Foundation, the Ono Medical Research Foundation, the Daiichi Sankyo Foundation of Life Science, the Japan Foundation for Applied Enzymology, the Takeda Science Foundation, the SENSHIN Medical Research Foundation, and the Terumo Foundation (to T.M.); by a Grants-in-Aid for Young Scientists (Start-up) (JSPS KAKENHI, Grant Number 26893080) and grants from the Uehara Memorial Foundation, Takeda Science Foundation, Kowa Life Science Foundation, Manpei Suzuki Diabetes Foundation, Kanae Foundation, Japan Heart Foundation Research Grant, the Senri Life Science Foundation, SENSHIN Medical Research Foundation, ONO Medical Research Foundation, Tsukada Grant for Niigata University Medical Research, the Nakajima Foundation, SUZUKEN Memorial Foundation, HOKUTO Corporation, Inamori Foundation, Mochida Memorial Foundation for Medical and Pharmaceutical Research, Banyu Foundation Research Grant, Grant for Basic Science Research Projects from The Sumitomo Foundation, and Grants-in-Aid for Encouragement of Young Scientists (A) (JSPS KAKENHI Grant Number 16H06244) (to I.S.); by a Grants-in-Aid for Encouragement of Young Scientists (B) (JSPS KAKENHI Grant Number 16K19531), Japan Heart Foundation, Dr. Hiroshi Irisawa & Dr. Aya Irisawa Memorial Research Grant, SENSHIN Medical Research Foundation grant, SUZUKEN Memorial Foundation, Takeda Science Foundation, ONO Medical Research Foundation, Uehara Memorial Foundation, and Research Foundation for Community Medicine (to Y.Y.); and by a Grant-in-Aid for Scientific Research (C) (JSPS KAKENHI Grant Number 15K09154) (to M.S.) and by a grant from Bourbon (to T.M., I.S., and Y.Y.).

Disclosures M.S., I.S., Y.Y., and T.M. disclose no conflict of interest.

References

1. Kusumanto YH, Van Weel V, Mulder NH, Smit AJ, Van den Dungen JJAM, Hooymans JMM, et al. Treatment with intramuscular vascular endothelial growth factor gene compared with placebo for patients with diabetes mellitus and critical limb ischemia: a double-blind randomized trial. Hum Gene Ther. 2006;17(6):683–91. doi:10.1089/Hum.2006.17.683.

2. Muona K, Makinen K, Hedman M, Manninen H, Yla-Herttuala S. 10-year safety follow-up in patients with local VEGF gene transfer to ischemic lower limb. Gene Ther. 2012;19(4):392–5. doi:10.1038/gt.2011.109.
3. Rajagopalan S, Mohler III ER, Lederman RJ, Mendelsohn FO, Saucedo JF, Goldman CK, et al. Regional angiogenesis with vascular endothelial growth factor in peripheral arterial disease: a phase II randomized, double-blind, controlled study of adenoviral delivery of vascular endothelial growth factor 121 in patients with disabling intermittent claudication. Circulation. 2003;108(16):1933–8. doi:10.1161/01.CIR.0000093398.16124.29.
4. Baumgartner I, Pieczek A, Manor O, Blair R, Kearney M, Walsh K, et al. Constitutive expression of phVEGF165 after intramuscular gene transfer promotes collateral vessel development in patients with critical limb ischemia. Circulation. 1998;97(12):1114–23.
5. Raval Z, Losordo DW. Cell therapy of peripheral arterial disease: from experimental findings to clinical trials. Circ Res. 2013;112(9):1288–302. doi:10.1161/CIRCRESAHA.113.300565.
6. Lederman RJ, Mendelsohn FO, Anderson RD, Saucedo JF, Tenaglia AN, Hermiller JB, et al. Therapeutic angiogenesis with recombinant fibroblast growth factor-2 for intermittent claudication (the TRAFFIC study): a randomised trial. Lancet. 2002;359(9323):2053–8.
7. Belch J, Hiatt WR, Baumgartner I, Driver IV, Nikol S, Norgren L, et al. Effect of fibroblast growth factor NV1FGF on amputation and death: a randomised placebo-controlled trial of gene therapy in critical limb ischaemia. Lancet. 2011;377(9781):1929–37. doi:10.1016/S0140-6736(11)60394-2.
8. Creager MA, Olin JW, Belch JJF, Moneta GL, Henry TD, Rajagopalan S, et al. Effect of hypoxia-inducible factor-1 alpha gene therapy on walking performance in patients with intermittent claudication. Circulation. 2011;124(16):1765–U148. doi:10.1161/Circulationaha.110.009407.
9. Tateishi-Yuyama E, Matsubara H, Murohara T, Ikeda U, Shintani S, Masaki H, et al. Therapeutic angiogenesis for patients with limb ischaemia by autologous transplantation of bone-marrow cells: a pilot study and a randomised controlled trial. Lancet. 2002;360(9331):427–35. doi:10.1016/S0140-6736(02)09670-8.
10. Huang PP, FengYang X, Li SZ, ChaoWen J, Zhang Y, Han ZC. Randomised comparison of G-CSF-mobilized peripheral blood mononuclear cells versus bone marrow-mononuclear cells for the treatment of patients with lower limb arteriosclerosis obliterans. Thromb Haemost. 2007;98(6):1335–42. doi:10.1160/TH07-02-0137.
11. Lara-Hernandez R, Lozano-Vilardell P, Blanes P, Torreguitart-Mirada N, Galmes A, Besalduch J. Safety and efficacy of therapeutic angiogenesis as a novel treatment in patients with critical limb ischemia. Ann Vasc Surg. 2010;24(2):287–94. doi:10.1016/j.avsg.2009.10.012.
12. Huang PP, Li SZ, Han MZ, Xiao ZJ, Yang RC, Han ZC. Autologous transplantation of granulocyte colony-stimulating factor-mobilized peripheral blood mononuclear cells improves critical limb ischemia in diabetes. Diabetes Care. 2005;28(9):2155–60. doi:10.2337/Diacare.28.9.2155.
13. Cooke JP, Losordo DW. Modulating the vascular response to limb ischemia angiogenic and cell therapies. Circ Res. 2015;116(9):1561–78. doi:10.1161/CIRCRESAHA.115.303565.
14. Kim SW, Kim H, Cho HJ, Lee JU, Levit R, Yoon YS. Human peripheral blood-derived CD31+ cells have robust angiogenic and vasculogenic properties and are effective for treating ischemic vascular disease. J Am Coll Cardiol. 2010;56(7):593–607. doi:10.1016/j.jacc.2010.01.070.
15. Tateno K, Minamino T, Toko H, Akazawa H, Shimizu N, Takeda S, et al. Critical roles of muscle-secreted angiogenic factors in therapeutic neovascularization. Circ Res. 2006;98(9):1194–202. doi:10.1161/01.RES.0000219901.13974.15.
16. Chang MC, Tsao CH, Huang WH, Chen PCH, Hung SC. Conditioned medium derived from mesenchymal stem cells overexpressing HPV16 E6E7 dramatically improves ischemic limb. J Mol Cell Cardiol. 2014;72:339–49. doi:10.1016/j.yjmcc.2014.04.012.
17. Kamata Y, Takahashi Y, Iwamoto M, Matsui K, Murakami Y, Muroi K, et al. Local implantation of autologous mononuclear cells from bone marrow and peripheral blood for treatment of ischaemic digits in patients with connective tissue diseases. Rheumatology. 2007;46(5):882–4. doi:10.1093/rheumatology/kel436.

18. Asahara T, Murohara T, Sullivan A, Silver M, van der Zee R, Li T, et al. Isolation of putative progenitor endothelial cells for angiogenesis. Science. 1997;275(5302):964–7.
19. Takahashi T, Kalka C, Masuda H, Chen D, Silver M, Kearney M, et al. Ischemia- and cytokine-induced mobilization of bone marrow-derived endothelial progenitor cells for neovascularization. Nat Med. 1999;5(4):434–8.
20. Dreger P, Haferlach T, Eckstein V, Jacobs S, Suttorp M, Loffler H, et al. G-Csf-mobilized peripheral-blood progenitor cells for allogeneic transplantation - safety, kinetics of mobilization, and composition of the graft. Br J Haematol. 1994;87(3):609–13. doi:10.1111/J.1365-2141.1994.Tb08321.X.
21. Hur J, Yoon CH, Kim HS, Choi JH, Kang HJ, Hwang KK, et al. Characterization of two types of endothelial progenitor cells and their different contributions to neovasculogenesis. ArterioscIerThrombVascBiol.2004;24(2):288–93.doi:10.1161/01.ATV.0000114236.77009.06.
22. Smadja DM, Bieche I, Silvestre JS, Germain S, Cornet A, Laurendeau I, et al. Bone morphogenetic proteins 2 and 4 are selectively expressed by late outgrowth endothelial progenitor cells and promote neoangiogenesis. Arterioscler Thromb Vasc Biol. 2008;28(12):2137–43. doi:10.1161/ATVBAHA.108.168815.
23. Ishida A, Ohya Y, Sakuda H, Ohshiro K, Higashiuesato Y, Nakaema M, et al. Autologous peripheral blood mononuclear cell implantation for patients with peripheral arterial disease improves limb ischemia. Circulation. 2005;69(10):1260–5. doi:10.1253/Circj.69.1260.
24. Ozturk A, Kucukardali Y, Tangi F, Erikci A, Uzun G, Bashekim C, et al. Therapeutical potential of autologous peripheral blood mononuclear cell transplantation in patients with type 2 diabetic critical limb ischemia. J Diabetes Complicat. 2012;26(1):29–33. doi:10.1016/j.jdiacomp.2011.11.007.
25. Mohammadzadeh L, Samedanifard SH, Keshavarzi A, Alimoghaddam K, Larijani B, Ghavamzadeh A, et al. Therapeutic outcomes of transplanting autologous granulocyte colony-stimulating factor-mobilised peripheral mononuclear cells in diabetic patients with critical limb ischaemia. Exp Clin Endocrinol Diabetes. 2013;121(1):48–53. doi:10.1055/s-0032-1311646.
26. Lenk K, Adams V, Lurz P, Erbs S, Linke A, Gielen S, et al. Therapeutical potential of blood-derived progenitor cells in patients with peripheral arterial occlusive disease and critical limb ischaemia. Eur Heart J. 2005;26(18):1903–9. doi:10.1093/eurheartj/ehi285.
27. Kawamoto A, Katayama M, Handa N, Kinoshita M, Takano H, Horii M, et al. Intramuscular transplantation of G-CSF-mobilized CD34(+) cells in patients with critical limb ischemia: a phase I/IIa, multicenter, single-blinded, dose-escalation clinical trial. Stem Cells. 2009;27(11):2857–64. doi:10.1002/stem.207.
28. Losordo DW, Kibbe MR, Mendelsohn F, Marston W, Driver VR, Sharafuddin M, et al. A randomized, controlled pilot study of autologous CD34+ cell therapy for critical limb ischemia. Circ Cardiovasc Interv. 2012;5(6):821–30. doi:10.1161/Circinterventions.112.968321.
29. Burt RK, Testori A, Oyama Y, Rodriguez HE, Yaung K, Villa M, et al. Autologous peripheral blood CD133+ cell implantation for limb salvage in patients with critical limb ischemia. Bone Marrow Transplant. 2010;45(1):111–6. doi:10.1038/bmt.2009.102.
30. Fan CL, Gao PJ, Che ZQ, Liu JJ, Wei J, Zhu DL. Therapeutic neovascularization by autologous transplantation with expanded endothelial progenitor cells from peripheral blood into ischemic hind limbs. Acta Pharmacol Sin. 2005;26(9):1069–75. doi:10.1111/j.1745-7254.2005.00168.x.
31. Newman PJ, Hillery CA, Albrecht R, Parise LV, Berndt MC, Mazurov AV, et al. Activation-dependent changes in human platelet PECAM-1: phosphorylation, cytoskeletal association, and surface membrane redistribution. J Cell Biol. 1992;119(1):239–46.
32. Baumann CI, Bailey AS, Li WM, Ferkowicz MJ, Yoder MC, Fleming WH. PECAM-1 is expressed on hematopoietic stem cells throughout ontogeny and identifies a population of erythroid progenitors. Blood. 2004;104(4):1010–6. doi:10.1182/blood-2004-03-0989.
33. Behrem S, Zarkovic K, Eskinja N, Jonjic N. Endoglin is a better marker than CD31 in evaluation of angiogenesis in glioblastoma. Croat Med J. 2005;46(3):417–22.
34. Nathan DM. Long-term complications of diabetes mellitus. N Engl J Med. 1993;328(23):1676–85. doi:10.1056/NEJM199306103282306.

35. Dubsky M, Jirkovska A, Bem R, Fejfarova V, Pagacova L, Sixta B, et al. Both autologous bone marrow mononuclear cell and peripheral blood progenitor cell therapies similarly improve ischaemia in patients with diabetic foot in comparison with control treatment. Diabetes Metab Res Rev. 2013;29(5):369–76. doi:10.1002/dmrr.2399.
36. Sun L, Wu L, Qiao Z, Yu J, Li L, Li S, et al. Analysis of possible factors relating to prognosis in autologous peripheral blood mononuclear cell transplantation for critical limb ischemia. Cytotherapy. 2014;16(8):1110–6. doi:10.1016/j.jcyt.2014.03.007.
37. Capla JM, Grogan RH, Callaghan MJ, Galiano RD, Tepper OM, Ceradini DJ, et al. Diabetes impairs endothelial progenitor cell-mediated blood vessel formation in response to hypoxia. Plast Reconstr Surg. 2007;119(1):59–70. doi:10.1097/01.prs.0000244830.16906.3f.
38. Chen YH, Lin SJ, Lin FY, TC W, Tsao CR, Huang PH, et al. High glucose impairs early and late endothelial progenitor cells by modifying nitric oxide-related but not oxidative stress-mediated mechanisms. Diabetes. 2007;56(6):1559–68. doi:10.2337/db06-1103.
39. Shimizu I, Yoshida Y, Suda M, Minamino T. DNA damage response and metabolic disease. Cell Metab. 2014;20(6):967–77. doi:10.1016/j.cmet.2014.10.008.
40. Shimizu I, Yoshida Y, Katsuno T, Tateno K, Okada S, Moriya J, et al. p53-induced adipose tissue inflammation is critically involved in the development of insulin resistance in heart failure. Cell Metab. 2012;15(1):51–64. doi:10.1016/j.cmet.2011.12.006. S1550-4131(11)00465-7 [pii]
41. Minamino T, Orimo M, Shimizu I, Kunieda T, Yokoyama M, Ito T, et al. A crucial role for adipose tissue p53 in the regulation of insulin resistance. Nat Med. 2009;15(9):1082–7. doi:10.1038/nm.2014.
42. Shimizu I, Yoshida Y, Moriya J, Nojima A, Uemura A, Kobayashi Y, et al. Semaphorin3E-induced inflammation contributes to insulin resistance in dietary obesity. Cell Metab. 2013;18(4):491–504. doi:10.1016/j.cmet.2013.09.001.
43. Sano M, Minamino T, Toko H, Miyauchi H, Orimo M, Qin Y, et al. p53-induced inhibition of Hif-1 causes cardiac dysfunction during pressure overload. Nature. 2007;446(7134):444–8. doi:10.1038/nature05602. nature05602 [pii]
44. Imanishi T, Moriwaki C, Hano T, Nishio I. Endothelial progenitor cell senescence is accelerated in both experimental hypertensive rats and patients with essential hypertension. J Hypertens. 2005;23(10):1831–7. doi:10.1097/01.Hjh.0000183524.73746.1b.
45. Michaud SE, Dussault S, Haddad P, Groleau J, Rivard A. Circulating endothelial progenitor cells from healthy smokers exhibit impaired functional activities. Atherosclerosis. 2006;187(2):423–32. doi:10.1016/j.atherosclerosis.2005.10.009.
46. de Nigris F, Balestrieri ML, Williams-Ignarro S, D'Armiento FP, Lerman LO, Byrns R, et al. Therapeutic effects of autologous bone marrow cells and metabolic intervention in the ischemic hindlimb of spontaneously hypertensive rats involve reduced cell senescence and CXCR4/Akt/eNOS pathways. J Cardiovasc Pharmacol. 2007;50(4):424–33. doi:10.1097/FJC.0b013e31812564e4.
47. Napoli C, Williams-Ignarro S, de Nigris F, de Rosa G, Lerman LO, Farzati B, et al. Beneficial effects of concurrent autologous bone marrow cell therapy and metabolic intervention in ischemia-induced angiogenesis in the mouse hindlimb. Proc Natl Acad Sci U S A. 2005;102(47):17202–6. doi:10.1073/pnas.0508534102.
48. Sica V, Williams-Ignarro S, de Nigris F, D'Armiento FP, Lerman LO, Balestrieri ML, et al. Autologous bone marrow cell therapy and metabolic intervention in ischemia-induced angiogenesis in the diabetic mouse hindlimb. Cell Cycle. 2006;5(24):2903–8. doi:10.4161/Cc.5.24.3568.
49. Schirmer SH, Millenaar DN, Werner C, Schuh L, Degen A, Bettink SI, et al. Exercise promotes collateral artery growth mediated by monocytic nitric oxide. Arterioscl Throm Vas. 2015;35(8):1862–71. doi:10.1161/ATVBAHA.115.305806.
50. Ott I, Keller U, Knoedler M, Gotze KS, Doss K, Fischer P, et al. Endothelial-like cells expanded from CD34(+) blood cells improve left ventricular function after experimental myocardial infarction. FASEB J. 2005;19(6):992–4. doi:10.1096/fj.04-3219fje.
51. Arai M, Misao Y, Nagai H, Kawasaki M, Nagashima K, Suzuki K, et al. Granulocyte colony-stimulating factor - a noninvasive regeneration therapy for treating atherosclerotic peripheral artery disease. Circulation. 2006;70(9):1093–8. doi:10.1253/Circj.70.1093.

52. Subramaniyam V, Waller EK, Murrow JR, Manatunga A, Lonial S, Kasirajan K, et al. Bone marrow mobilization with granulocyte macrophage colony-stimulating factor improves endothelial dysfunction and exercise capacity in patients with peripheral arterial disease. Am Heart J. 2009;158(1):53–U5. doi:10.1016/j.ahj.2009.04.014.
53. van Royen N, Schirmer SH, Atasever B, Behrens CYH, Ubbink D, Buschmann EE, et al. START trial - a pilot study on STimulation of ARTeriogenesis using subcutaneous application of granulocyte-macrophage colony-stimulating factor as a new treatment for peripheral vascular disease. Circulation. 2005;112(7):1040–6. doi:10.1161/Circulationaha.104.529552.
54. Poole J, Mavromatis K, Binongo JN, Khan A, Li Q, Khayata M, et al. Effect of progenitor cell mobilization with granulocyte-macrophage colony-stimulating factor in patients with peripheral artery disease: a randomized clinical trial. JAMA. 2013;310(24):2631–9. doi:10.1001/jama.2013.282540.
55. Jujo K, Hamada H, Iwakura A, Thorne T, Sekiguchi H, Clarke T, et al. CXCR4 blockade augments bone marrow progenitor cell recruitment to the neovasculature and reduces mortality after myocardial infarction. Proc Natl Acad Sci U S A. 2010;107(24):11008–13. doi:10.1073/pnas.0914248107.
56. Hattori K, Dias S, Heissig B, Hackett NR, Lyden D, Tateno M, et al. Vascular endothelial growth factor and angiopoietin-1 stimulate postnatal hematopoiesis by recruitment of vasculogenic and hematopoietic stem cells. J Exp Med. 2001;193(9):1005–14. doi:10.1084/Jem.193.9.1005.
57. Bosch-Marce M, Okuyama H, Wesley JB, Sarkar K, Kimura H, Liu YV, et al. Effects of aging and hypoxia-inducible factor-1 activity on angiogenic cell mobilization and recovery of perfusion after limb ischemia. Circ Res. 2007;101(12):1310–8. doi:10.1161/CIRCRESAHA.107.153346.
58. Rafii S, Heissig B, Hattori K. Efficient mobilization and recruitment of marrow-derived endothelial and hematopoietic stem cells by adenoviral vectors expressing angiogenic factors. Gene Ther. 2002;9(10):631–41. doi:10.1038/sj.gt.3301723.
59. Yau TM, Kim C, Li G, Zhang Y, Weisel RD, Li RK. Maximizing ventricular function with multimodal cell-based gene therapy. Circulation. 2005;112(9 Suppl):I123–8. doi:10.1161/CIRCULATIONAHA.104.525147.
60. GH S, Sun YF, YX L, Shuai XX, Liao YH, Liu QY, et al. Hepatocyte growth factor gene-modified bone marrow-derived mesenchymal stem cells transplantation promotes angiogenesis in a rat model of hindlimb ischemia. J Huazhong Univ Sci Technolog Med Sci. 2013;33(4):511–9. doi:10.1007/s11596-013-1151-6.
61. Yin T, He SS, Su C, Chen XC, Zhang DM, Wan Y, et al. Genetically modified human placenta-derived mesenchymal stem cells with FGF-2 and PDGF-BB enhance neovascularization in a model of hindlimb ischemia. Mol Med Rep. 2015;12(4):5093–9. doi:10.3892/mmr.2015.4089.
62. Kawamoto A, Murayama T, Kusano K, Ii M, Tkebuchava T, Shintani S, et al. Synergistic effect of bone marrow mobilization and vascular endothelial growth factor-2 gene therapy in myocardial ischemia. Circulation. 2004;110(11):1398–405. doi:10.1161/01.CIR.0000141563.71410.64.
63. Rowe GC, Raghuram S, Jang C, Nagy JA, Patten IS, Goyal A, et al. PGC-1 alpha induces SPP1 to activate macrophages and orchestrate functional angiogenesis in skeletal muscle. Circ Res. 2014;115(5):504–17. doi:10.1161/CIRCRESAHA.115.303829.
64. Zhu HM, Sun AJ, Zou YZ, Ge JB. Inducible metabolic adaptation promotes mesenchymal stem cell therapy for ischemia a hypoxia-induced and glycogen-based energy prestorage strategy. Arterioscler Thromb Vasc Biol. 2014;34(4):870–6. doi:10.1161/ATVBAHA.114.303194.
65. Kudo T, Hosoyama T, Samura M, Katsura S, Nishimoto A, Kugimiya N, et al. Hypoxic preconditioning reinforces cellular functions of autologous peripheral blood-derived cells in rabbit hindlimb ischemia model. Biochem Biophys Res Commun. 2014;444(3):370–5. doi:10.1016/j.bbrc.2014.01.054.
66. Iida K, Luo H, Hagisawa K, Akima T, Shah PK, Naqvi TZ, et al. Noninvasive low-frequency ultrasound energy causes vasodilation in humans. J Am Coll Cardiol. 2006;48(3):532–7. doi:10.1016/j.jacc.2006.03.046.

67. Atar S, Siegel RJ, Akel R, Ye Y, Lin Y, Modi SA, et al. Ultrasound at 27 kHz increases tissue expression and activity of nitric oxide synthases in acute limb ischemia in rabbits. Ultrasound Med Biol. 2007;33(9):1483–8. doi:10.1016/j.ultrasmedbio.2007.03.008.
68. Belcik JT, Mott BH, Xie A, Zhao Y, Kim S, Lindner NJ et al. Augmentation of limb perfusion and reversal of tissue ischemia produced by ultrasound-mediated microbubble cavitation. Circ Cardiovasc Imaging. 2015;8(4). doi:10.1161/CIRCIMAGING.114.002979.
69. Song J, Qi M, Kaul S, Price RJ. Stimulation of arteriogenesis in skeletal muscle by microbubble destruction with ultrasound. Circulation. 2002;106(12):1550–5.
70. Enomoto S, Yoshiyama M, Omura T, Matsumoto R, Kusuyama T, Nishiya D, et al. Microbubble destruction with ultrasound augments neovascularisation by bone marrow cell transplantation in rat hind limb ischaemia. Heart. 2006;92(4):515–20. doi:10.1136/hrt.2005.064162.
71. Kobulnik J, Kuliszewski MA, Stewart DJ, Lindner JR, Leong-Poi H. Comparison of gene delivery techniques for therapeutic angiogenesis ultrasound-mediated destruction of carrier microbubbles versus direct intramuscular injection. J Am Coll Cardiol. 2009;54(18):1735–42. doi:10.1016/j.jacc.2009.07.023.
72. Brenner W, Aicher A, Eckey T, Massoudi S, Zuhayra M, Koehl U, et al. In-111-labeled CD34+ hematopoietic progenitor cells in a rat myocardial infarction model. J Nucl Med. 2004;45(3):512–8.
73. Aicher A, Brenner W, Zuhayra M, Badorff C, Massoudi S, Assmus B, et al. Assessment of the tissue distribution of transplanted human endothelial progenitor cells by radioactive labeling. Circulation. 2003;107(16):2134–9. doi:10.1161/01.CIR.0000062649.63838.C9.
74. Hofmann M, Wollert KC, Meyer GP, Menke A, Arseniev L, Hertenstein B, et al. Monitoring of bone marrow cell homing into the infarcted human myocardium. Circulation. 2005;111(17):2198–202. doi:10.1161/01.CIR.0000163546.27639.AA.
75. Dedobbeleer C, Blocklet D, Toungouz M, Lambermont M, Unger P, Degaute JP, et al. Myocardial homing and coronary endothelial function after autologous blood CD34(+) progenitor cells intracoronary injection in the chronic phase of myocardial infarction. J Cardiovasc Pharmacol. 2009;53(6):480–5.
76. Tongers J, Webber MJ, Vaughan EE, Sleep E, Renault MA, Roncalli JG, et al. Enhanced potency of cell-based therapy for ischemic tissue repair using an injectable bioactive epitope presenting nanofiber support matrix. J Mol Cell Cardiol. 2014;74:231–9. doi:10.1016/j.yjmcc.2014.05.017.
77. Dash BC, Thomas D, Monaghan M, Carroll O, Chen XZ, Woodhouse K, et al. An injectable elastin-based gene delivery platform for dose-dependent modulation of angiogenesis and inflammation for critical limb ischemia. Biomaterials. 2015;65:126–39. doi:10.1016/j.biomaterials.2015.06.037.

Chapter 4
Endothelial Progenitor Cells for Ischemic Diseases

Takayuki Asahara and Haruchika Masuda

Abstract Endothelial progenitor cells (EPCs) are believed to home to sites of neovascularization, where they contribute to vascular regeneration by forming a structural component of capillaries and by secreting angiogenic factors, thereby enhancing vascular and blood flow recovery in ischemic tissue. This therapeutic strategy has been effective in animal models of ischemia, and we and other clinical trials have demonstrated that it was safe and feasible for treatment of critical ischemic limb and cardiovascular diseases. However, the decline of EPCs in the peripheral blood and evidence that several disease states reduced EPC number and/or function have prompted the development of several strategies to overcome these limitations, including the administration of genetically modified EPCs that overexpress angiogenic growth factors. To optimize therapeutic outcomes, investigators must keep refining methods of EPC purification, expansion, and administration and to develop techniques that overcome the intrinsic decline and phenotypic deficiencies of EPCs. In this chapter, we have illustrated EPC biology and the therapeutic potential of EPCs for vascular regeneration demonstrating our data of clinical study.

Keywords Endothelial progenitor cell • Cell therapy • Ischemia • Angiogenesis • Regeneration

4.1 Introduction

Endothelial progenitor cells (EPCs) have been isolated from adult human peripheral blood (PB) [1], to accumulate in active angiogenic foci and participate in neovascularization following ischemic insults [2, 3], exhibiting common stem/progenitor cell characteristics. The evidence that BM-derived EPCs home to sites of neovascularization differentiating into endothelial cells (ECs) in situ is consistent with

T. Asahara, M.D., Ph.D. (✉) • H. Masuda, M.D., Ph.D.
Department of Regenerative Medicine Science, Tokai University School of Medicine,
143 Shimokasuya, Isehara, Kanagawa 259-1193, Japan
e-mail: asa777@is.icc.u-tokai.ac.jp

Y. Higashi, T. Murohara (eds.), *Therapeutic Angiogenesis*,
DOI 10.1007/978-981-10-2744-4_4

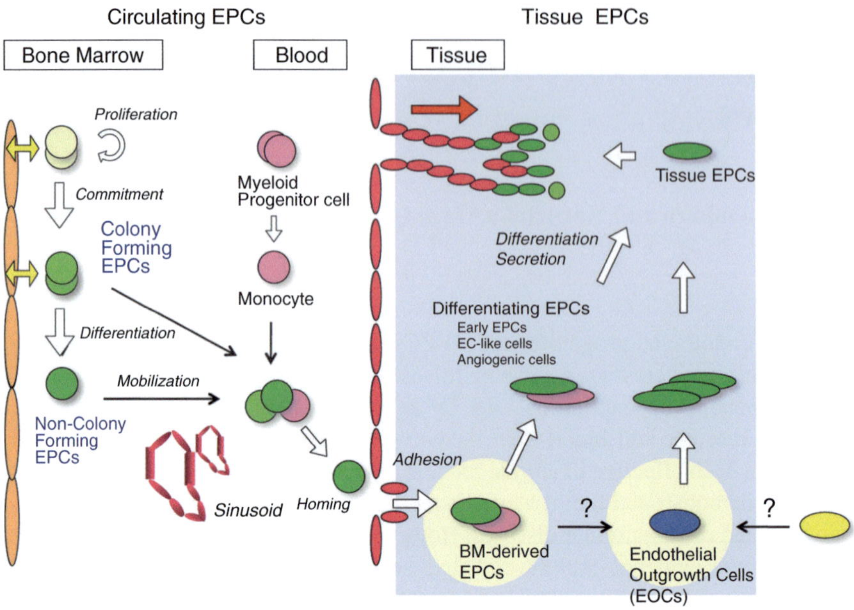

Fig. 4.1 Kinetics of circulating EPCs and tissue EPCs. The relationship among EPCs in the bone marrow (BM), blood, and organ tissues, and their differentiation cascade is represented in the figure. *HPP* high proliferative, *LPP* low proliferative, *ECFC* endothelial colony-forming cell

"vasculogenesis," a critical paradigm well described in embryonic neovascularization but recently proposed in adults in whom a reservoir of stem/progenitor cells contribute to postnatal vascular formation (Fig. 4.1). The discovery of EPCs has therefore drastically changed our understanding of adult blood vessel formation specifically in ischemic tissue. The following issue highlights the potential utility of EPCs for therapeutic angio-/vasculogenesis in ischemic diseases, updating the notion of EPC biology.

4.2 Biological Characteristics and Definition of EPCs

Endothelial progenitor cells or "EPCs" were originally described as blood-bound cells with the ability to differentiate into the endothelial lineage [4]. Believed to be a "progenitor cell," EPCs were thought to be able to reside in their immature state and upon the encounter of appropriate stimuli to migrate, proliferate, or differentiate into a more mature lineage, capable of either direct contribution to or at least support of regenerative processes, namely, the regeneration of the injured cardiovascular system.

EPCs are currently believed to be represented by the following hallmarks: (1) the ability for endothelial lineage commitment and the acquisition of an EC-specific or "EC-equivalent" phenotype; (2) initial immatureness, while preserving the competence to differentiate, indicated by a primitive progenitor cell phenotype and the (partial) lack of mature EC markers; and (3) the presence of pro-angiogenic and vasculogenic properties, with a strong biological activity toward neovascular formation resulting in functional recovery and regeneration of the injured vascular system. Besides these general hallmarks, EPCs can be distinguished and subdivided into various categories.

4.2.1 *Tissue EPCs vs. Circulating EPCs*

Based on their in vivo classification, one can distinguish between "tissue EPCs" and "circulating EPCs." Tissue EPCs are characterized by their adhesive nature and the fact that they can be isolated directly from organ tissues, representing either EPCs in the wake of differentiation originating from the circulation, so-called "homed-down" circulatory EPCs, endothelial outgrowth cells (EOCs) of a yet to be defined origin, or cells of the endothelial lineage which are directly derived from organ-based stem and progenitor cells such as cardiac stem cells [5], neural stem cells [6], myogenic stem cells, or mesenchymal stem cells [7] (Fig. 4.1). On the other hand, circulating EPCs are cellular components of blood which can be isolated from peripheral blood (PB), umbilical cord blood (UCB), bone marrow (BM), and organs or organ blood vessels. Circulating EPCs emerge as floating, nonadhesive cells present in and moving throughout the circulatory system. A "suspended, non-attaching blood cell state" is therefore most characteristic for circulating EPCs which can mobilize and be recruited from preservative and educational niches in the BM into the bloodstream and home to sites of ischemic and/or vascular distress, contributing to the regeneration of the target tissue by transforming into adhesive EPCs.

4.2.2 *Hematopoietic EPCs vs. Non-hematopoietic EPCs*

Circulating EPCs can be subdivided into two main categories, hematopoietic lineage EPCs (h-EPCs) and non-hematopoietic lineage EPCs (nh-EPCs) (Fig. 4.1). The h-EPCs originate from BM and represent a pro-vasculogenic subpopulation of hematopoietic stem cells (HSCs). The h-EPCs can enter circulation upon stimulation as cellular components of blood, compromising a possibly heterogeneous cell population, represented by, for example, colony-forming EPCs, non-colony-forming "differentiating" EPCs, myeloid EPCs or angiogenic cells, etc. The nh-EPCs are not HSC-derived cells, which can be isolated from blood or tissue samples via the help of adhesive cell culture techniques and distinguished by their rather obvious EC-like phenotype. The origin of nh-EPCs remains to be clarified, but they are generally thought to be derived from non-hematopoietic tissue prone lineage stem cells or organ blood vessels.

The h-EPCs can be further subdivided into three distinct classes. The first class is represented by EPCs which can be classified as direct descendant of HSCs, which can form immature hematopoietic-like EPC colonies and commit into circulating EC-like cells. The second class is represented by myeloid cells derived from myeloid progenitors, already committed to the myeloid lineage but still capable to differentiate into EC-like cells, mimicking an endothelial cell phenotype. The third type is represented by cells, loosely termed circulating angiogenic cells (CACs), which can give rise to EC-like cells and contribute to neovascularization mainly by the secretion of pro-angiogenic growth factors. The characterization and identification of HSC-derived EPCs are tightly linked to and associated with the methods and markers already applied in the hematopoietic field. EPCs and HSCs can both be isolated using antibodies against various cell surface markers, including membrane receptors like CD34, CD133, Flk-1/KDR, CXCR4, and CD105 (Endoglin) for human samples [1, 8–13] and receptors like c-Kit [14], Sca-1 [15, 16], and CD34 [14, 17] in combination with Flk-1 (VEGFR2) in case of mouse samples. Nevertheless, the identification of a unique combination of receptors specific and selective for primary EPCs, enabling an unambiguous distinction between EPCs and HSCs, is still missing. The introduction of a definitive assay system capable of clearly distinguishing between EPCs and HSCs, thus enabling the identification of the long sought precise primary EPC phenotype, is highly anticipated but still missing.

4.2.3 Colony-Forming EPCs vs. Non-colony-Forming EPCs

4.2.3.1 Colony-Forming EPCs

A novel recently developed EPC colony-forming assay (EPC-CFA) system, capable to address and overcome most of the abovementioned limitations of the classical assay systems, is challenging several of the predominant classical opinions about EPCs and enabling an until now missing differential hierarchic view on EPCs. We recently reported one of the first examples of such an assay system, initially designed to work with mouse samples. C-Kit$^+$/Sca-1$^+$/lineage-negative (KSL) cells were used as a putative murine hematopoietic EPC-enriched cell population, allowing the identification of two clearly distinguishable types of colonies (small and large colonies) that in turn correspond to two distinct EPC populations, primitive (small) and definitive (large) EPCs, respectively [18–20] (Fig. 4.2). The concept of an EPC-CFA was recently introduced and further developed for analysis of human EPC samples (Masuda et al., unpublished data). The EPC-CFA enables hereby not only the EPC colony formation analysis of single and/or bulk cells from EPC-enriched arbitrary fractions or nonselected cell populations but allows also the cell fate analysis of primary and/or suspension culture cultivated single and/or bulk cells. It can further be easily combined with a classical HPC colony assay system, thus allowing a direct and comprehensive elucidation of the differences and similarities between EPCs and HPCs via the clarification of the cell fate of each cell type. The use of such an EPC-CFA allows not only the elucidation of a possible but so far elusive

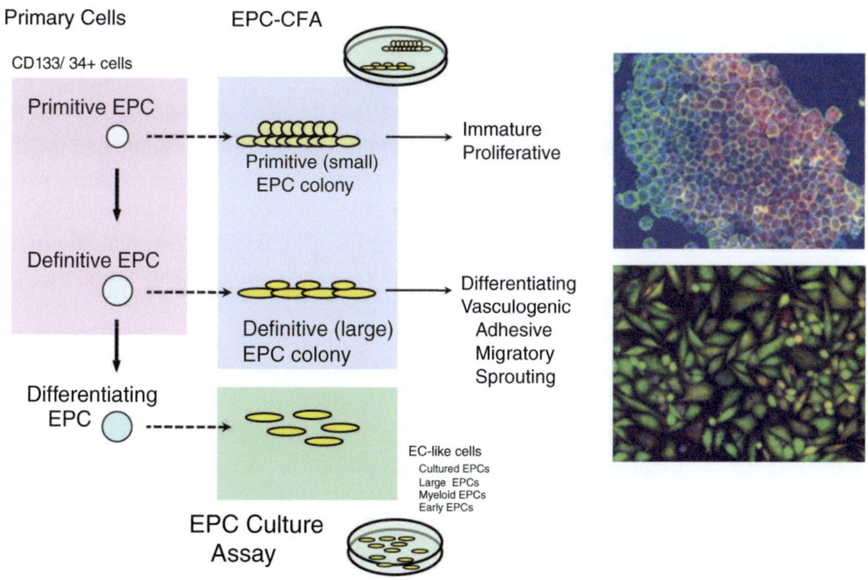

Fig. 4.2 Differentiation cascade of blood EPCs and in vitro EPC assay system. (1) Heterogeneous cell populations including myeloid cell, lymphoid cell, and EPC aggregates are assessed by Hill's colony assay system, (2) relatively purified EPC-rich cell populations including primitive (small) EPCs and definitive (large) EPCs are assessed by EPC-CFA, and (3) small/large EPCs, monocytic EPCs, and angiogenic monocyte/macrophages are assessed by culture EPC assay system

differentiation hierarchy of EPCs but can be further used to identify and characterize the parameters associated with proliferation, commitment, and differentiation of EPCs in vitro and in vivo.

Indeed, the application of EPC-CFA on human CD34+ or CD133+ stem/progenitor cells enabled the identification of small and large distinct colony types each derived from a single cell, small EPCs, and large EPCs, respectively (Fig. 4.2). Small EPCs showed a higher rate of proliferative activity with a higher number of cells being in the S-phase, when compared to large EPCs. Interestingly, large EPCs showed a significantly higher rate of vasculogenic activity with overall increased potential for cell adhesion and tube-like structure formation in vitro as well as a high in vivo de novo blood vessel-forming activity following transplantation of these cells into a murine ischemic hindlimb model, as compared to small EPCs. In contrast to small EPCs, large EPCs did not form secondary colonies but gave rise to isolated endothelial cell (EC)-like cells when reseeded. Due to the observed in vitro (by FACS analysis) and in vivo characteristics of these colony types, small EPCs were further characterized and believed to represent "primitive EPCs," a highly immature and proliferative population of cells, compared to large EPCs which are believed to represent "definitive EPCs," cells prone to differentiate and promote vasculogenesis.

4.2.3.2 Non-colony-Forming EPCs

The widely used "classical EPC culture assay" systems are characterized by the appearance of adhesive endothelial lineage(-like) cells upon conditioning of PB- or BM-derived mononuclear cells with endothelial growth factor-supplemented media [21–23]. These overall reproducible and standardized assay systems were used for the characterization of a wide range of EPCs, ranging from "cultured EPCs" [21, 23–26], "EC-like cells" [27], and "early EPCs" [26, 28, 29] to so-called circulating angiogenic cells [30, 31], which generally do not form colonies under conventional endothelial differentiation conditions (Fig. 4.2).

Cultured EPCs are often called "EC-like cells" due to the expression of certain endothelial features such as (1) the expression of certain endothelial lineage marker genes/proteins, like CD31, Flk-1/KDR, Flt-1, VE-cadherin, Tie-2, and vWF; (2) an EC-like bioactivity, characterized by their capacity to migrate toward an angiogenic growth factor gradient or to support the formation of or incorporate into tube-like structures; and (3) their direct/indirect contribution to the formation of new blood vessels in ischemic tissues after in vivo transplantation. Other characteristics of these cells cover also non-endothelial features like (1) hematopoietic cell marker expression, e.g., CD45 or CD14 up to 2 weeks in culture, (2) loss of EC monolayer formation, and (3) reduction of their proliferative activity in culture similar to cultured human endothelial cells, e.g., HUVEC [21, 24, 26]. The obvious discrepancies between differentiating EPCs and differentiated ECs characterized by a lack of certain endothelial specific markers and properties of EC-like cells and the obvious diminished EPC differentiation capacity into totally differentiated EC phenotype in vitro have been discussed for years and still remain to be clarified.

4.3 EPC Mobilization and Kinetics in Peripheral Blood

As described previously, tissue trauma causes mobilization of hematopoietic cells as well as pluripotent stem or progenitor cells from the hematopoietic system [32]. Consistent with the notion that EPCs and HSCs share common surface angigens, our recent data has shown that mobilization of BM-derived EPCs constitutes a natural response to tissue ischemia. The murine BMT model also provided direct evidence of enhanced BM-derived EPC incorporation into foci of corneal neovascularization following the development of hindlimb ischemia [22], indicating that circulating EPCs are mobilized endogenously in response to tissue ischemia and can incorporate into neovascular foci to promote tissue repair. These results in animals were recently confirmed by human studies illustrating EPC mobilization in patients following burns, CABG, or acute myocardial infarction [33].

In the pathophysiological events that require neovascularization in vivo, a variety of cytokines, growth factors, or hormones released from the jeopardized tissue affect BM remotely and cause EPC mobilization from BM. For instance, granulocyte macrophage colony-stimulating factor (GM-CSF) is well known to stimulate hematopoietic progenitor cells and myeloid lineage cells but has recently been shown to exert a potent stimulatory effect on EPC kinetics. The delivery of this cytokine induced EPC mobilization and enhanced neovascularization in severely ischemic tissues and de novo corneal vascularization [22]. Vascular endothelial growth factor (VEGF), critical for angio-/vasculogenesis in the embryo [34–36], has also been shown to be an important stimulus of adult EPC kinetics recently. Our studies performed first in mice [2] and subsequently in patients undergoing VEGF gene transfer for limb or myocardial ischemia [37] revealed a previously unappreciated mechanism by which VEGF contributes to neovascularization in part by mobilizing BM-derived EPCs. Similar modulation of EPC kinetics has been observed in response to other hematopoietic stimulators such as granulocyte colony-stimulating factor (G-CSF) and stromal cell-derived factor-1 (SDF-1) [38], growth factors such as platelet-derived growth factor-CC (PDGF-CC) [39], brain-derived neurotrophic factor (BDNF) [40], insulin-like growth factor 2 (IGF-2) [41], placental growth factor (PlGF) [42], and hormones such as estrogen [16] and erythropoietin [17] (Fig. 4.1). The distinct mechanism by which EPCs are mobilized to the peripheral circulation remains unknown but may mimic aspects of embryonic development.

EPC mobilization has recently been implicated not only by natural hematopoietic or angiogenic stimulants but also by pharmacological agents. For instance, 3-hydroxy-3-methylglutaryl-coenzyme A (HMG-CoA) reductase inhibitors (statins) are known to rapidly activate Akt signaling in ECs, thereby stimulating EC bioactivity in vitro and enhancing angiogenesis in vivo [43]. Recent studies by Dimmeler et al. and our laboratory have demonstrated a novel function of statins by mobilizing BM-derived EPCs through the stimulation of the Akt signaling pathway [21, 44–46]. Also, emerging evidences indicated that AMD3100 which is an antagonist of CXCR4 receptor for SDF-1 exhibited therapeutic effects on wound healing [47], diabetic neuropathy [48], and cardiovascular diseases [49, 50]. One concrete mechanism for these favorable effects is due to EPC mobilization by blocking SDF-1/CXCR4 axis in EPC retention in BM; however, since high-dose AMD3100 blocks the SDF-1/CXCR4 axis not only in BM but also inhibits circulating or mobilized EPCs' homing capacity to sites of ischemia or target, the therapeutic dose of AMD3100 should be determined carefully before any application for EPC mobilizer.

Therefore, these newly appreciated role of statins and that of AMD3100 suggest that they can be beneficial in treating various forms of vascular diseases expecting endogenous EPC contribution.

4.4 Role of EPCs in Postnatal Neovascularization

Postnatal neovascularization was originally recognized to be constituted by the mechanism of "angiogenesis," which is a new vessel formation, operated by in situ proliferation and migration of preexisting ECs as previously described [51]. However, the discovery of EPCs resulted in the addition of the new mechanism for vascular formation in adults, "vasculogenesis," which is frequently observed during embryogenesis. "Vasculogenesis" is de novo vessel formation by in situ incorporation, differentiation, migration, and/or proliferation of BM-derived EPCs [2]. The incorporation of BM-derived EPCs into foci of physiological and pathological neovascularization has been demonstrated in various animal experiments. One well-established model that allows us to detect BM-derived EPCs utilizes wild-type mice with BM cells transplanted from transgenic mice in which LacZ expresses under the regulation of an EC lineage-specific promoter, Flk-1 or Tie-2 (Flk-1/LacZ/BMT, Tie-2/LacZ/BMT). Using these mice, Flk-1- or Tie-2-expressing endothelial lineage cells derived from BM (EPCs) have been shown to localize to vessels during tumor growth, wound healing, skeletal and cardiac ischemia, corneal neovascularization, and endometrial remodeling following hormone-induced ovulation [2] (Figs. 4.1 and 4.2). On the other hand, tissue-specific stem/progenitor cells with the potency of differentiation into myocytes or ECs were also isolated in skeletal muscle tissue in murine hindlimb later on, although the origin of the cells remains to be cleared [52]. This finding suggests that the origin of EPCs may not be limited to BM, e.g., tissue-specific stem/progenitor cells possibly provide "in situ EPCs" as other sources of EPCs than BM. Regardless of the origin of EPCs, they certainly play a significant role contributing to neovascularization directly via vasculogenesis in the tissue.

Apart from the established role of EPCs in neovascularization, namely, "direct participation in neovasculature via vasculogenesis," recruited EPCs to the jeopardized tissue that requires vessel regeneration do not always participate in the neovasculature but rather stay in interstitial tissue along with neovascularization (Fig. 4.3). These

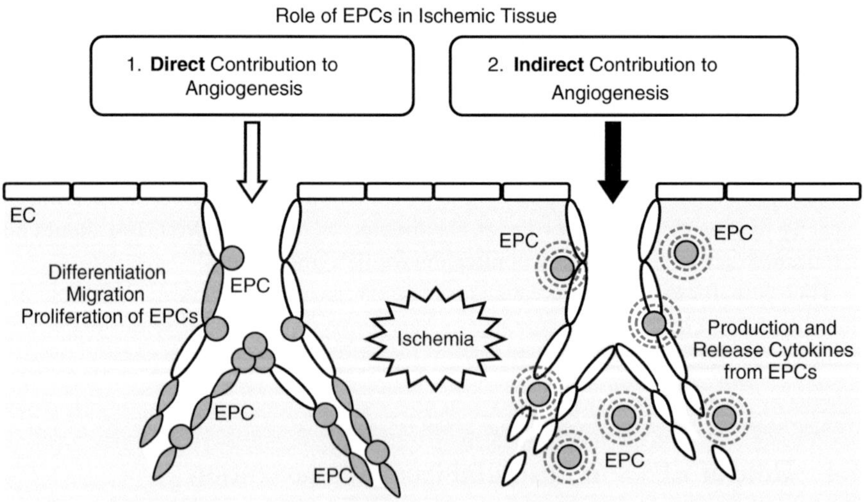

Fig. 4.3 Two different roles of EPCs in neovascularization. In the case of new vessel formation, one role of EPCs is the direct participation of EPCs in neovascularization accompanying preexisting EC proliferation and migration (*left* in the figure). The other role of EPCs is the indirect effect on angiogenesis with the production and release of pro-angiogenic cytokines/growth factors from recruited EPCs. These EPCs remain in the site without participating in the neovasculature, exhibiting so-called paracrine effect (*right* in the figure)

"resting" EPCs in the tissue produce a variety of cytokines/growth factors, specifically pro-angiogenic ones, and promote preexisting EC proliferation and migration resulting in angiogenesis. This paracrine effect of EPCs represents indirect contribution to neovascularization. As far as we and others confirm the cytokines/growth factors produced from EPCs, EPCs will release VEGF, hepatocyte growth factor (HGF), angiopoietin-1 (Ang-1), endothelial nitric oxide synthase (eNOS), inducible nitric oxide synthase (iNOS), SDF-1α, and insulin-like growth factor-1 (IGF-1), etc. Both VEGF and HGF promote EC proliferation leading to angiogenesis, and Ang-1 may play a role for stabilizing prematured vessels in ischemic tissue. Nitric oxide (NO) synthase, by ether eNOS or iNOS, maintains tissue blood perfusion in microcirculating systems acting as a vasodilator. Since little eNOS expression is observed in cardiac capillaries except for ECs in coronary arteries, "imported" eNOS produced from the recruited EPCs is thought to be a major source of eNOS and important in short-term ischemia, specifically ischemia-reperfusion injury [53]. iNOS is also produced from the recruited EPCs; however, the expression is prominent only when the tissue hypoxia is sustained for a long time, i.e., in the case of chronic myocardial ischemia rather than ischemia-reperfusion injury. BM-derived cell eNOS or iNOS deletion results in the exacerbation of myocardial infarction induced by ischemia-reperfusion injury or permanent vessel occlusion, respectively, suggesting that "imported" NOS is crucial for preventing ischemic myocardium depending on the type of ischemic injury [53]. SDF-1α released from recruited EPCs further recruits more EPCs triggering a chain reaction. On the other hand, EPCs will prevent cardiac apoptosis caused by ischemia via a production of IGF-1, a potent antiapoptotic factor, activating the Akt signaling pathway. Thus, EPCs demonstrate tissue-protective effects producing favorable factors, namely, "indirect contribution to neovascularization in ischemic tissue."

4.5 EPC-Based Therapeutic Angiogenesis

Since the discovery of EPCs in 1997, we immediately focused on the regenerative potential of stem/progenitor cells as well as the unique characteristics. In vitro, stem/progenitor cells have the capability of self-renewal and differentiation into organ-specific cell types. In vivo, these cells are then directed by the appropriate milieu that allows them to differentiate and reconstitute target organs. The novel therapeutic strategy for ischemic diseases, EPC transplantation, may therefore be an epoch as a cell therapy involving the classic paradigm of angiogenesis developed by Folkman and colleagues.

4.5.1 EPC Transplantation in Experimental Animals

We and others indicated that cell therapy with culture-expanded EPCs can successfully promote neovascularization in ischemic tissue, even when administered as "sole therapy," i.e., in the absence of angiogenic growth factors. Such a "supply-side" version of therapeutic neovascularization in which the substrate (EPCs/ECs) rather than ligand (growth factor) comprises the therapeutic tool was first reported by intravenously transplanting human EPCs to immunodeficient mice with hindlimb ischemia [24]. These findings provided a novel insight that exogenously administered EPCs restored impaired neovascularization in a mouse ischemic hindlimb model. A similar study in which human EPCs were transplanted in a myocardial ischemia model of nude rat demonstrated that transplanted EPCs were localized to the area of neovascularization with the differentiation into mature ECs. These findings were consistent with preserved left ventricular (LV) function and reduced infraction size [54]. Another study in which human cord blood-derived EPCs were transplanted in an ischemic hindlimb model of nude rats also demonstrated similar findings with enhanced neovascularization in ischemic tissue [25].

Recently, other investigators have explored the therapeutic potential of CD34+ cells as an EPC-enriched fraction. Shatteman et al. transplanted freshly isolated human CD34+ cells into diabetic nude mice with hindlimb ischemia and showed a blood flow recovery in the ischemic limb [55]. Also, Kocher et al. attempted intravenous infusion of freshly isolated human CD34+ cells into nude rats with myocardial ischemia and observed preservation of LV function consistent with the inhibition of cardiac apoptosis [56]. CD34+ cell dose-dependent contribution to LV functional recovery and neovascularization in ischemic myocardium has been demonstrated [57]. The major mechanism for these therapeutic effects is attributed by the paracrine effect rather than direct contribution to neovascularization of CD34+ cells due to its less ability of transdifferentiation to cardiovascular lineage cells. In order to promote the direct contribution of CD34+ cells to therapeutic angiogenesis, we and others focused on one of the morphogens in embryonic stage, sonic hedgehog (SHh), and tested the effect of pretreated CD34+ cells with SHh comparing with non-treated CD34+ cells in mouse ischemic disease models [58, 59]. The SHh treatment significantly upregulated gene expressions of angiogenic growth factor and vascular-related marker in vitro and exhibited high therapeutic outcomes with increased vascularity including endothelial and smooth muscle cell differentiation of the locally transplanted CD34+ cells (Fig. 4.3).

4.5.2 EPC Transplantation in Clinical Trials

Therapeutic effects of EPCs seen in animal models on ischemic diseases, utilizing a broad range of cells which are all believed to consist of or contain to a certain extent EPCs and/or pro-vasculogenic/angiogenic cell populations, are evaluated in numerous clinical trials (Table 4.1). Excellent in-depth reviews summarizing the cells, conditions, as well as the therapeutic outcome, efficacy, and safety of the applied strategies are available [60–62].

We have recently reported a phase I/II clinical trial regarding intramuscular transplantation of autologous and G-CSF-mobilized CD34+ cells in patients with intractable critical limb ischemia (CLI) [63]. The first-in man trial was conducted as

Table 4.1 Clinical trials for ischemic diseases with EPCs

Trial name/ author	Disease type	Number of patients (T/C)	EPC type	Study design	Outcome	Reference
TOPCARE-AMI	AMI	30/29	PB-/BM-derived cultured EPCs	RT	Effective	[85]
Bartunek et al.	AMI	19/16	CD133	RT	Effective	[86]
Li et al.	AMI	35/35	Gm-PB-CD34	Cohort	Effective	[87]
Tatsumi et al.	AMI	36/18	PB-MNC	Cohort	Effective	[88]
Dobert et al.	AMI	11/15	PB-/BM-derived cultured EPCs	Cohort	Effective	[89]
Stamm et al.	RMI	46/9	CD133	NRT	Effective	[90, 91]
Ahmadi et al.	RMI	18/9	CD133	NRT	Effective	[92]
Balogh et al.	RMI	8/18	CD34	NRT	Inconclusive	[93]
Erbs et al.	OMI	13/13	PB-derived cultured EPCs	RT	Effective	[94]
Assmus et al.	OMI	24/23	PB-derived cultured EPCs	RCT	Ineffective	[95]
Boyle et al.	OMI	5/0	Gm-PB-CD34	NRT	NA	[96]
Losordo et al.	AP	18/6	Gm-PB-CD34	RT	Safe and feasible	[61]
Lara-Hernandez et al.	CLI	28/0	Gm-PB-CD34	Cohort	Effective	[97]
EPOCH-CLI	CLI	17/0	Gm-PB-CD34	Cohort	Effective	[63]
Kuroda et al.	NUF	4/0	Gm-PB-CD34	Cohort	Effective	Ongoing in Kobe, Japan

AMI acute myocardial infarction, *OMI* old myocardial infarction, *RMI* recent myocardial infarction, *AP* angina pectoris, *CLI* critical limb ischemia, *NUF* nonunion fracture, *PB* peripheral blood, *BM* bone marrow, *Gm* G-CSF (granulocyte colony-stimulating factor) mobilized, *T/C* treatment/control, *RT* randomized trial, *NRT* nonrandomized trial, *NA* not available

a prospective, multicenter, single-blinded, and dose-escalation study since 2003 in our institute. G-CSF was used to efficiently mobilize BM-EPCs to PB, and the mobilized CD34+ cells were isolated as EPC-enriched fraction.

In all subjects, for the primary endpoint, the efficacy score at week 12 was a positive value indicating improvement of lower limb ischemia after the cell therapy. In addition, both subjective and objective parameters of lower limb ischemia such as toe brachial pressure index (TBPI), transcutaneous partial oxygen pressure ($TcPO_2$), total walking distance (TWD), pain-free walking distance (PFWD), Wong-Baker's pain rating scale, and the ulcer size significantly (Fig. 4.4) and serially improved after transplantation of CD34+ cells. Because this was not a randomized, controlled study, possibility of the placebo effect after CD34+ cell transplantation needs to be evaluated in the large-scale future trial. As for the safety evaluation, neither death

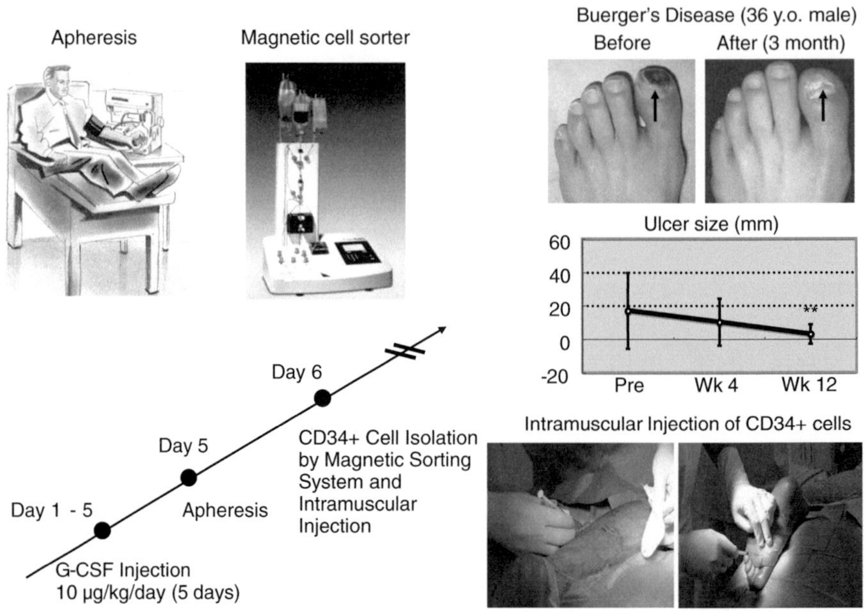

Fig. 4.4 Representative case of autologous CD34+ cell transplantation therapy for CLI in Buerger's disease. A 36-year-old male patient who had toe necrosis due to microcirculation failure received CD34+ cell injection at 40 sites in ischemic limb under lumbar anesthesia, and the necrosis was significantly improved with blood flow recovery with reduced skin ulcer size 3 months after the treatment. Quantitative analysis for skin ulcer size exhibits significant improvement of toe necrosis (*graph*). $^{**}P < 0.01$ vs. Pre (*baseline*). The improvement could be maintained for more than 1 year without recurrence

nor life-threatening adverse events were observed in this study. No severe adverse event, for which relation to a series of cell therapy could not be denied, was also observed. Although mild to moderate adverse events were frequent, these events were transient and expected. No malignant tumor was also clinically identified during the study period.

In addition to CD34, CD133 is a surface marker of early EPC phenotype. A recent clinical study by Burt et al. showed the safety and feasibility of autologous, G-CSF-mobilized CD133+ cell implantation into lower extremity muscles of nine patients with CLI including a patient with Buerger's disease [64]. PB-MNCs were collected by leukapheresis after G-CSF mobilization (10 μg/kg/day for 4–5 days), and CD133+ stem cells were selected using a magnetic separation system. There were no major complications from either leukapheresis or cell injection. The patient with Buerger's disease underwent the procedure twice. After the procedure, rest pain resolved rapidly by day 2, and seven of nine patients including a case of Buerger's disease were able to avoid limb amputation during the 1-year follow-up.

Although these studies were small-sized, nonrandomized trials, these initial results suggest the potential effectiveness of the purified EPC population in CLI patients.

As for the EPC therapy for coronary artery disease, Losordo et al. recently reported a phase II, randomized, placebo-controlled, and dose-ranging clinical trial for 167 patients with refractory angina. Autologous CD34+ cells isolated from G-CSF-mobilized apheresis products were intramyocardially injected into ischemic myocardium under the guidance of NOGA endomyocardial mapping. Six and 12 months later, angina counts and changes in exercise time significantly improved in CD34+ cell group than placebo group [65]. These promising outcomes also support the clinical usefulness of EPC transplantation for reduction of tissue ischemia.

4.5.3 Problems in EPC Transplantation

Our animal studies [24] suggest that heterologous EPC transplantation requires systemic injection of 0.5–2.0 × 10^4 human EPCs/g body weight of the recipient animal to achieve satisfactory improvement of hindlimb ischemia. In general, cultured EPCs obtained from healthy human volunteers yield 5.0 × 10^6 cells per 100 ml of peripheral blood on day 7. Based on these data in human, a blood volume of as much as 12 L will be necessary to obtain enough number of EPCs to treat patients who have critical ischemic hindlimb. Therefore, the background factors in clinical patients such as aging [66], diabetes [23, 67], hypercholesterolemia [23], hypertension [23, 68], and smoking [69, 70] that may reduce the number of circulating/BM-EPCs and the function will cause major limitations of primary EPC transplantation. In reality, most of the patients who are going to undergo EPC therapy for the ischemic diseases more or less have background diseases as described above. Considering autologous EPC therapy, certain technical improvements that may help to overcome the malfunction of EPCs should include (1) local delivery of EPCs, (2) endogenous EPC mobilization, i.e., cytokine/growth factor supplements to promote BM-derived EPC mobilization [2, 22], (3) enrichment procedures, i.e., leukapheresis or BM aspiration, (4) enhancement of EPC functions by gene transduction, or (5) culture expansion of EPCs from self-renewable primitive stem/progenitor cells in BM or other sources. Unless the quality and quantity of autologous EPCs are obtained by the technical improvements as described above, allogenic EPCs derived from umbilical cord blood or culture expanded from human embryonic stem cells [25, 71] may be another alternative source supplying EPCs.

4.6 EPC as a Biomarker for Ischemic Diseases

Previous clinical studies reported that the number of circulating EPCs defined with cell surface markers, CD34+, CD34+/KDR+, CD133+/KDR, or CD34/CD133/KDR+, inversely correlated with the severity of cardiovascular diseases including congestive heart failure [33, 72–76] (Table 4.2). Colony-forming activity of EPCs analyzed by Hill's method [77] has also been known to correlate with

Table 4.2 Correlation between circulating EPC number/function and cardiovascular diseases

Subject disease	Control disease	*N*	EPC type	EPC number/ function	Reference
AMI (day 7 after onset)	AMI (day of onset)	16	CD34+	2X↑/CFA↑	[33]
Non-ST↑AMI (with collaterals)	Non-ST↑AMI (without collaterals)	20	CD133+/KDR+	2X↑/CFA→	[73]
CAD (cardiac event+)	CAD (cardiac event−)	77	CD34+/KDR+	0.5X↓/(−)	[74]
CAD (cardiac event−)	CAD (cardiac event+)	519	CD34+/KDR+	1.5-2X↑/(−)	[76]
Congestive HF	Healthy volunteer	46	CD34+/CD133+/ KDR+	Mild HF: 3-4X↑/ CFA↑ Severe HF: 0.7–0.5X↓/CFA↓	[75]
Eisenmenger syndrome (with pulmonary HT)	Healthy volunteer	96	CD34+, CD34+/ CD133+, KDR+	0.3–0.5X↓/CFA↓	[72]

AMI acute myocardial infarction, *CAD* coronary artery disease, *HF* heart failure, *HT* hypertension, *CFA* colony-forming activity

the number of circulating EPCs and used for the assessment of EPC function; however, Hill's colony assay is recognized as just a method for detecting EPC aggregation.

Thus, we have recently developed a novel EPC colony-forming assay (EPC-CFA) system, which is capable of addressing and overcoming most of the limitations of the classical assay systems, challenging several of the predominant classical opinions about EPCs, and enabling an until now missing differential hierarchic view on EPCs. We have reported one of the first examples of such an assay system, initially designed to work with mouse samples. C-Kit^{+}/Sca-1^{+}/lineage-negative (KSL) cells were used as a putative murine hematopoietic EPC-enriched cell population, allowing the identification of two clearly distinguishable types of colonies (small and large colonies) that in turn correspond to two distinct EPC populations, primitive (small) and definitive (large) EPCs, respectively [18–20] (Fig. 4.3). The concept of an EPC-CFA was recently introduced and further developed for analysis of human EPC samples [78]. The EPC-CFA enables hereby not only the EPC colony formation analysis of single and/or bulk cells from EPC-enriched arbitrary fractions or nonselected cell populations but allows also the cell fate analysis of primary and/or suspension culture cultivated single and/or bulk cells. It can further be easily combined with a classical HPC colony assay system, thus allowing a direct and comprehensive elucidation of the differences and similarities between EPCs and HPCs via the clarification of the cell fate of each cell type. The use of such an EPC-CFA allows not only the elucidation of a possible but so far elusive differentiation hierarchy of EPCs but can be further used to identify and characterize the parameters associated with proliferation, commitment, and differentiation of EPCs in vitro and in vivo [78].

Indeed, application of EPC-CFA on human CD34+ or CD133+ stem/progenitor cells enabled the identification of small and large distinct colony types each derived from a single cell, small EPCs, and large EPCs, respectively. Small EPCs showed a higher rate of proliferative activity with a higher number of cells being in the S-phase, when compared to large EPCs. Interestingly, large EPCs showed a significantly higher rate of vasculogenic activity with overall increased potential for cell adhesion and tube-like structure formation in vitro as well as a high in vivo de novo blood vessel-forming activity following transplantation of these cells into a murine ischemic hindlimb model, as compared to small EPCs. In contrast to small EPCs, large EPCs did not form secondary colonies but gave rise to isolated endothelial cell (EC)-like cells when reseeded. Due to the observed in vitro (by FACS analysis) and in vivo characteristics of these colony types, small EPCs were further characterized and believed to represent "primitive EPCs," a highly immature and proliferative population of cells, compared to large EPCs which are believed to represent "definitive EPCs," cells prone to differentiate and promote vasculogenesis.

The advantage of these assessments for the number and colony-forming activity of circulating EPCs is a convenient tool for clinical application in terms of a medical regulatory feasibility of sampling from blood cells by antibody targeting isolation and a potent effectiveness on ischemic diseases through vasculogenic and angiogenic mechanisms by primary cells.

4.7 Future Strategy with EPC Transplantation

The possible and feasible strategy that may recover potential EPC dysfunction in ischemic disorders should be considered, given the findings that EPC function and mobilization may be impaired in certain diseases. One of the strategies, genetic modification of EPCs to overexpress angiogenic growth factors, will enhance signaling activity of the angiogenic response and reactivate the bioactivity and/or extend the life span of EPCs.

We have recently shown for the first time that gene-modified EPCs rescue impaired neovascularization in an animal model of limb ischemia [79]. Transplantation of heterologous EPCs transduced with adenovirus encoding human VEGF165 improved neovascularization and blood flow recovery, reducing the limb necrosis and auto-amputation rate in comparison with controls. The dose of EPCs needed to achieve limb salvage in these in vivo experiments was 30 times less than that required in the previous experiments involving unmodified EPCs [24]. Other investigators have also demonstrated the therapeutic efficacy of genetically engineered EPCs with a variety of target genes such as adrenomedullin (AM) [80], eNOS [81], tissue plasminogen activator (tPA) [82], and integrin-like kinase (ILK) [83] in animal models. Thus, genetic modification might overcome the potential problems in the patients' EPCs for EPC transplantation therapy in ischemic diseases as so-called second-generation EPC therapy. Also, combining EPC cell therapy with gene (i.e., VEGF) therapy [84] may be another option to address the limited number and function of EPCs that can be isolated from peripheral blood in patients.

4.8 Summary

There is accumulating evidence that BM-derived EPCs have characteristics similar to those of angioblasts demonstrating the potential to promote postnatal vasculogenesis in adults, and clinical applications of EPCs in regenerative medicine are now ongoing. To acquire optimal quality and quantity of EPCs, however, several issues remain to be addressed, such as the development of a more efficient method of EPC purification and expansion, the methods of administration, and background disease-induced dysfunction or senescence in EPCs in patients. Alternatively, in the case of impossible utility of autologous BM-derived EPCs in patients with impaired BM function, appreciable EPCs isolated from umbilical cord blood or differentiated from tissue-specific stem/progenitor or embryonic stem cells need to be optimized for EPC therapy. However, the unlimited potential of EPCs along with the emerging concepts of autologous cell therapy with gene modification suggests that they may soon reach clinical fruition.

References

1. Asahara T, Murohara T, Sullivan A, Silver M, van der Zee R, Li T, Witzenbichler B, Schatteman G, Isner JM. Isolation of putative progenitor endothelial cells for angiogenesis. Science. 1997;275:964–7.
2. Asahara T, Masuda H, Takahashi T, Kalka C, Pastore C, Silver M, Kearne M, Magner M, Isner JM. Bone marrow origin of endothelial progenitor cells responsible for postnatal vasculogenesis in physiological and pathological neovascularization. Circ Res. 1999;85:221–8.
3. Asahara T, Takahashi T, Masuda H, Kalka C, Chen D, Iwaguro H, Inai Y, Silver M, Isner JM. VEGF contributes to postnatal neovascularization by mobilizing bone marrow-derived endothelial progenitor cells. EMBO J. 1999;18:3964–72.
4. Lyden D, Hattori K, Dias S, Costa C, Blaikie P, Butros L, Chadburn A, Heissig B, Marks W, Witte L, Wu Y, Hicklin D, Zhu Z, Hackett NR, Crystal RG, Moore MA, Hajjar KA, Manova K, Benezra R, Rafii S. Impaired recruitment of bone-marrow-derived endothelial and hematopoietic precursor cells blocks tumor angiogenesis and growth. Nat Med. 2001;7:1194–201.
5. Beltrami AP, Barlucchi L, Torella D, Baker M, Limana F, Chimenti S, Kasahara H, Rota M, Musso E, Urbanek K, Leri A, Kajstura J, Nadal-Ginard B, Anversa P. Adult cardiac stem cells are multipotent and support myocardial regeneration. Cell. 2003;114:763–76.
6. Ii M, Nishimura H, Sekiguchi H, Kamei N, Yokoyama A, Horii M, Asahara T. Concurrent vasculogenesis and neurogenesis from adult neural stem cells. Circ Res. 2009;105:860–8.

7. Kovacic JC, Boehm M. Resident vascular progenitor cells: an emerging role for non-terminally differentiated vessel-resident cells in vascular biology. Stem Cell Res. 2009;2:2–15.
8. Quirici N, Soligo D, Caneva L, Servida F, Bossolasco P, Deliliers GL. Differentiation and expansion of endothelial cells from human bone marrow CD133(+) cells. Br J Haematol. 2001;115:186–94.
9. Peichev M, Naiyer AJ, Pereira D, Zhu Z, Lane WJ, Williams M, Oz MC, Hicklin DJ, Witte L, Moore MA, Rafii S. Expression of VEGFR-2 and AC133 by circulating human CD34(+) cells identifies a population of functional endothelial precursors. Blood. 2000;95:952–8.
10. Handgretinger R, Gordon PR, Leimig T, Chen X, Buhring HJ, Niethammer D, Kuci S. Biology and plasticity of CD133+ hematopoietic stem cells. Ann N Y Acad Sci. 2003;996:141–51.
11. Gehling UM, Ergun S, Schumacher U, Wagener C, Pantel K, Otte M, Schuch G, Schafhausen P, Mende T, Kilic N, Kluge K, Schafer B, Hossfeld DK, Fiedler W. In vitro differentiation of endothelial cells from AC133-positive progenitor cells. Blood. 2000;95:3106–12.
12. Friedrich EB, Walenta K, Scharlau J, Nickenig G, Werner N. CD34−/CD133+/VEGFR-2+ endothelial progenitor cell subpopulation with potent vasoregenerative capacities. Circ Res. 2006;98:e20–5.
13. Eggermann J, Kliche S, Jarmy G, Hoffmann K, Mayr-Beyrle U, Debatin KM, Waltenberger J, Beltinger C. Endothelial progenitor cell culture and differentiation in vitro: a methodological comparison using human umbilical cord blood. Cardiovasc Res. 2003;58:478–86.
14. Patschan D, Krupincza K, Patschan S, Zhang Z, Hamby C, Goligorsky MS. Dynamics of mobilization and homing of endothelial progenitor cells after acute renal ischemia: modulation by ischemic preconditioning. Am J Physiol Renal Physiol. 2006;291:F176–85.
15. Hamada H, Kim MK, Iwakura A, Ii M, Thorne T, Qin G, Asai J, Tsutsumi Y, Sekiguchi H, Silver M, Wecker A, Bord E, Zhu Y, Kishore R, Losordo DW. Estrogen receptors alpha and beta mediate contribution of bone marrow-derived endothelial progenitor cells to functional recovery after myocardial infarction. Circulation. 2006;114:2261–70.
16. Iwakura A, Luedemann C, Shastry S, Hanley A, Kearney M, Aikawa R, Isner JM, Asahara T, Losordo DW. Estrogen-mediated, endothelial nitric oxide synthase-dependent mobilization of bone marrow-derived endothelial progenitor cells contributes to reendothelialization after arterial injury. Circulation. 2003;108:3115–21.
17. Heeschen C, Aicher A, Lehmann R, Fichtlscherer S, Vasa M, Urbich C, Mildner-Rihm C, Martin H, Zeiher AM, Dimmeler S. Erythropoietin is a potent physiologic stimulus for endothelial progenitor cell mobilization. Blood. 2003;102:1340–6.
18. Kwon SM, Eguchi M, Wada M, Iwami Y, Hozumi K, Iwaguro H, Masuda H, Kawamoto A, Asahara T. Specific Jagged-1 signal from bone marrow microenvironment is required for endothelial progenitor cell development for neovascularization. Circulation. 2008;118:157–65.
19. Tanaka K, Sata M. Contribution of circulating vascular progenitors in lesion formation and vascular healing: lessons from animal models. Curr Opin Lipidol. 2008;19:498–504.
20. Kamei N, Kwon SM, Alev C, Ishikawa M, Yokoyama A, Nakanishi K, Yamada K, Horii M, Nishimura H, Takaki S, Kawamoto A, Ii M, Akimaru H, Tanaka N, Nishikawa S, Ochi M, Asahara T. Lnk deletion reinforces the function of bone marrow progenitors in promoting neovascularization and astrogliosis following spinal cord injury. Stem Cells. 2010;28:365–75.
21. Dimmeler S, Aicher A, Vasa M, Mildner-Rihm C, Adler K, Tiemann M, Rutten H, Fichtlscherer S, Martin H, Zeiher AM. HMG-CoA reductase inhibitors (statins) increase endothelial progenitor cells via the PI 3-kinase/Akt pathway. J Clin Invest. 2001;108:391–7.
22. Takahashi T, Kalka C, Masuda H, Chen D, Silver M, Kearney M, Magner M, Isner JM, Asahara T. Ischemia- and cytokine-induced mobilization of bone marrow-derived endothelial progenitor cells for neovascularization. Nat Med. 1999;5:434–8.
23. Vasa M, Fichtlscherer S, Aicher A, Adler K, Urbich C, Martin H, Zeiher AM, Dimmeler S. Number and migratory activity of circulating endothelial progenitor cells inversely correlate with risk factors for coronary artery disease. Circ Res. 2001;89:E1–7.

24. Kalka C, Masuda H, Takahashi T, Kalka-Moll WM, Silver M, Kearney M, Li T, Isner JM, Asahara T. Transplantation of ex vivo expanded endothelial progenitor cells for therapeutic neovascularization. Proc Natl Acad Sci U S A. 2000;97:3422–7.
25. Murohara T, Ikeda H, Duan J, Shintani S, Sasaki K, Eguchi H, Onitsuka I, Matsui K, Imaizumi T. Transplanted cord blood-derived endothelial precursor cells augment postnatal neovascularization. J Clin Invest. 2000;105:1527–36.
26. Sharpe 3rd EE, Teleron AA, Li B, Price J, Sands MS, Alford K, Young PP. The origin and in vivo significance of murine and human culture-expanded endothelial progenitor cells. Am J Pathol. 2006;168:1710–21.
27. Timmermans F, Plum J, Yoder MC, Ingram DA, Vandekerckhove B, Case J. Endothelial progenitor cells: identity defined? J Cell Mol Med. 2009;13:87–102.
28. Gulati R, Jevremovic D, Peterson TE, Chatterjee S, Shah V, Vile RG, Simari RD. Diverse origin and function of cells with endothelial phenotype obtained from adult human blood. Circ Res. 2003;93:1023–5.
29. Hur J, Yang HM, Yoon CH, Lee CS, Park KW, Kim JH, Kim TY, Kim JY, Kang HJ, Chae IH, Oh BH, Park YB, Kim HS. Identification of a novel role of T cells in postnatal vasculogenesis: characterization of endothelial progenitor cell colonies. Circulation. 2007;116:1671–82.
30. Rehman J, Li J, Orschell CM, March KL. Peripheral blood "endothelial progenitor cells" are derived from monocyte/macrophages and secrete angiogenic growth factors. Circulation. 2003;107:1164–9.
31. Shepherd RM, Capoccia BJ, Devine SM, Dipersio J, Trinkaus KM, Ingram D, Link DC. Angiogenic cells can be rapidly mobilized and efficiently harvested from the blood following treatment with AMD3100. Blood. 2006;108:3662–7.
32. Grzelak I, Olszewski WL, Zaleska M, Ziolkowska A, Durlik M, Lagiewska B, Muszynski M, Rowinski W. Surgical trauma evokes a rise in the frequency of hematopoietic progenitor cells and cytokine levels in blood circulation. Eur Surg Res. 1998;30:198–204.
33. Shintani S, Murohara T, Ikeda H, Ueno T, Honma T, Katoh A, Sasaki K, Shimada T, Oike Y, Imaizumi T. Mobilization of endothelial progenitor cells in patients with acute myocardial infarction. Circulation. 2001;103:2776–9.
34. Ferrara N, Carver-Moore K, Chen H, Dowd M, Lu L, O'Shea KS, Powell-Braxton L, Hillan KJ, Moore MW. Heterozygous embryonic lethality induced by targeted inactivation of the VEGF gene. Nature. 1996;380:439–42.
35. Carmeliet P, Ferreira V, Breier G, Pollefeyt S, Kieckens L, Gertsenstein M, Fahrig M, Vandenhoeck A, Harpal K, Eberhardt C, Declercq C, Pawling J, Moons L, Collen D, Risau W, Nagy A. Abnormal blood vessel development and lethality in embryos lacking a single VEGF allele. Nature. 1996;380:435–9.
36. Shalaby F, Rossant J, Yamaguchi TP, Gertsenstein M, Wu XF, Breitman ML, Schuh AC. Failure of blood-island formation and vasculogenesis in Flk-1-deficient mice. Nature. 1995;376:62–6.
37. Kalka C, Masuda H, Takahashi T, Gordon R, Tepper O, Gravereaux E, Pieczek A, Iwaguro H, Hayashi SI, Isner JM, Asahara T. Vascular endothelial growth factor(165) gene transfer augments circulating endothelial progenitor cells in human subjects. Circ Res. 2000;86:1198–202.
38. Moore MA, Hattori K, Heissig B, Shieh JH, Dias S, Crystal RG, Rafii S. Mobilization of endothelial and hematopoietic stem and progenitor cells by adenovector-mediated elevation of serum levels of SDF-1, VEGF, and angiopoietin-1. Ann N Y Acad Sci. 2001;938:36–45. discussion 45–7
39. Li X, Tjwa M, Moons L, Fons P, Noel A, Ny A, Zhou JM, Lennartsson J, Li H, Luttun A, Ponten A, Devy L, Bouche A, Oh H, Manderveld A, Blacher S, Communi D, Savi P, Bono F, Dewerchin M, Foidart JM, Autiero M, Herbert JM, Collen D, Heldin CH, Eriksson U, Carmeliet P. Revascularization of ischemic tissues by PDGF-CC via effects on endothelial cells and their progenitors. J Clin Invest. 2005;115:118–27.
40. Kermani P, Rafii D, Jin DK, Whitlock P, Schaffer W, Chiang A, Vincent L, Friedrich M, Shido K, Hackett NR, Crystal RG, Rafii S, Hempstead BL. Neurotrophins promote revascularization

by local recruitment of TrkB+ endothelial cells and systemic mobilization of hematopoietic progenitors. J Clin Invest. 2005;115:653–63.
41. Maeng YS, Choi HJ, Kwon JY, Park YW, Choi KS, Min JK, Kim YH, Suh PG, Kang KS, Won MH, Kim YM, Kwon YG. Endothelial progenitor cell homing: prominent role of the IGF2-IGF2R-PLCbeta2 axis. Blood. 2009;113:233–43.
42. Hattori K, Heissig B, Wu Y, Dias S, Tejada R, Ferris B, Hicklin DJ, Zhu Z, Bohlen P, Witte L, Hendrikx J, Hackett NR, Crystal RG, Moore MA, Werb Z, Lyden D, Rafii S. Placental growth factor reconstitutes hematopoiesis by recruiting VEGFR1(+) stem cells from bone-marrow microenvironment. Nat Med. 2002;8:841–9.
43. Kureishi Y, Luo Z, Shiojima I, Bialik A, Fulton D, Lefer DJ, Sessa WC, Walsh K. The HMG-CoA reductase inhibitor simvastatin activates the protein kinase Akt and promotes angiogenesis in normocholesterolemic animals. Nat Med. 2000;6:1004–10.
44. Vasa M, Fichtlscherer S, Adler K, Aicher A, Martin H, Zeiher AM, Dimmeler S. Increase in circulating endothelial progenitor cells by statin therapy in patients with stable coronary artery disease. Circulation. 2001;103:2885–90.
45. Urbich C, Dernbach E, Zeiher AM, Dimmeler S. Double-edged role of statins in angiogenesis signaling. Circ Res. 2002;90:737–44.
46. Llevadot J, Murasawa S, Kureishi Y, Uchida S, Masuda H, Kawamoto A, Walsh K, Isner JM, Asahara T. HMG-CoA reductase inhibitor mobilizes bone marrow--derived endothelial progenitor cells. J Clin Invest. 2001;108:399–405.
47. Nishimura Y, Ii M, Qin G, Hamada H, Asai J, Takenaka H, Sekiguchi H, Renault MA, Jujo K, Katoh N, Kishimoto S, Ito A, Kamide C, Kenny J, Millay M, Misener S, Thorne T, Losordo DW. CXCR4 antagonist AMD3100 accelerates impaired wound healing in diabetic mice. J Invest Dermatol. 2012;132:711–20.
48. Kim BJ, Lee JK, Schuchman EH, Jin HK, Bae JS. Synergistic vasculogenic effects of AMD3100 and stromal-cell-derived factor-1alpha in vasa nervorum of the sciatic nerve of mice with diabetic peripheral neuropathy. Cell Tissue Res. 2013;354:395–407.
49. Jujo K, Ii M, Sekiguchi H, Klyachko E, Misener S, Tanaka T, Tongers J, Roncalli J, Renault MA, Thorne T, Ito A, Clarke T, Kamide C, Tsurumi Y, Hagiwara N, Qin G, Asahi M, Losordo DW. CXC-chemokine receptor 4 antagonist AMD3100 promotes cardiac functional recovery after ischemia/reperfusion injury via endothelial nitric oxide synthase-dependent mechanism. Circulation. 2013;127:63–73.
50. Jujo K, Hamada H, Iwakura A, Thorne T, Sekiguchi H, Clarke T, Ito A, Misener S, Tanaka T, Klyachko E, Kobayashi K, Tongers J, Roncalli J, Tsurumi Y, Hagiwara N, Losordo DW. CXCR4 blockade augments bone marrow progenitor cell recruitment to the neovasculature and reduces mortality after myocardial infarction. Proc Natl Acad Sci U S A. 2010;107:11008–13.
51. Folkman J, Shing Y. Angiogenesis. J Biol Chem. 1992;267:10931–4.
52. Tamaki T, Akatsuka A, Ando K, Nakamura Y, Matsuzawa H, Hotta T, Roy RR, Edgerton VR. Identification of myogenic-endothelial progenitor cells in the interstitial spaces of skeletal muscle. J Cell Biol. 2002;157:571–7.
53. Ii M, Nishimura H, Iwakura A, Wecker A, Eaton E, Asahara T, Losordo DW. Endothelial progenitor cells are rapidly recruited to myocardium and mediate protective effect of ischemic preconditioning via "imported" nitric oxide synthase activity. Circulation. 2005;111:1114–20.
54. Kawamoto A, Gwon HC, Iwaguro H, Yamaguchi JI, Uchida S, Masuda H, Silver M, Ma H, Kearney M, Isner JM, Asahara T. Therapeutic potential of ex vivo expanded endothelial progenitor cells for myocardial ischemia. Circulation. 2001;103:634–7.
55. Schatteman GC, Hanlon HD, Jiao C, Dodds SG, Christy BA. Blood-derived angioblasts accelerate blood-flow restoration in diabetic mice. J Clin Invest. 2000;106:571–8.
56. Kocher AA, Schuster MD, Szabolcs MJ, Takuma S, Burkhoff D, Wang J, Homma S, Edwards NM, Itescu S. Neovascularization of ischemic myocardium by human bone-marrow-derived angioblasts prevents cardiomyocyte apoptosis, reduces remodeling and improves cardiac function. Nat Med. 2001;7:430–6.

57. Iwasaki H, Kawamoto A, Ishikawa M, Oyamada A, Nakamori S, Nishimura H, Sadamoto K, Horii M, Matsumoto T, Murasawa S, Shibata T, Suehiro S, Asahara T. Dose-dependent contribution of CD34-positive cell transplantation to concurrent vasculogenesis and cardiomyogenesis for functional regenerative recovery after myocardial infarction. Circulation. 2006;113:1311–25.
58. Kanaya K, Masaaki I, Okazaki T, Nakamura T, Horii-Komatsu M, Alev C, Akimaru H, Kawamoto A, Akashi H, Tanaka H, Asahi M, Asahara T. Sonic hedgehog signaling regulates vascular differentiation and function in human CD34 positive cells: vasculogenic CD34(+) cells with sonic hedgehog. Stem Cell Res. 2015;14:165–76.
59. Mackie AR, Klyachko E, Thorne T, Schultz KM, Millay M, Ito A, Kamide CE, Liu T, Gupta R, Sahoo S, Misener S, Kishore R, Losordo DW. Sonic hedgehog-modified human CD34+ cells preserve cardiac function after acute myocardial infarction. Circ Res. 2012;111:312–21.
60. Losordo DW, Kishore R. A big promise from the very small identification of circulating embryonic stem-like pluripotent cells in patients with acute myocardial infarction. J Am Coll Cardiol. 2009;53:10–2.
61. Losordo DW, Schatz RA, White CJ, Udelson JE, Veereshwarayya V, Durgin M, Poh KK, Weinstein R, Kearney M, Chaudhry M, Burg A, Eaton L, Heyd L, Thorne T, Shturman L, Hoffmeister P, Story K, Zak V, Dowling D, Traverse JH, Olson RE, Flanagan J, Sodano D, Murayama T, Kawamoto A, Kusano KF, Wollins J, Welt F, Shah P, Soukas P, Asahara T, Henry TD. Intramyocardial transplantation of autologous CD34+ stem cells for intractable angina: a phase I/IIa double-blind, randomized controlled trial. Circulation. 2007;115:3165–72.
62. Jujo K, Ii M, Losordo DW. Endothelial progenitor cells in neovascularization of infarcted myocardium. J Mol Cell Cardiol. 2008;45:530–44.
63. Kawamoto A, Katayama M, Handa N, Kinoshita M, Takano H, Horii M, Sadamoto K, Yokoyama A, Yamanaka T, Onodera R, Kuroda A, Baba R, Kaneko Y, Tsukie T, Kurimoto Y, Okada Y, Kihara Y, Morioka S, Fukushima M, Asahara T. Intramuscular transplantation of G-CSF-mobilized CD34(+) cells in patients with critical limb ischemia: a phase I/IIa, multicenter, single-blinded, dose-escalation clinical trial. Stem Cells. 2009;27:2857–64.
64. Burt RK, Testori A, Oyama Y, Rodriguez HE, Yaung K, Villa M, Bucha JM, Milanetti F, Sheehan J, Rajamannan N, Pearce WH. Autologous peripheral blood CD133+ cell implantation for limb salvage in patients with critical limb ischemia. Bone Marrow Transplant. 2010;45:111–6.
65. Losordo DW, Henry TD, Davidson C, Sup Lee J, Costa MA, Bass T, Mendelsohn F, Fortuin FD, Pepine CJ, Traverse JH, Amrani D, Ewenstein BM, Riedel N, Story K, Barker K, Povsic TJ, Harrington RA, Schatz RA. Intramyocardial, autologous CD34+ cell therapy for refractory angina. Circ Res. 2011;109:428–36.
66. Heiss C, Keymel S, Niesler U, Ziemann J, Kelm M, Kalka C. Impaired progenitor cell activity in age-related endothelial dysfunction. J Am Coll Cardiol. 2005;45:1441–8.
67. Ii M, Takenaka H, Asai J, Ibusuki K, Mizukami Y, Maruyama K, Yoon YS, Wecker A, Luedemann C, Eaton E, Silver M, Thorne T, Losordo DW. Endothelial progenitor thrombospondin-1 mediates diabetes-induced delay in reendothelialization following arterial injury. Circ Res. 2006;98:697–704.
68. Imanishi T, Moriwaki C, Hano T, Nishio I. Endothelial progenitor cell senescence is accelerated in both experimental hypertensive rats and patients with essential hypertension. J Hypertens. 2005;23:1831–7.
69. Kondo T, Hayashi M, Takeshita K, Numaguchi Y, Kobayashi K, Iino S, Inden Y, Murohara T. Smoking cessation rapidly increases circulating progenitor cells in peripheral blood in chronic smokers. Arterioscler Thromb Vasc Biol. 2004;24:1442–7.
70. Michaud SE, Dussault S, Haddad P, Groleau J, Rivard A. Circulating endothelial progenitor cells from healthy smokers exhibit impaired functional activities. Atherosclerosis. 2006;187:423–32.
71. Levenberg S, Golub JS, Amit M, Itskovitz-Eldor J, Langer R. Endothelial cells derived from human embryonic stem cells. Proc Natl Acad Sci U S A. 2002;99:4391–6.

72. Diller GP, van Eijl S, Okonko DO, Howard LS, Ali O, Thum T, Wort SJ, Bedard E, Gibbs JS, Bauersachs J, Hobbs AJ, Wilkins MR, Gatzoulis MA, Wharton J. Circulating endothelial progenitor cells in patients with Eisenmenger syndrome and idiopathic pulmonary arterial hypertension. Circulation. 2008;117:3020–30.
73. Lev EI, Kleiman NS, Birnbaum Y, Harris D, Korbling M, Estrov Z. Circulating endothelial progenitor cells and coronary collaterals in patients with non-ST segment elevation myocardial infarction. J Vasc Res. 2005;42:408–14.
74. Schmidt-Lucke C, Rossig L, Fichtlscherer S, Vasa M, Britten M, Kamper U, Dimmeler S, Zeiher AM. Reduced number of circulating endothelial progenitor cells predicts future cardiovascular events: proof of concept for the clinical importance of endogenous vascular repair. Circulation. 2005;111:2981–7.
75. Valgimigli M, Rigolin GM, Fucili A, Porta MD, Soukhomovskaia O, Malagutti P, Bugli AM, Bragotti LZ, Francolini G, Mauro E, Castoldi G, Ferrari R. CD34+ and endothelial progenitor cells in patients with various degrees of congestive heart failure. Circulation. 2004;110: 1209–12.
76. Werner N, Kosiol S, Schiegl T, Ahlers P, Walenta K, Link A, Bohm M, Nickenig G. Circulating endothelial progenitor cells and cardiovascular outcomes. N Engl J Med. 2005;353: 999–1007.
77. Hill JM, Zalos G, Halcox JP, Schenke WH, Waclawiw MA, Quyyumi AA, Finkel T. Circulating endothelial progenitor cells, vascular function, and cardiovascular risk. N Engl J Med. 2003;348:593–600.
78. Asahara T, Kawamoto A, Masuda H. Concise review: circulating endothelial progenitor cells for vascular medicine. Stem Cells. 2011;29:1650–5.
79. Iwaguro H, Yamaguchi J, Kalka C, Murasawa S, Masuda H, Hayashi S, Silver M, Li T, Isner JM, Asahara T. Endothelial progenitor cell vascular endothelial growth factor gene transfer for vascular regeneration. Circulation. 2002;105:732–8.
80. Nagaya N, Kangawa K, Kanda M, Uematsu M, Horio T, Fukuyama N, Hino J, Harada-Shiba M, Okumura H, Tabata Y, Mochizuki N, Chiba Y, Nishioka K, Miyatake K, Asahara T, Hara H, Mori H. Hybrid cell-gene therapy for pulmonary hypertension based on phagocytosing action of endothelial progenitor cells. Circulation. 2003;108:889–95.
81. Kong D, Melo LG, Gnecchi M, Zhang L, Mostoslavsky G, Liew CC, Pratt RE, Dzau VJ. Cytokine-induced mobilization of circulating endothelial progenitor cells enhances repair of injured arteries. Circulation. 2004;110:2039–46.
82. Griese DP, Achatz S, Batzlsperger CA, Strauch UG, Grumbeck B, Weil J, Riegger GA. Vascular gene delivery of anticoagulants by transplantation of retrovirally-transduced endothelial progenitor cells. Cardiovasc Res. 2003;58:469–77.
83. Cho HJ, Youn SW, Cheon SI, Kim TY, Hur J, Zhang SY, Lee SP, Park KW, Lee MM, Choi YS, Park YB, Kim HS. Regulation of endothelial cell and endothelial progenitor cell survival and vasculogenesis by integrin-linked kinase. Arterioscler Thromb Vasc Biol. 2005;25:1154–60.
84. Kawamoto A, Murayama T, Kusano K, Ii M, Tkebuchava T, Shintani S, Iwakura A, Johnson I, von Samson P, Hanley A, Gavin M, Curry C, Silver M, Ma H, Kearney M, Losordo DW. Synergistic effect of bone marrow mobilization and vascular endothelial growth factor-2 gene therapy in myocardial ischemia. Circulation. 2004;110:1398–405.
85. Schachinger V, Assmus B, Britten MB, Honold J, Lehmann R, Teupe C, Abolmaali ND, Vogl TJ, Hofmann WK, Martin H, Dimmeler S, Zeiher AM. Transplantation of progenitor cells and regeneration enhancement in acute myocardial infarction: final one-year results of the TOPCARE-AMI trial. J Am Coll Cardiol. 2004;44:1690–9.
86. Bartunek J, Vanderheyden M, Vandekerckhove B, Mansour S, De Bruyne B, De Bondt P, Van Haute I, Lootens N, Heyndrickx G, Wijns W. Intracoronary injection of CD133-positive enriched bone marrow progenitor cells promotes cardiac recovery after recent myocardial infarction: feasibility and safety. Circulation. 2005;112:I178–83.
87. Li CJ, Gao RL, Yang YJ, Hu FH, Yang WX, Song LF, Ruan YM, Qiao SB, Chen JL, Qin XW, Chen ZJ. Effects of intracoronary transplantation of autologous bone marrow mononuclear

cells or endothelial progenitor cells in mini-swine model of myocardial ischemia-reperfusion. Zhonghua Xin Xue Guan Bing Za Zhi. 2007;35:936–9.

88. Tatsumi T, Ashihara E, Yasui T, Matsunaga S, Kido A, Sasada Y, Nishikawa S, Hadase M, Koide M, Nakamura R, Irie H, Ito K, Matsui A, Matsui H, Katamura M, Kusuoka S, Matoba S, Okayama S, Horii M, Uemura S, Shimazaki C, Tsuji H, Saito Y, Matsubara H. Intracoronary transplantation of non-expanded peripheral blood-derived mononuclear cells promotes improvement of cardiac function in patients with acute myocardial infarction. Circ J. 2007;71:1199–207.
89. Dobert N, Britten M, Assmus B, Berner U, Menzel C, Lehmann R, Hamscho N, Schachinger V, Dimmeler S, Zeiher AM, Grunwald F. Transplantation of progenitor cells after reperfused acute myocardial infarction: evaluation of perfusion and myocardial viability with FDG-PET and thallium SPECT. Eur J Nucl Med Mol Imaging. 2004;31:1146–51.
90. Stamm C, Westphal B, Kleine HD, Petzsch M, Kittner C, Klinge H, Schumichen C, Nienaber CA, Freund M, Steinhoff G. Autologous bone-marrow stem-cell transplantation for myocardial regeneration. Lancet. 2003;361:45–6.
91. Stamm C, Kleine HD, Choi YH, Dunkelmann S, Lauffs JA, Lorenzen B, David A, Liebold A, Nienaber C, Zurakowski D, Freund M, Steinhoff G. Intramyocardial delivery of CD133+ bone marrow cells and coronary artery bypass grafting for chronic ischemic heart disease: safety and efficacy studies. J Thorac Cardiovasc Surg. 2007;133:717–25.
92. Ahmadi H, Baharvand H, Ashtiani SK, Soleimani M, Sadeghian H, Ardekani JM, Mehrjerdi NZ, Kouhkan A, Namiri M, Madani-Civi M, Fattahi F, Shahverdi A, Dizaji AV. Safety analysis and improved cardiac function following local autologous transplantation of CD133(+) enriched bone marrow cells after myocardial infarction. Curr Neurovasc Res. 2007;4:153–60.
93. Balogh L, Czuriga I, Hunyadi J, Galuska L, Kristof E, Edes I. Effects of autologous bone marrow derived CD34+ stem cells on the left ventricular function following myocardial infarction. Orv Hetil. 2007;148:243–9.
94. Erbs S, Linke A, Adams V, Lenk K, Thiele H, Diederich KW, Emmrich F, Kluge R, Kendziorra K, Sabri O, Schuler G, Hambrecht R. Transplantation of blood-derived progenitor cells after recanalization of chronic coronary artery occlusion: first randomized and placebo-controlled study. Circ Res. 2005;97:756–62.
95. Assmus B, Honold J, Schachinger V, Britten MB, Fischer-Rasokat U, Lehmann R, Teupe C, Pistorius K, Martin H, Abolmaali ND, Tonn T, Dimmeler S, Zeiher AM. Transcoronary transplantation of progenitor cells after myocardial infarction. N Engl J Med. 2006;355:1222–32.
96. Boyle AJ, Whitbourn R, Schlicht S, Krum H, Kocher A, Nandurkar H, Bergmann S, Daniell M, O'Day J, Skerrett D, Haylock D, Gilbert RE, Itescu S. Intra-coronary high-dose CD34+ stem cells in patients with chronic ischemic heart disease: a 12-month follow-up. Int J Cardiol. 2006;109:21–7.
97. Lara-Hernandez R, Lozano-Vilardell P, Blanes P, Torreguitart-Mirada N, Galmes A, Besalduch J. Safety and efficacy of therapeutic angiogenesis as a novel treatment in patients with critical limb ischemia. Ann Vasc Surg. 2010;24:287–94.

Chapter 5
Therapeutic Angiogenesis with Adipose Tissue-Derived Regenerative Cells

Toyoaki Murohara and Kazuhisa Kondo

Abstract Therapeutic angiogenesis is an important strategy to treat tissues against severe ischemic diseases in patients with no other treatment option including endovascular intervention or bypass surgery. Recent studies indicated some possibilities of cell transplantation-mediated angiogenesis using autologous bone marrow cells, $CD34^{+}$ or $CD133^{+}$ stem cells, and peripheral blood mononuclear cells. Subcutaneous adipose tissues can be harvested by relatively easy and less invasive techniques. Recent studies indicated that adipose tissue contains mesenchymal progenitor cells that can give rise to several lineage cells. Moreover, these mesenchymal progenitor cells can release a variety of angiogenic growth factors including vascular endothelial growth factor, hepatocyte growth factor, and chemokine stromal cell-derived factor. The combination of these biological properties of adipose-derived regenerative cells (ADRCs) implicates that autologous adipose tissue would be a useful cell source for therapeutic angiogenesis.

Keywords Angiogenesis • Adipose-derived regenerative cells • Endothelial cells • Peripheral artery occlusive disease

5.1 Introduction

When tissue is exposed to severe ischemia, new blood vessels develop into the ischemic foci to prevent further cellular damage and necrosis [1]. However, this endogenous protective mechanism (i.e., collateral vessel formation and capillary angiogenesis) is often hampered by coexisting morbidity including atherosclerotic risk factors such as diabetes, hypercholesterolemia, smoking, etc. [2]. For example, it is frequent for diabetic patients to lose their extremities after suffering from severe peripheral artery obstructive disease (PAOD) because of insufficient development

T. Murohara, M.D., Ph.D. (✉) • K. Kondo
Department of Cardiology, Nagoya University Graduate School of Medicine,
65 Tsurumai, Showa-ku, Nagoya 466-8550, Japan
e-mail: murohara@med.nagoya-u.ac.jp

Y. Higashi, T. Murohara (eds.), *Therapeutic Angiogenesis*,
DOI 10.1007/978-981-10-2744-4_5

of collateral vessels/angiogenesis. Therefore, the strategy called "therapeutic angiogenesis" is a very important means to salvage tissues against critical ischemia [1].

Subcutaneous adipose tissue can be harvested by relatively easy and less invasive technology by an established liposuction method. Furthermore, recent studies have indicated that adipose tissues contain progenitor cells that can give rise to several lineage cells including the fat, bone, cartilage, muscle, etc. [3, 4]. These progenitor cells are termed as adipose-derived stem/progenitor cells (ASCs or ADSCs) or adipose-derived regenerative cells (ADRCs), and these terminologies including adipose-derived stromal vascular fraction (SVF) are essentially considered as an identical cell fraction [5]. ADSCs/ADRCs can also release multiple angiogenesis-related growth factors including vascular endothelial growth factor (VEGF), hepatocyte growth factor (HGF), and chemokine stromal cell-derived factor-1 (SDF-1) [6–8]. The combination of these biological properties (i.e., progenitor cell supply and growth factor release) suggests that autologous subcutaneous adipose tissue will be a good candidate for a cell source of therapeutic angiogenesis [6–13] (Table 5.1).

Table 5.1 Animal models of therapeutic angiogenesis by adipose-derived cells

Study	Year	Cells and ischemic models	Efficacy	Proposed mechanisms
Rehman et al. [6]	2004	Human ASCs to mouse hind limb	Increased angiogenesis	Cytokines (VEGF, HGF, TGFβ)
Miranville et al. [9]	2004	Human SVF to mouse hind limb	Increased angiogenesis	CD34+/CD31- cells, EC differentiation
Planat-Benard et al. [10]	2004	Human SVF to mouse hind limb	Increased angiogenesis	EC differentiation
Nakagami et al. [7]	2005	Mouse ADSC to mouse hind limb	Increased angiogenesis	Cytokines (VEGF, HGF)
Moon et al. [11]	2006	Human ADSC to mouse hind limb	Increased angiogenesis	Possibly cytokines (n.i.)
Sumi et al. [12]	2007	Mouse SVF to mouse hind limb	Increased angiogenesis	EC, VSMC differentiation
Cai et al. [13]	2008	Human ASC to rat MI	Increased angiogenesis	Possibly cytokines (n.i.)
Kondo et al. [8]	2009	Mouse ADRC to mouse hind limb	Increased angiogenesis	EPC recruitment, cytokines (SDF-1, VEGF)
Harada et al. [57]	2013	Rat ADRC to mouse hind limb	Increased angiogenesis	Cytokines (VEGF, bFGF, HGF)
Hao et al. [56]	2014	Rabbit ADRC to rabbit hind limb	Increased angiogenesis	Cytokines (HGF, VEGF)
Kosaraju et al. [62]	2016	Mouse ADRC to mouse wound	Increased angiogenesis	EPC recruitment

All hind limb model is the ischemic hind limb model

ASCs adipose-derived stromal cells or adipose-derived stem cells, *SVF* stromal vascular fraction, *ADSCs* adipose-derived stem/progenitor cells, *ADRCs* adipose-derived regenerative cells, *MI* myocardial infarction, *VEGF* vascular endothelial growth factor, *HGF* hepatocyte growth factor, *TGF* transforming growth factor, *EC* endothelial cell, *EPCs* endothelial progenitor cells, *VSMC* vascular smooth muscle cell, *SDF* stromal cell-derived factor, *n.i.* not identified

5.2 Therapeutic Angiogenesis: Growth Factor and Cytokine Therapy

Past Professors Judah Folkman and Jeffrey Isner, great pioneers for angiogenesis research, had initiated the concept of therapeutic angiogenesis in the early 1970s [14–16]. Folkman and co-workers observed that pathological angiogenesis is essential for the growth and metastasis of solid tumors. His original idea that the suppression of tumor angiogenesis would be effective against tumor growth had been developed into a completely new paradigm of anti-angiogenic therapy against neoplastic diseases, tumor dormancy therapy [15].

After the identification of angiogenic growth factors including VEGF and basic fibroblast growth factor (bFGF), investigators in the cardiovascular field started testing their hypothesis that facilitating angiogenesis with externally treated growth factor genes and/or proteins would improve tissue blood perfusion and function in critically ischemic diseases [17–19]. A number of preclinical data supported the safety and feasibility of clinical application of therapeutic angiogenesis using growth factor genes or proteins. Thereafter, abundant clinical trials have been performed to conduct therapeutic angiogenesis with genes and cytokines [20]. Initial uncontrolled small clinical trials showed successful results; however, newer clinical studies with randomized placebo-controlled trials failed to show sufficient improvement of angiogenesis, function, or symptoms [20]. These findings suggest that the clinical trial of therapeutic angiogenesis using single factor may have a limited efficacy. This will be explained by the fact that the expression of more than 750 genes was either upregulated or downregulated more than twofold by acute ischemic insult in vivo [21], and such dramatic changes may not easily be overcome by a cytokine monotherapy (i.e., single gene or protein). So far, only small numbers of studies using HGF or basic fibroblast growth factor (bFGF) have been shown to reveal promising data, but these may be because a number of cytokines or transcription factors are expressed by downstream to these cytokine-mediated signal transductions [22–24].

5.3 Therapeutic Angiogenesis: Cell Therapy

Meantime, a pioneering work conducted by Asahara, Isner, and co-workers regarding the identification of putative endothelial progenitor cells (EPCs) in adult human peripheral blood was published in 1997 [25]. Since EPCs and hematopoietic stem cells (HSCs) share several cell surface antigens such as Flk-1, Tie-2, CD31, and CD34, EPCs are believed to derive from so-called hemangioblasts or closely related cell class existing in the bone marrow in adults [26]. Mobilized EPCs from the bone marrow circulate in the peripheral blood and participate in new blood vessel formation and/or re-endothelialization on the injured arterial wall [25]. Although transplantation of culture-expanded EPCs could successfully augment angiogenesis and tissue blood supply in experimental animal models with ischemia [27], this procedure has not been developed into clinical trials because of difficulties in culture expansion techniques and limited number of cells to implant.

Instead, based on the fact that patients' own bone marrow can be obtained by relatively easy way and that fresh bone marrow contains HSCs and EPCs [26], implantation of autologous bone marrow mononuclear cells has been investigated in both preclinical and clinical studies [28–31]. Adult bone marrow has been shown to contain EPCs that mobilize into peripheral circulation after stimulation by ischemia or by cytokines such as VEGF and SDF-1 [32–34]. We previously demonstrated that implantation of autologous bone marrow mononuclear cells into ischemic skeletal muscles successfully augmented angiogenesis and collateral vessel formation in both animal studies and human trials [28, 29, 31, 35]. Especially in the clinical study called the "TACT trial," we have shown that more than 80% of Buerger's disease patients and 50% of PAD patients were well responsive to this therapeutic procedure [36, 37].

However, there are several limiting factors for this procedure. One is that the procedure itself is sometime too invasive for patients with critical limb ischemia. In the TACT protocol, approximately $1–3 \times 10^9$ bone marrow mononuclear cells were required for cell transplantation therapy in one limb of a patient, which contains $1–3 \times 10^7$ $CD34^+$ cells and needs about ~800 mL of fresh bone marrow aspirated from the ileum [31, 37]. Moreover, for this entire protocol, patients must be under the general anesthesia. Secondly, mechanism of angiogenesis by bone marrow mononuclear cell implantation had been studied by investigators, and these studies suggested that differentiation of bone marrow mononuclear cells into endothelial cells was rarely seen within the implanted skeletal muscle tissues. However, cytokines and chemokines released from implanted mononuclear cells would stimulate preexisting endothelial cells to sprout out. Kamihata and co-workers showed that cytokines such as VEGF and bFGF released from implanted bone marrow mononuclear cells were likely main players for angiogenesis by stimulating "resident" endothelial cells [29]. Tateno and co-workers showed that inflammatory cytokine IL-1 released from implanted mononuclear cells once stimulates immature host skeletal myocytes to release VEGF which then stimulated the angiogenic sprout of resident endothelial cells [38]. Thus, the initially presented direct differentiation mechanism of bone marrow mononuclear cells into mature endothelial cells is now skeptical. Thirdly, although the safety and efficiency of the TACT protocol have been established, we recently reported that patients with severe end-stage PAOD had poor responses to the TACT procedure [35]. For example, patients with diabetes or chronic hemodialysis due to end-stage renal disease (ESRD) showed limited responses to the TACT procedure [35]. Moreover, recent studies indicated that patients with severe PAOD, ischemic heart disease, and/or multiple coronary risk factors had diminished functions of EPCs and poor responses to angiogenic cell therapy [39–42]. These results collectively suggest that even fresh autologous bone marrow mononuclear cells were isolated and the function of stem/progenitor cell per se is already reduced and is too weak to induce sufficient angiogenesis after implantation into the "host" ischemic tissues [39]. It has been demonstrated that the efficacy of the TACT procedure was limited in patients with poorly controlled diabetes and ESRD [35]. In such patients, the number and functions of EPCs are also reduced and are difficult to be isolated. Also, we previously demonstrated that the number and function of circulating EPCs were markedly reduced in chronic smokers even without other risk factors [43]. And this reduction was persistent and was not easily restored by the treatment with HMG-CoA reductase inhibitor statins that have been shown to stimulate mobilization of EPCs from the bone marrow [44, 45].

5.4 Adipose-Derived Stem/Progenitor Cells or Regenerative Cells

Since bone marrow aspiration is invasive for patients with severe ischemic diseases, less invasive techniques for isolating cells for angiogenic therapy have been investigated. One of the most attractive sources for isolating cells for therapeutic angiogenesis would be autologous subcutaneous adipose tissue [5, 46] (Table 5.1). Adipose tissue mainly comprises two classes of cells. One is mature adipocytes (MAs) forming major part of adipose tissue volume, and the other is stromal cells termed as stromal vascular fraction (SVF). Several studies have demonstrated that SVF contains multipotent mesenchymal stem/progenitor cells that could differentiate into various lineages including fibroblasts, adipocytes, pericytes, osteoblasts, chondrocytes, and myocytes [3, 4]. Recently, it has been reported that adipose tissues contain multipotent mesenchymal stem cells termed adipose-derived stem/progenitor cells (ASCs or ADSCs) or adipose-derived regenerative cells (ADRCs) that have an ability to regenerate damaged tissues. ADRCs could be isolated from even a small amount of human subcutaneous adipose tissues and culture expandable [4]. Therefore, autologous ADRCs could be isolated with minimal invasiveness such as a liposuction method [47].

5.5 Therapeutic Angiogenesis Using Adipose-Derived Regenerative Cells: A New Paradigm

Although some studies reported that adipose-derived cells could be able to differentiate into ECs or EPCs [9, 10, 12], in our recent investigation, we could not confirm ADRCs to differentiate into the endothelial lineage [8]. One study used low-serum medium supplemented with VEGF and insulin-like growth factor-1 (IGF-1) to direct ADRCs to ECs or EPCs [9]. The difference of such culture conditions might have affected the differentiation capacity of ADRCs into ECs or other lineages [48]. Nevertheless, a recent study failed to show that human ADRCs differentiated into ECs even under culture with endothelial-specific medium EGM-2 [49]. In addition in vivo study, we could not confirm differentiation of ADRCs into ECs. Implanted GFP-positive ADRCs expressed no endothelial makers, but implanted ADRCs were positive for CD140b, also known as PDGF receptor B and a myofibroblast-specific marker, and co-localized with vessel-like pericytes [8]. These results suggest that

ADRCs might not have the capability to differentiate into EPCs or mature ECs but pericytes or smooth muscle lineage cells. Similarly, several recent studies showed that ADRCs differentiated into pericytes in vitro and in vivo [49–51].

Although our study and previous studies failed to show the ADRCs or ASCs differentiation into the EC lineage, it has been shown that ADRCs can secrete multiple angiogenic growth factors or cytokines such as vascular endothelial growth factor (VEGF) and hepatocyte growth factor (HGF) [6, 7]. Thus, it seems that autologous adipose tissue-derived cells may be useful for therapeutic angiogenesis via growth factor-dependent mechanisms. Nakagami and co-workers and Sumi and co-workers reported that implantation of adipose-derived mesenchymal cells did induce angiogenesis via a secretion of angiogenic growth factors [7, 12]. We also found that implantation of ADRCs significantly augmented angiogenesis in a mouse model of severe hind limb ischemia (Fig. 5.1) [8]. Our data indicated that implantation of ADRCs induced angiogenesis not only by an endothelial differentiation of ADRCs but also by released chemokines such as SDF-1 [8]. Interestingly, we found that implantation of mature adipocytes (MAs) into skeletal muscles even worsened angiogenesis compared to saline-injected control mice. We found that ADRCs expressed SDF-1 and the abundance of mRNA and protein expression was significantly greater in ADRCs than in MAs [8]. Recent studies indicated that MAs release other adipocytokines including TNF-α and IL-6. These deleterious inflammatory cytokines might have negatively affected angiogenesis by MA implantation observed in our recent study [6]. SDF-1 is a member of CXC chemokines originally isolated from murine BM stromal cells [52]. CXCR4 is the receptor for SDF-1 and is a co-receptor for HIV type 1 infection [53]. The SDF-1/CXCR4 axis regulates multiple physiological processes including embryonic development and organ homeostasis. Interestingly, SDF-1 is considered as one of the key regulators of EPCs trafficking from BM into PB as well [34]. Thus, SDF-1 has been shown to augment neovascularization by acceleration of EPC recruitment into ischemic foci [34, 54]. In addition, VEGF is one of the powerful angiogenic cytokines that can

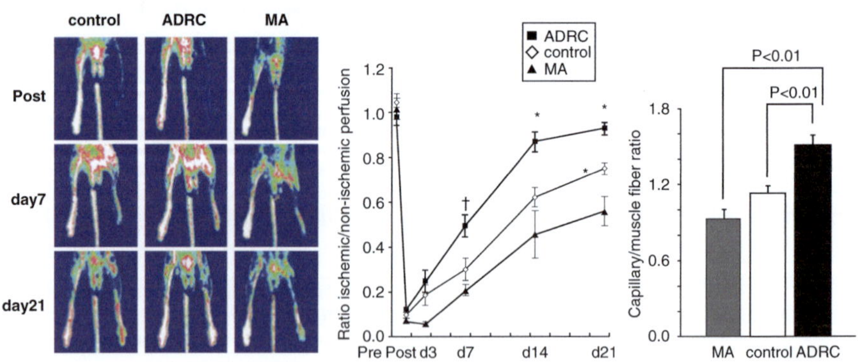

Fig. 5.1 Implantation of ADRCs augmented ischemia-induced angiogenesis. Mouse ADCRs were isolated and injected into the skeletal muscles of a mouse hind limb ischemia model. Implantation of ADRCs significantly augmented angiogenesis, tissue capillary density, and blood perfusion in the ischemic tissues. Mature adipocyte (MA) injection did not affect or slightly suppressed the angiogenesis. Adopted from [8]

also mobilize EPCs from BM and inhibit EPC apoptosis [55]. Interestingly, in the mouse ischemic hind limb model, VEGF-A-mediated angiogenesis partly depends on the activation of the SDF-1/CXCR4 pathway [54]. Taken together, chemokine SDF-1 likely plays a pivotal role for the ADRC-mediated angiogenesis [8]. In fact, therapeutic efficacies and mobilization of EPCs of ADRC implantation were markedly suppressed by injection of an anti-SDF-1 neutralizing mAb in our study [8].

5.6 Modification of ADRCs for the Enhancement of Angiogenic Efficacy

After the initial reports of therapeutic angiogenesis using ADRCs, there have been several unique methods and findings published regarding the modification of ADRC functions. We recently compared the efficacy of angiogenesis between bone marrow mononuclear cells and ADRCs in a rabbit model of hind limb ischemia [56]. Overall efficacy of therapeutic angiogenesis was comparable between the bone marrow mononuclear cells and ADRCs, but the cytokine-releasing capacity was much greater in ADRCs (Fig. 5.2) [56]. Interestingly, Harada and

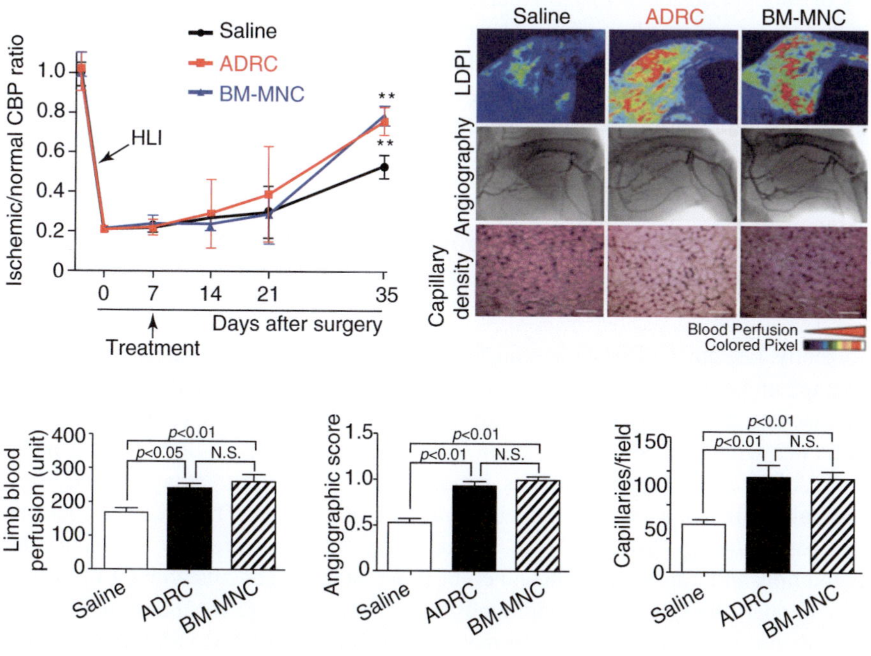

Fig. 5.2 Implanted autologous ADRCs and BM-MNCs equally induced angiogenesis. Implantation of autologous ADRCs and bone marrow mononuclear cells (BM-MNCs) equally augmented angiogenesis, blood perfusion, and tissue capillary density in a rabbit model of unilateral hind limb ischemia. Adopted from [56]

co-workers showed that transplantation of freshly isolated ADRCs (fADRCs) significantly improved blood perfusion, capillary density, and production of several angiogenic factors in transplanted ischemic limbs compared with the culture-expanded ADRC (cADRC)-injected group. fADRCs revealed significantly higher expression levels of angiogenic factors than cADRCs [57]. Thus, fADRCs have an effective angiogenic capacity, and they would be more valuable as a source for cell-based therapeutic angiogenesis compared to cADRCs [57]. Bhang and co-workers demonstrated that ADRCs delivered intramuscularly into ischemic hind limbs in combination with nanosphere-mediated bFGF delivery enhanced limb survival and blood perfusion, as well as survival of the transplanted ADRCs and secretion of angiogenic growth factors (i.e., VEGF, HGF, and bFGF) [58]. Horikoshi-Ishihara and co-workers showed that coadministration of ADRCs and control-released bFGF enhanced angiogenesis in a murine ischemic hind limb model. bFGF coadministration induced the release of HGF, VEGF, and TGF-β1 [59]. Delle Monache and co-workers found that hypoxia-conditioned ADRCs released leptin that enhanced Flt-1 and Tie-1, angiogenic growth factor receptors, and thereby enhanced angiogenic responses [60]. Kishimoto and co-workers recently showed that a combination of heparin/protamine micro-/nanoparticle (LH/P-MP)-conjugated ADRCs further enhanced angiogenesis as compared to ADRCs alone in mouse ischemic hind limb model [61]. Kosaraju and co-workers demonstrated that ADRC-seeded hydrogels enhanced recruitment and functionality of endogenous bone marrow-derived progenitor cells and improved angiogenesis and wound healing [62]. These studies suggest that coadministration of ADRCs and growth factors such as bFGF with tissue-retaining compounds is a promising strategy to further enhance the ADRC-mediated therapeutic angiogenesis. This field warrants further investigation in both basic and clinical studies.

5.7 ADRC Therapy for Angiogenesis: Unanswered Questions

Recently, Lee and colleagues conducted a clinical trial to test the safety and efficacy of implanting autologous ADRCs into the ischemic limbs of patients with Buerger's disease and diabetes [63]. Their study involved 15 patients with critical limb ischemia (12 with Buerger's disease and 3 with diabetic foot). ADRCs were isolated from subcutaneous adipose tissues, expanded in culture until passage 3, and intramuscularly injected into the ischemic muscles of patients. They showed that this cell-based therapy using ADRCs might be feasible and effective for increasing blood flow and improving pain and clinical outcomes in patients with CLI [63].

Although the efficacy of autologous ADRC implantation for therapeutic angiogenesis has been established in animal and human studies, it is still unknown how many cells should be used for optimal angiogenesis. In the case of bone marrow mononuclear cells, it has been reported that the injection of too many cells resulted in adverse effects in an animal model [64]. The optimum dosage of ADRC implantation will be necessary to be determined. Second question is what the optimal cell isolation method is? Recently, freshly isolated ADRCs by means of collagenase digestion followed by an automatic centrifuge system using a specific machine are useful for regeneration of adipose tissues for reconstructive surgeries [47, 65]. Such technique does not need either cell cultivation or an established cell processing center. If two classes of ADRCs obtained by either culture method or fresh isolation method have equal potency in terms of releasing angiogenic growth factors, it will be useful to employ such machine for cell-mediated therapeutic angiogenesis. Thirdly, either ADRC number or functions may be affected by concomitant patients' conditions or atherogenic risk factors. For example, SDF-1 secretion is reduced in type 2 diabetes [66], and the function and number of progenitors are generally reduced in patients with multiple risk factors including diabetes, hypertension, aging, and smoking [40–43]. Such pathological conditions are commonly complicated in patients with PAOD and/or Buerger's disease. It is still unknown whether ADRC number and functions are influenced by the presence of such risk factors as well. Before applying this new technique to therapeutic angiogenesis, it will be needed to elucidate these points.

References

1. Isner JM, Asahara T. Angiogenesis and vasculogenesis as therapeutic strategies for postnatal neovascularization. J Clin Invest. 1999;103:1231–6.
2. Zeiher AM, Drexler H, Saurbier B, Just H. Endothelium-mediated coronary blood flow modulation in humans: effects of age, atherosclerosis, hypercholesterolemia, and hypertension. J Clin Invest. 1993;92:652–62.
3. Zuk PA, Zhu M, Mizuno H, Huang J, Futrell JW, Katz AJ, Benhaim P, Lorenz HP, Hedrick MH. Multilineage cells from human adipose tissue: implications for cell-based therapies. Tissue Eng. 2001;7:211–28.
4. Miyazaki T, Kitagawa Y, Toriyama K, Toriyama K, Kobori M, Torii S. Isolation of two human fibroblastic cell populations with multiple but distinct potential of mesenchymal differentiation by ceiling culture of mature fat cells from subcutaneous adipose tissue. Differentiation. 2005;73:69–78.
5. Murohara T. Autologous adipose tissue as a new source of progeni- tor cells for therapeutic angiogenesis. J Cardiol. 2009;53:155–63.
6. Rehman J, Traktuev D, Li J, Merfeld-Clauss S, Temm-Grove CJ, Bovenkerk JE, Pell CL, Johnstone BH, Considine RV, March KL. Secretion of angiogenic and antiapoptotic factors by human adipose stromal cells. Circulation. 2004;109:1292–8.
7. Nakagami H, Maeda K, Morishita R, Iguchi S, Nishikawa T, Takami Y, Kikuchi Y, Saito Y, Tamai K, Ogihara T, Kaneda Y. Novel autologous cell therapy in ischemic limb disease through growth factor secretion by cultured adipose tissue-derived stromal cells. Arterioscler Thromb Vasc Biol. 2005;25:2542–7.
8. Kondo K, Shintani S, Shibata R, Murakami H, Murakami R, Imaizumi T, Kitagawa S, Murohara T. Implantation of adipose-derived regenerative cells enhances ischemia-induced angiogenesis. Arterioscler Thromb Vasc Biol. 2009;29:61–6.
9. Miranville A, Heeschen C, Sengenes C, Curat CA, Busse R, Bouloumié A. Improvement of postnatal neovascularization by human adipose tissue-derived stem cells. Circulation. 2004;110:349–55.

10. Planat-Benard V, Silvestre JS, Cousin B, André M, Nibbelink M, Tamarat R, Clergue M, Manneville C, Saillan-Barreau C, Duriez M, Tedgui A, Levy B, Pénicaud L, Casteilla L. Plasticity of human adipose lineage cells toward endothelial cells: physiological and therapeutic perspectives. Circulation. 2004;109:656–63.
11. Moon MH, Kim SY, Kim YJ, Kim SJ, Lee JB, Bae YC, Sung SM, Jung JS. Human adipose tissue-derived mesenchymal stem cells improve postnatal neovascularization in a mouse model of hindlimb ischemia. Cell Physiol Biochem. 2006;17:279–90.
12. Sumi M, Sata M, Toya N, Yanaga K, Ohki T, Nagai R. Transplantation of adipose stromal cells, but not mature adipocytes, augments ischemia-induced angiogenesis. Life Sci. 2007;80:559–65.
13. Cai L, Johnstone BH, Cook TG, Tan J, Fishbein MC, Chen PS, March KL. Human adipose tissue-derived stem cells induce angiogenesis and nerve sprouting following myocardial infarction, in conjunction with potent preservation of cardiac function. Stem Cells. 2009;27:230–7.
14. Folkman J. Tumor angiogenesis: therapeutic implications. N Engl J Med. 1971;285:1182–6.
15. Folkman J. Angiogenesis in cancer, vascular, rheumatoid and other disease. Nat Med. 1995;1:27–30.
16. Isner JM. Myocardial gene therapy. Nature. 2002;415:234–9.
17. Yanagisawa-Miwa A, Uchida Y, Nakamura F, Tomaru T, Kido H, Kamijo T, Sugimoto T, Kaji K, Utsuyama M, Kurashima C, et al. Salvage of infarcted myocardium by angiogenic action of basic fibroblast growth factor. Science. 1992;257:1401–3.
18. Takeshita S, Zheng LP, Brogi E, Kearney M, LQ P, Bunting S, et al. Therapeutic angiogenesis. A single intraarterial bolus of vascular endothelial growth factor augments revascularization in a rabbit ischemic hind limb model. J Clin Invest. 1994;93:662–70.
19. Morishita R, Nakamura S, Hayashi S, Taniyama Y, Moriguchi A, Nagano T, Taiji M, Noguchi H, Takeshita S, Matsumoto K, Nakamura T, Higaki J, Ogihara T. Therapeutic angiogenesis induced by human recombinant hepatocyte growth factor in rabbit hind limb ischemia model as cytokine supplement therapy. Hypertension. 1999;33:1379–84.
20. Losordo DW, Dimmeler S. Therapeutic angiogenesis and vasculogenesis for ischemic disease. Part I: angiogenic cytokines. Circulation. 2004;109:2487–91.
21. Lee CW, Stabile E, Kinnaird T, Shou M, Devaney JM, Epstein SE, Burnett MS. Temporal patterns of gene expression after acute hindlimb ischemia in mice: insights into the genomic program for collateral vessel development. J Am Coll Cardiol. 2004;43:474–82.
22. Powell RJ, Simons M, Mendelsohn FO, Daniel G, Henry TD, Koga M, Morishita R, Annex BH. Results of a double-blind, placebo-controlled study to assess the safety of intramuscular injection of hepatocyte growth factor plasmid to improve limb perfusion in patients with critical limb ischemia. Circulation. 2008;118:58–65.
23. van Belle E. Potentiated angiogenic effect of scatter factor/hepatocyte growth factor via induction of vascular endothelial cell growth factor: the case for paracrine amplification of angiogenesis. Circulation. 1998;97:381–90.
24. Fujii T, Yonemitsu Y, Onimaru M, Inoue M, Hasegawa M, Kuwano H, Sueishi K. VEGF function for upregulation of endogenous PlGF expression during FGF-2-mediated therapeutic angiogenesis. Atherosclerosis. 2008;200:51–7.
25. Asahara T, Murohara T, Sullivan A, Silver M, van der Zee R, Li T, Witzenbichler B, Schatteman G, Isner JM. Isolation of putative progenitor endothelial cells for angiogenesis. Science. 1997;275:964–7.
26. Asahara T, Masuda H, Takahashi T, Kalka C, Pastore C, Silver M, et al. Bone marrow origin of endothelial progenitor cells responsible for postnatal vasculogenesis in physiological and pathological neovascularization. Circ Res. 1999;85:221–8.
27. Kalka C, Masuda H, Takahashi T, Kalka-Moll WM, Silver M, Kearney M, Li T, Isner JM, Asahara T. Transplantation of ex vivo expanded endothelial progenitor cells for therapeutic neovascularization. Proc Natl Acad Sci U S A. 2000;97:3422–7.
28. Shintani S, Murohara T, Ikeda H, Ueno T, Sasaki K, Duan J, et al. Augmentation of postnatal neovascularization with autologous bone marrow transplantation. Circulation. 2001;103:897–903.

29. Kamihata H, Matsubara H, Nishiue T, Fujiyama S, Tsutsumi Y, Ozono R, et al. Implantation of bone marrow mononuclear cells into ischemic myocardium enhances collateral perfusion and regional function via side supply of angioblasts, angiogenic ligands, and cytokines. Circulation. 2001;104:1046–52.
30. Losordo DW, Dimmeler S. Therapeutic angiogenesis and vasculogenesis for ischemic disease: part II: cell-based therapies. Circulation. 2004;109:2692–7.
31. Tateishi-Yuyama E, Matsubara H, Murohara T, Ikeda U, Shintani S, Masaki H, et al. Therapeutic angiogenesis for patients with limb ischaemia by autologous transplantation of bone-marrow cells: a pilot study and a randomized controlled trial. Lancet. 2002;360:427–35.
32. Takahashi T, Kalka C, Masuda H, Chen D, Silver M, Kearney M, et al. Ischemia- and cytokine-induced mobilization of bone marrow-derived endothelial progenitor cells for neovascularization. Nat Med. 1999;5(4):434–8.
33. Kalka C, Tehrani H, Laudenberg B, et al. VEGF gene transfer mobilizes endothelial progenitor cells in patients with inoperable coronary disease. Ann Thorac Surg. 2000;70:829–34.
34. Yamaguchi J, Kusano KF, Masuo O, Kawamoto A, Silver M, Murasawa S, Bosch-Marce M, Masuda H, Losordo DW, Isner JM, Asahara T. Stromal cell-derived factor-1 effects on ex vivo expanded endothelial progenitor cell recruitment for ischemic neovascularization. Circulation. 2003;107:1322–8.
35. Kajiguchi M, Kondo T, Izawa H, Kobayashi K, Yamamoto K, Shintani S, Numaguchi Y, Naoe T, Takamatsu J, Komori K, Murohara T. Safety and efficacy of autologous progenitor cell transplantation for therapeutic angiogenesis in patients with critical limb ischemia. Circ J. 2007;71:196–201.
36. Saito Y, Sasaki K, Katsuda Y, Murohara T, Takeshita Y, Okazaki T, Arima K, Katsuki Y, Shintani S, Shimada T, Akashi H, Ikeda H, Imaizumi T. Effect of autologous bone marrow cell transplantation on ischemic ulcer in patients with Burger's disease. Circ J. 2007;71:1187–92.
37. Matoba S, Tatsumi T, Murohara T, Imaizumi T, Katsuda Y, Ito M, Saito Y, Uemura S, Suzuki H, Fukumoto S, Yamamoto Y, Onodera R, Teramukai S, Fukushima M, Matsubara H, Follow-up Study Investigators TACT. Long-term clinical outcome after intramuscular implantation of bone marrow mononuclear cells (therapeutic angiogenesis by cell transplantation [TACT] trial) in patients with chronic limb ischemia. Am Heart J. 2008;156:1010–8.
38. Tateno K, Minamino T, Toko H, Akazawa H, Shimizu N, Takeda S, Kunieda T, Miyauchi H, Oyama T, Matsuura K, Nishi J, Kobayashi Y, Nagai T, Kuwabara Y, Iwakura Y, Nomura F, Saito Y, Komuro I. Critical roles of muscle-secreted angiogenic factors in therapeutic neovascularization. Circ Res. 2006;98:1194–202.
39. Heeschen C, Lehmann R, Honold J, Assmus B, Aicher A, Walter DH, et al. Profoundly reduced neovascularization capacity of bone marrow mononuclear cells derived from patients with chronic ischemic heart disease. Circulation. 2004;109:1615–22.
40. Vasa M, Fichtlscherer S, Adler K, et al. Increase in circulating endothelial progenitor cells by statin therapy in patients with stable coronary artery disease. Circulation. 2001;103:2885–90.
41. Hill JM, Zalos G, Halcox JP, Schenke WH, Waclawiw MA, Quyyumi AA, Finkel T. Circulating endothelial progenitor cells, vascular function, and cardiovascular risk. N Engl J Med. 2003;348:593–600.
42. Choi JH, Kim KL, Huh W, Kim B, Byun J, Suh W, et al. Decreased number and impaired angiogenic function of endothelial progenitor cells in patients with chronic renal failure. Arterioscler Thromb Vasc Biol. 2004;24:1246–52.
43. Kondo T, Hayashi M, Takeshita K, Numaguchi Y, Kobayashi K, Iino S, et al. Smoking cessation rapidly increases circulating progenitor cells in peripheral blood in chronic smokers. Arterioscler Thromb Vasc Biol. 2004;24:1442–7.
44. Yoshida O, Kondo T, Sugiura T, Okumura K, Murohara T. Pitavastatin ameliorates impaired endothelial function in chronic smokers by reducing oxidative stress. Circ J. 2007;71(Suppl I):417. abstract
45. Walter DH, Zeiher AM, Dimmeler S. Effects of statins on endothelium and their contribution to neovascularization by mobilization of endothelial progenitor cells. Coron Artery Dis. 2004;15:235–42.

46. Takahashi M. Adipose tissue -an alternative source for therapeutic angiogenesis. Circ J. 2012;76:1597–8.
47. Fraser JK, Wulur I, Alfonso Z, Hedrick MH. Fat tissue: an underappreciated source of stem cells for biotechnology. Trends Biotechnol. 2006;24:150–4.
48. Suga H, Shigeura T, Matsumoto D, Inoue K, Kato H, Aoi N, Murase S, Sato K, Gonda K, Koshima I, Yoshimura K. Rapid expansion of human adipose-derived stromal cells preserving multipotency. Cytotherapy. 2007;9:738–45.
49. Traktuev D, Merfeld-Clauss S, Li J, Kolonin M, Arap W, Pasqualini R, Johnstone B, March K. A population of multipotent CD34-positive adipose stromal cells share pericyte and mesenchymal surface markers, reside in a periendothelial location, and stabilize endothelial networks. Circ Res. 2008;102:77–85.
50. Zannettino ACW, Paton S, Arthur A, Khor F, Itescu S, Gimble JM, Gronthos S. Multipotential human adipose-derived stromal stem cells exhibit a perivascilar phenotype in vitro and vivo. J Cell Physiol. 2008;214:413–21.
51. Amos P, Shang H, Bailey A, Taylor A, Katz A, Peirce S. IFATS series: the role of human adipose-derived stromal cells in inflammatory microvascular remodeling and evidence of a perivascular phenotype. Stem Cells. 2008;26:2682–90.
52. Nagasawa T, Nakajima T, Tachibana K, Iizasa H, Bleul CC, Yoshie O, Matsushima K, Yoshida N, Springer TA, Kishimoto T. Molecular cloning and characterization of a murine pre-B-cell growth-stimulating factor/stromal cell-derived factor 1 receptor, a murine homolog of the human immunodeficiency virus 1 entry coreceptor fusin. Proc Natl Acad Sci U S A. 1996;93:14726–9.
53. Bleul CC, Farzan M, Choe H, Parolin C, Clark-Lewis I, Sodroski J, Springer TA. The lymphocyte chemoattractant SDF-1 is a ligand for LESTR/fusin and blocks HIV-1 entry. Nature. 1996;382:829–33.
54. Hiasa K, Ishibashi M, Ohtani K, Inoue S, Zhao Q, Kitamoto S, Sata M, Ichiki T, Takeshita A, Egashira K. Gene transfer of stromal cell-derived factor-1alpha enhances ischemic vasculogenesis and angiogenesis via vascular endothelial growth factor/endothelial nitric oxide synthase-related pathway: next-generation chemokine therapy for therapeutic neovascularization. Circulation. 2004;109:2454–61.
55. Asahara T, Takahashi T, Masuda H, Kalka C, Chen D, Iwaguro H, Inai Y, Silver M, Isner JM. VEGF contributes to postnatal neovascularization by mobilizing bone marrow-derived endothelial progenitor cells. EMBO J. 1999;18:3964–72.
56. Hao CN, Shintani S, Shimizu Y, Kondo K, Ishii M, Wu H, Murohara T. Therapeutic angiogenesis by autologous adipose-derived regenerative cells: comparison with bone marrow mononuclear cells. Am J Physiol Heart Circ Physiol. 2014;307:H869–79.
57. Harada Y, Yamamoto Y, Tsujimoto S, Matsugami H, Yoshida A, Hisatome I. Transplantation of freshly isolated adipose tissue-derived regenerative cells enhances angiogenesis in a murine model of hind limb ischemia. Biomed Res. 2013;34:23–9.
58. Bhang SH, Cho SW, Lim JM, Kang JM, Lee TJ, Yang HS, Song YS, Park MH, Kim HS, Yoo KJ, Jang Y, Langer R, Anderson DG, Kim BS. Locally delivered growth factor enhances the angiogenic efficacy of adipose-derived stromal cells transplanted to ischemic limbs. Stem Cells. 2009;27:1976–86.
59. Horikoshi-Ishihara H, Tobita M, Tajima S, Tanaka R, Oshita T, Tabata Y, Mizuno H. Coadministration of adipose-derived stem cells and control-released basic fibroblast growth factor facilitates angiogenesis in a murine ischemic hind limb model. J Vasc Surg. 2015; doi:10.1016/j.jvs.2015.09.054.
60. Delle Monache S, Calgani A, Sanità P, Zazzeroni F, Gentile Warshauer E, Giuliani A, Amicucci G, Angelucci A. Adipose-derived stem cells sustain prolonged angiogenesis through leptin secretion. Growth Factors. 2016;30:1–10.
61. Kishimoto S, Inoue K, Nakamura S, Hattori H, Ishihara M, Sakuma M, Toyoda S, Iwaguro H, Taguchi I, Inoue T, Yoshida K. Low-molecular weight heparin protamine complex augmented the potential of adipose-derived stromal cells to ameliorate limb ischemia. Atherosclerosis. 2016;249:132–9.

62. Kosaraju R, Rennert RC, Maan ZN, Duscher D, Barrera J, Whittam AJ, Januszyk M, Rajadas J, Rodrigues M, Gurtner GC. Adipose-derived stem cell-seeded hydrogels increase endogenous progenitor cell recruitment and neovascularization in wounds. Tissue Eng Part A. 2016;22:295–305.
63. Lee HC, An SG, Lee HW, Park JS, Cha KS, Hong TJ, et al. Safety and effect of adipose tissue-derived stem cell implantation in patients with critical limb ischemia: a pilot study. Circ J. 2012;76:1750–60.
64. Kawamoto A, Iwasaki H, Kusano K, Murayama T, Oyamada A, Silver M, Hulbert C, Gavin M, Hanley A, Ma H, Kearney M, Zak V, Asahara T, Losordo DW. CD34-positive cells exhibit increased potency and safety for therapeutic neovascularization after myocardial infarction compared with total mononuclear cells. Circulation. 2006;114:2163–9.
65. Moseley TA, Zhu M, Hedrick MH. Adipose-derived stem and progenitor cells as fillers in plastic and reconstructive surgery. Plast Reconstr Surg. 2006;118(3 Suppl):121S–8S.
66. Gallagher KA, Liu ZJ, Xiao M, Chen H, Goldstein LJ, Buerk DG, Nedeau A, Thom SR, Velazquez OC. Diabetic impairments in NO-mediated endothelial progenitor cell mobilization and homing are reversed by hyperoxia and SDF-1 alpha. J Clin Invest. 2007;117:1249–59.

Chapter 6
Cell Therapy for Ischemic Heart Disease

Hiroshi Kurazumi, Tohru Hosoyama, and Kimikazu Hamano

Abstract Myocardial infarction causes the loss of many cardiomyocytes, leading to dysfunction of the pumping function of the heart and resulting in heart failure. Progress in medical therapies and coronary artery revascularization has improved the prognosis of coronary artery disease and heart failure, but heart failure remains one of the main causes of mortality in the Western world.

In recent decades, basic research studies have revealed that intrinsic stem cells can differentiate into new cardiomyocytes and that the human heart has cardiac progenitor cells and self-renewal potential. Therefore, cell-based myocardial repair is an alternative to conventional therapy for ischemic cardiomyopathy.

Cell-based myocardial repair is achieved by transplanting stem cells into the failing heart to regenerate the organ. Various cell types, including bone marrow-derived cells (BMCs), skeletal myoblasts, endothelial progenitor cells (EPCs), mesenchymal stem cells (MSCs), cardiac stem/progenitor cells (CPCs), embryonic stem cells (ESCs), and induced pluripotent stem cells (iPSCs), have been tested in animal models. Various delivery methods, including intramyocardial, intracoronary arterial, and cell sheet applications, have also been tested.

Many clinical trials investigating stem cell therapy in patients suffering from heart failure have been performed as a consequence of the positive effect of this therapy in animal models. In clinical trials, cell therapy improves the ejection fraction and myocardial perfusion and decreases the scar size while increasing the amount of viable heart tissue. However, these effects only modestly contribute to clinical outcomes. Although past clinical trials have not achieved dramatic improvement in cardiac function, cell therapy remains an increasingly promising therapy.

Keywords Cell therapy • Regenerative therapy • Ischemic heart disease

H. Kurazumi (✉)
Department of Surgery, Saiseikai Yamaguchi Hospital, 2-11 Midori-chyo, Yamaguchi, Yamaguchi 753-8517, Japan
e-mail: kurazumi@yamaguchi-u.ac.jp

T. Hosoyama • K. Hamano
Department of Surgery and Clinical Science, Yamaguchi University Graduate School of Medicine, 1-1-1 Minami-Kogushi, Ube, Yamaguchi 755-8505, Japan

Y. Higashi, T. Murohara (eds.), *Therapeutic Angiogenesis*,
DOI 10.1007/978-981-10-2744-4_6

6.1 History

Ischemic heart disease (IHD) develops through atherosclerotic degeneration, resulting in coronary artery stenosis or occlusion. Myocardial infarction and subsequent ischemic cardiomyopathy result in significant mortality and morbidity worldwide [1].

Percutaneous coronary intervention (PCI) and coronary artery bypass grafting (CABG) were developed to revascularize the ischemic heart, and they remain the gold standard therapy for IHD. Many medications, including diuretic agents, angiotensin converting enzyme inhibitors, and β-blockers, are well-established treatments for heart failure from IHD. These agents prevent congestive heart failure and remodeling of the diseased heart. However, ischemic heart disease patients with severely deteriorated heart function are not adequately treated with these common therapies. Approximately 45% of patients with coronary artery disease develop congestive heart failure despite progress in coronary revascularization and medical therapy [2]. More aggressive interventions, such as cardiac transplantation and the use of a left ventricular assist device, are the only treatments available to compensate for the cardiomyocyte loss. While a few patients are treated with a left ventricular assist device or heart transplantation, the majority cannot afford these luxuries due to a limited number of donor hearts and the associated high cost of medical care [3].

Various novel therapies for IHD have been developed and investigated in the last three decades. Examples include physical therapy with transmyocardial laser revascularization, which was introduced in the 1980s, and gene therapy using various growth factors, which was introduced in the 1990s [4, 5].

In 2000, some groups reported that local transplantation of bone marrow cells with primitive cells facilitates in vivo angiogenesis in the ischemic heart and hind limb [6]. Subsequently, cell therapy for ischemic heart disease was proposed as a potential novel treatment strategy. In 2001, we performed the first clinical trial examining bone marrow cell transplantation to treat ischemic heart disease during CABG, demonstrating the feasibility of cell therapy, albeit in a small number of patients [7]. A variety of cell delivery methods and stem/progenitor cells were then translated from bench to bedside for cell-based myocardial regeneration. To date, many randomized controlled studies have been designed and completed using various cells and delivery methods during various disease phases (acute or chronic). Most of these studies were conducted in European countries and the United States. A recent meta-analysis revealed the obvious efficacy of cell-based myocardial repair [8–11].

6.2 The Mechanism of Cell-Based Myocardial Repair

The mechanism of cell-based myocardial repair is very complicated, and the underlying details remained to be unclear. A possible mechanism of cell therapy for heart failure is reviewed in this paragraph. Schematic drawings of cell-based myocardial repair for ischemic heart diseases are shown in Fig. 6.1.

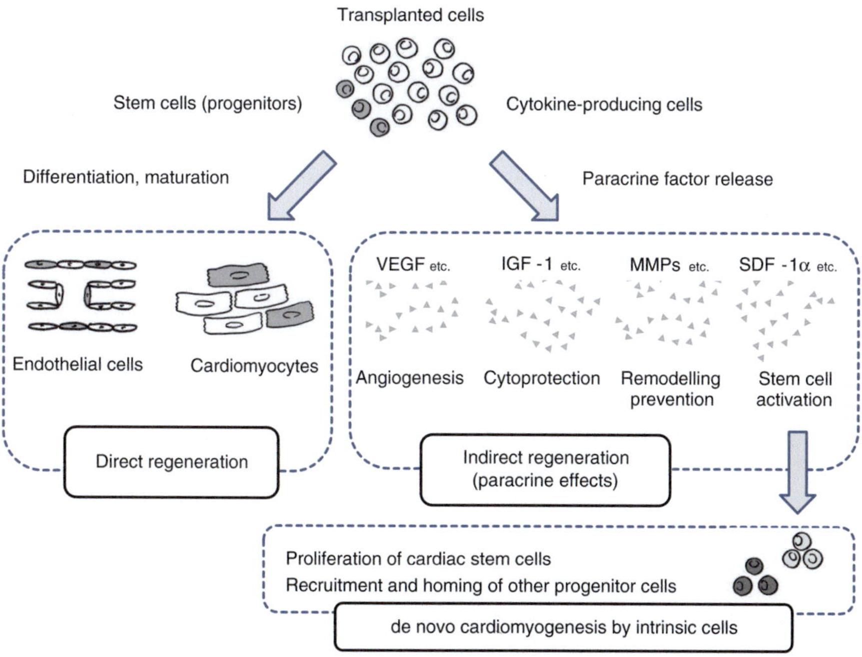

Fig. 6.1 Mechanism of cell-based myocardial repair

6.2.1 Direct Regeneration

It was first supposed that primitive cells transplanted into diseased myocardium differentiate into mature myocardium, contributing to the amelioration of cardiac function. This process is called "direct regeneration." However, most of implanted cells were lost due to washing out from injected area and cell death from ischemia, anoikis, and inflammation [12], and it is thought that very few numbers of stem/progenitor cells might be able to directly contribute to vascular structures and heart regeneration after transplantation. Also, there are no clinical evidences that transplanted stem/progenitor cells differentiate into new mature myocardium so far.

6.2.2 Indirect Regeneration (Paracrine Effects)

To date, it is thought that the major mechanism of cell-based myocardial repair is "indirect regeneration" due to paracrine mechanism. The engrafted stem/progenitor cells secrete various soluble factors affecting regeneration-mediated cardiac events such as protection of cardiomyocyte death, cardiac contraction, neovessel formation, and stem cell activation (described in Sect. 6.2.3). These effects resulted in an amelioration of cardiac functions and prevention of cardiac remodeling [13, 14]. For example, soluble factors such as HGF and bFGF from engrafted stem/progenitor cells inhibit apoptosis of cardiomyocytes, subsequently increasing survived cardiomyocytes in failing hearts. Cardiac stem/progenitor cells also secrete proangiogenic factors such as VEGF in addition to cell mobilization factors such as SDF-1 [15]. Finally, it has been reported that intramyocardial injection of bone marrow-derived mesenchymal stem cells activated resident cardiac stem/progenitor cells, suggesting paracrine effects of graft cells on stem cell stimulation [16]. Taken together, these findings clearly indicate paracrine manners of graft cells in cell-based myocardial repair of failing hearts.

6.2.3 De Novo Cardiomyogenesis by Intrinsic Cells

Beginning with the finding that intramyocardial c-kit-positive cells could provide de novo cardiomyocytes, it has been believed that postnatal heart can be regenerated by providing new cardiomyocytes from resident stem/progenitor cells after injuries [17]. However, intrinsic regenerative potential of postnatal heart is not sufficient to improve cardiac functions in severe injuries, even if endogenous cardiac stem/progenitor cells such as c-kit-positive cells provide de novo cardiomyocytes. In other words, if it can be enhanced regenerative capacity of resident cardiac stem/progenitor cells, postnatal heart might be regenerated much more efficiently. Riley and colleagues recently reported that thymosin-β4, an actin-binding monomeric peptide, activates epicardial-derived cardiac progenitor cells and induces cardiac regeneration after injury [18]. However, effect of thymosin-β4 on activation of epicardial cardiac progenitor cells is limited in postnatal heart, and it should be administrated prior to heart injury.

As another resource of de novo cardiomyocytes in postnatal heart, preexisting cardiomyocytes in the heart are considered, because human cardiomyocyte can undergo mitosis and maintain a capacity to regenerate after injury [19]. However, dividing capacity of preexisting cardiomyocytes is limited, and a radioactive isotope study demonstrated that human cardiomyocytes gradually renew 1% of the cell population annually at the age of 25 years to 0.45% at the age of 75 years [20]. Because recent studies reported the possibility that microRNAs and non-glycosylated growth factor can accelerate mitosis of postnatal cardiomyocytes, intrinsic cardiomyocytes might become strong tool for cell-based myocardial repair [21, 22].

6.3 Indication for Cell Therapy

Because surgical or catheter coronary revascularization is a confirmed therapy for ischemic heart disease, cell therapy for ischemic heart disease should be an additional therapeutic option after coronary revascularization (PCI or CABG) and standard medical therapy.

Major indication for cell therapy for ischemic heart disease is as follows:

(a) Acute myocardial infarction

In acute myocardial infarction patients, stem/progenitor cells are delivered from the culprit coronary artery after revascularization by PCI to reduce the infarction area.

(b) Chronic heart failure due to ischemic cardiomyopathy

In chronic phase of coronary artery disease, stem/progenitor cells are delivered to ischemic myocardium because coronary arteries are too small to perform revascularization.

(c) Refractory angina

Patients who have untreatable angina despite maximal conventional therapies (coronary revascularization or medical therapy) are the candidates of cell therapy.

6.4 Application Approaches

The application approaches are one of the major concerns for cell-based myocardial repair. Various stem cell delivery methods have been developed. To date, the best mode of delivery has not been determined. In this paragraph, a possible delivery mode is introduced. The schematic drawings of the application approaches are provided in Fig. 6.2.

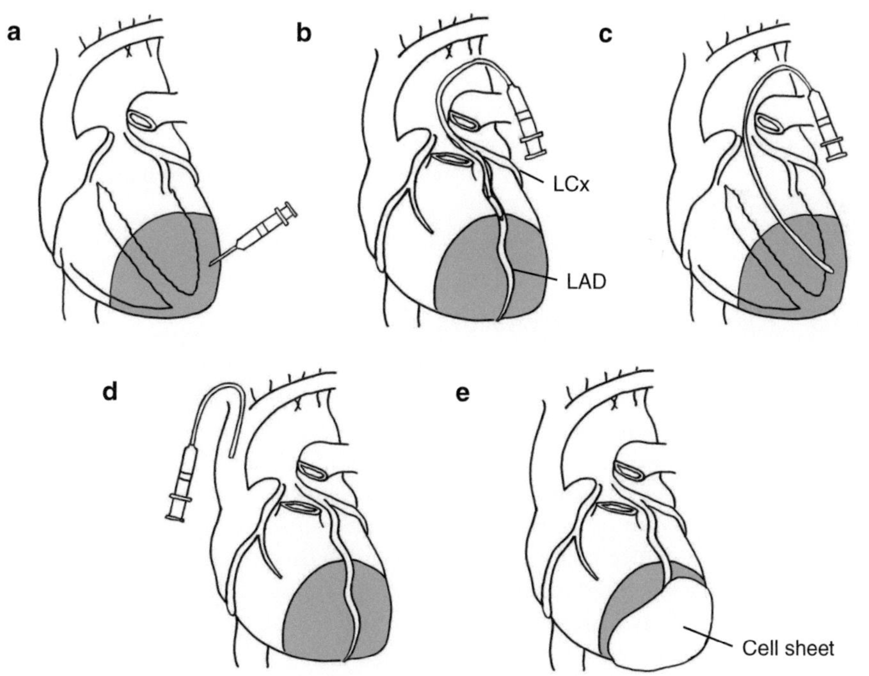

Fig. 6.2 Routes of application. (**a**) Epicardial intramyocardial application. (**b**) Intracoronary arterial application. (**c**) Endocardial intramyocardial application. (**d**) Intravenous application. (**e**) Cell sheet application. *LAD* left anterior descending artery, *LCx* left circumflex artery

6.4.1 Epicardial Intramyocardial Application

This is the most classical cell delivery method. Cells diluted in osmolaric media are injected into the target region of the myocardium through the epicardium using a handheld needle and syringes during open heart surgery [7]. After completion of the surgical revascularization, the cells are injected into the border zone of the infarct area or the area that have not been grafted arteries because of a lack of graftable coronary arteries. This method can be applied successfully during off-pump and conventional CABG. This method usually requires general anesthesia and sternotomy or thoracotomy. The advantage of this method is the direct injection of cells into the targeted area. However, a number of transplanted cells are lost via leakage from puncture holes; they are squeezed out by the beating myocardium. Potential adverse effects are left ventricular perforation or bleeding from the myocardium, but serious adverse effects were not reported in a previous clinical trial.

6.4.2 Intracoronary Arterial Application

This is a catheter-based technique that is performed using a percutaneous approach, which is less invasive than the surgical approach of direct intramyocardial injection. Intracoronary artery transplantation, using a balloon catheter, has the advantage that cells can be directly injected into a target region of the myocardium, thus preserving the blood flow and oxygen supply [23]. This procedure may be favorable for cell survival and engraftment; however, the cells cannot be delivered to regions with lower blood flow, which may represent an ischemic area of the myocardium.

6.4.3 Endocardial Intramyocardial Application

A catheter is inserted retrograde into the left ventricle across the aortic valve via a percutaneous, femoral arterial approach. Stem cells are transendocardially transplanted into the myocardium via a needle built into a catheter. This method is less invasive than direct surgical injection, which was described previously in Sect. 6.4.1. A number of transplanted cells are considered to be lost into the ventricle from puncture holes as well as from the direct surgical intramyocardial injection. This method requires deftness and equipment for electromechanically mapping the myocardium to define the target ischemic area. The potential adverse effects are ventricular arrhythmias and cardiac tamponade [24]. Cell therapy using this delivery method can be safely repeated for serial treatment [25].

6.4.4 Intravenous Application

Stem cells are supplied to the systemic circulation intravenously. This method is indicated for patients with acute myocardial infarction [26]. This is the least invasive procedure because it does not require general anesthesia, sternotomy, thoracotomy, or arterial puncture. However, in an animal model of acute myocardial infarction, the transplanted cells mainly accumulate in the lungs and spleen [27]. Improved accumulation of the stem cells in the injured heart cannot be expected at this time.

6.4.5 Cell Sheet Application

Tissue engineering technologies have been used to develop cell delivery methods for regenerative therapy in various organs. Using this approach, stem cells are expanded in temperature-responsive culture dishes and harvested as sheets. Cell sheets are delivered onto the surface of the damaged heart via sternotomy or thoracotomy. Improved cell survival and engraftment, resulting in a positive effect on cardiac function, have been reported [28].

The details of cell sheet myocardial repair are described in Chap. 6.

6.5 Cell Sources

Cell sources are another major concern for cell-based myocardial repair. Various cell sources have been investigated in previous animal models and clinical trials. To date, the best mode of delivery has not been revealed. In this paragraph, possible cell sources for cell-based myocardial repair are briefly introduced with current status in clinical settings. The characteristics of each cell sources are summarized in Table 6.1.

6.5.1 Bone Marrow-Derived Cells

Bone marrow-derived cells (BMCs) have been often used in a variety of clinical settings. Although some reports have shown possible cardiac differentiation ability of BMCs, it is used in hope that BMCs can contribute to therapeutic angiogenesis in ischemic heart [29]. Indeed, transplantation of BMCs into the patients with myocardial infarction improved coronary blood flow resulted from angiogenesis in ischemic hearts [30, 31]. However, another studies reported no evidences that transplantation of BMCs produces any functional improvements in patients with

Table. 6.1 Lists of cell source

Cell types	Angio-genesis	Differentiate to cardio-myocyte	Auto-logous source	Cell supply	Transplan-tation immunity	Safety	Clinical trial
Bone marrow-derived cells	○	Limited	○	◎	○	◎	◎
Endothelial progenitor cells (EPCs)	○	×	○	○	○	◎	○
Skeletal myoblasts	○	×	○	◎	○	△[a]	○
Mesen-chymal stem cells (MSCs)	○	Limited	○	○	○	○	○
Adipose-derived cells	○	Limited	○	○	○	○	○
Umbilical cord blood cells	○	Unclear	×	×	×	○	×
Cardiac progenitor cells	○	○	○	○	○	○	○
Embryonic stem cells (ES cells)	○	○	×	×	×	Unclear	×
Induced pluripotent stem cells (iPS cells)	○	○	○	○	○	Unclear	×

[a]Concern for arrhythmia

myocardial infarction [32, 33]. For this point, it is also pointed out that different cell delivery methods in each trials might result in a discrepancy in outcomes of BMC transplantation. Thus, therapeutic benefits of BMC transplantation into ischemic heart diseases are still controversial, and it should be followed up for therapeutic availability of BMCs for ischemic heart diseases in future trials.

6.5.2 *Endothelial Progenitor Cells*

As another cellular tool for therapeutic angiogenesis, endothelial progenitor cells (EPCs) are considered. EPCs, which are characterized by CD34 cell surface marker, can be isolated from peripheral blood and also from bone marrow [34]. Because

EPCs can give rise to vascular cells but not to cardiomyocytes, clinical application of EPC transplantation is expected to increase the capillary density and subsequently improve the microcirculation around the transplanted sites in infarcted heart [35]. A clinical trial reported that intramyocardial injection of autologous EPCs improves angina refractory heart disease and exercise tolerance [36].

6.5.3 Skeletal Myoblasts

Skeletal myoblasts, which are easily isolated from skeletal muscle tissue, were first cells to be clinically investigated cell-based myocardial repair [37–39]. However, clinical trial using autologous skeletal myoblasts (Myoblast Autologous Grafting in Ischemic Cardiomyopathy (MAGIC)) resulted in disappointments in efficacy and increased incidence of arrhythmias, declining an importance of skeletal myoblasts as valuable tool in cell therapy for heart failure patients [40]. On the other hand, its therapeutic availability for heart failure was re-demonstrated using cell sheet technology in patient with dilated cardiomyopathy, indicating that skeletal myoblasts are still valuable therapeutic tools when cells are delivered onto the surface of failing heart [41].

6.5.4 Mesenchymal Stem Cells

Mesenchymal stem cells (MSCs) are multipotent stromal cells and can differentiate into a wide variety of cell types including cardiomyocytes [42, 43]. MSCs can be isolated from multiple tissues such as bone marrow, skeletal muscle, and adipose tissue and can be easily expanded in vitro. Although MSCs induce cardiac repair in infarcted heart after transplantation, the role of MSCs in cardiac regeneration is to supply not only new cardiomyocytes by trans-differentiation but also angiogenic factors by paracrine fashion [44].

In randomized clinical trial, which is Reinfusion of Enriched Progenitor cells And Infarct Remodeling in Acute Myocardial Infarction (REPAIR-AMI), transplantation of autologous bone marrow-derived MSCs induced improved outcomes and ventricular function in patients with acute myocardial infarction at 2 years after transplantation [45]. In contrast to this trial, however, recent two trials using MSCs, which are Timing In Myocardial infarction Evaluation (TIME) and PercutaneOus StEm cell Injection Delivery effects On Neomyogenesis (POSEIDON), did not show any improvement in ventricular function of heart failure patients after intracoronary or transendocardial delivery of autologous bone marrow-derived MSCs [46, 47]. Thus, whether MSCs are valuable and effective therapeutic tool for ischemic heart diseases have to be decided from accumulated scientific knowledges in future investigations.

6.5.5 Cardiac Stem/Progenitor Cells

As a result of further investigations of cardiac stem cells and cell cultures, cell sources for stem cell therapy started to become obtainable from endogenous cardiac progenitor cells in the 2010s. Cardiac precursor cells were expanded from heart biopsy pieces and proposed to be a new cell source. These cells are expected to possess greater cardiogenic potency than other stem cells. An in vivo study has shown that these cardiac progenitor cells have superior paracrine potency and myocardial repair efficacy compared with other available stem/progenitor cells [48]. These cell sources are unlikely to stimulate immune rejection, but techniques for expanding the cells in culture are needed.

Two prominent clinical trials for cell therapy using this new cell source have been performed in the United States. In the study "Stem Cell Infusion in Patients with Ischemic cardiOmyopathy (SCIPIO)," autologous c-kit-positive cells, which have high cardiogenic potential, were isolated from the atrial samples during CABG and then administered by intracoronary infusion. This therapy resulted in an increase of left ventricular ejection fraction (LVEF) and reduction of infarct size, suggesting that c-kit-positive cardiac stem/progenitor cells are promising cell type in cell therapy for ischemic heart diseases [49]. On the other hand, CArdiosphere-Derived aUtologous stem CElls to reverse ventricUlar dySfunction (CADUCEUS) trial was performed using different types of cardiac stem/progenitor cells. In this trial, cardiosphere-derived cells isolated from endomyocardial biopsy specimens were used and administered to heart failure patients by intracoronary infusion [50]. In initial 6 months after transplantation, autologous graft CDCs induced reduction of scar mass, increase of viable heart mass, and regional contractility of infarcted heart. And, these therapeutic effects on infarcted heart were continued to at least 12 months after cell transplantation [51]. Taken together, cardiac stem/progenitor cells may be a practical choice in clinical settings. However, reliability of the SCIPIO trial was recently being under investigation by the institution of trial implementation because of concerns about integrity of data from the trial [52]. So, we should carefully follow up the investigation for this clinical trial.

6.5.6 Embryonic Stem Cells

Embryonic stem cells (ESCs) are derived from the inner cell masses of blastocysts, an early-stage embryo, and are able to differentiate into three germ layers: endoderm, ectoderm, and mesoderm [53–55]. Because of its pluripotent property, ESCs can provide any types of cells in specific culture conditions, being an unlimited source of desired cell types including cardiomyocytes [56]. Considering that cardiac stem/progenitor cells in adult heart have numerical problem to provide sufficient numbers of de novo cardiomyocytes, it is no doubt that ECSs are valuable resource in cell therapy for heart failure. Indeed, billions of human cardiomyocytes, which are theoretically required order for clinical use, can be prepared from ESCs, and preclinical study using nonhuman primates resulted in cardiac regeneration and an improvement of ventricular functions by transplantation of human ESC-derived cardiomyocytes into infarcted hearts [57]. Because ESC-derived cardiomyocytes are generally thought to be an immature fetal cardiomyocytes, an occurrence of arrhythmia has been concerned after transplantation into adult heart. However, the study on nonhuman primate models showed no obvious signs of arrhythmias after transplantation, suggesting that ESC-derived cardiomyocytes were matured in adult heart environment and synchronously functioned with preexisting mature cardiomyocytes in monkey heart. Although tumorigenesis of graft cells is required to be carefully monitored hereafter, this study encouraged us to apply ESC-derived cardiomyocytes for clinical settings.

6.5.7 Induced Pluripotent Stem Cells

As mentioned above, ESCs are promising tool for cell therapy for ischemic heart diseases. However, several limitations in ethics and required immunosuppression after transplantation are still remained to apply clinical settings. For this situation, it is expected that induced pluripotent stem cells (iPSCs) may break major limitations surrounding ESCs and boost an application of pluripotent stem cells for clinical settings.

iPSCs, originally established from mouse fibroblasts in 2006 and then from human fibroblasts in 2007, possess pluripotency and rich expansion property in similar to ESCs [58–60]. iPSCs can be generated from a wide variety of somatic cells by induction of pluripotent genes [61], indicating that large numbers of autologous pluripotent stem cells and its derivatives such as cardiomyocytes can be developed [62]. Human iPSC-derived cardiomyocytes showed similar functional characteristics to ESC-derived ESCs and improved left ventricular function with reduction of infarcted area in large animal models of myocardial infarction [63, 64]. Because clinical trial using autologous iPSC derivatives have just started from 2014 in Japan for patient with age-related macular degeneration, it is expected in future that iPSC-derived cardiomyocytes will be applied to patients with ischemic heart diseases. Establishing protocols to efficiently generate high-quality and high-quantity autologous iPSCs and cardiomyocyte derivatives within therapeutic time frame remains a challenge for clinical application.

6.5.8 Other Cell Sources

Umbilical cord blood and adipose-derived cells contain stem/progenitor cells with cellular characteristics that are quite distinct from bone marrow and adult peripheral blood. Human umbilical cord blood cells have regenerative effects in the injured hearts of severe combined immunodeficiency (SCID) mice [65]. Adipose-derived stem cells are also available as a cell source for cell therapy, and their safety and efficacy have been reported in a clinical trial [66].

6.6 Timing of Cell Therapy

The optimal timing of cell therapy after acute myocardial infarction (AMI) has been investigated in several trials using bone marrow-derived cells with catheter-based applications [33, 46]. In a recent systematic review, cardiac parameters (LVEF, left ventricular end-systolic volume (LVESV), and left ventricular end-diastolic volume (LVEDV)) were significantly improved when stem cells were transplanted between 3 and 10 days after AMI, whereas the infarct area was significantly smaller when the stem cells were injected within 48 h [10]. For acute myocardial infarction, stem cells should be delivered at least 10 days after onset.

6.7 Effectiveness and Adverse Effects

Many clinical trials investigating stem cell therapy have been reported. Because of the variability in study design, cell types, delivery method, and patient characteristics, the efficacy of cell therapy is discordant. Several meta-analyses have been performed based on the previous clinical trials to investigate the overall efficacy of cell therapy. Most cell therapy trials have been conducted using bone marrow-derived cells. Therefore, the reports describing overall efficacy are mostly based on therapy using bone marrow-derived cells. In this paragraph, potential effectiveness and adverse effect of cell-based myocardial repair are briefly described.

6.7.1 Effectiveness

The left ventricular ejection fraction is significantly increased in patients treated with cell therapy. There is an approximately 3% increase in the ejection fraction with cell therapy using bone marrow-derived cells [9]. This 3% increase is small, but it is important because most candidates for cell therapy are often maximally treated with medical therapy and coronary revascularization and have no options for treatment. Additionally, a dose-dependent contribution to cardiac recovery has been revealed [67]. A cell number less than 50 million of bone marrow-derived cells is thought to be ineffective for cardiac repair [9, 10]. Significant amelioration of the LVEF and LVESV has been noted with the administration of more than 100 million cells. Cell therapy for ischemic heart disease may improve myocardial perfusion, cardiac remodeling, and exercise tolerance [10, 36, 51, 59, 66]. Cell therapy for ischemic heart disease may decrease all-cause mortality, recurrent myocardial infarction, and ventricular arrhythmia [8, 11, 26, 36].

6.7.2 Adverse Effects

The potential adverse effects of cell therapy are tumor formation, ectopic calcification, arrhythmia, myocardial infarction (especially intracoronary application), and ventricle rupture (especially intramyocardial application). Although ventricular tachycardia has been documented after skeletal myoblast transplantation in clinical studies [68], other critical adverse effects have not been reported. However, most clinical trials comprised a small number of patients with a short follow-up period. Thus, a larger sample size with a long-term follow-up is needed to confirm the efficacy of cell therapy for ischemic heart disease.

6.8 Conclusions

Although cell therapy has complex mechanisms and previous clinical trials have not demonstrated dramatic improvement in cardiac function, cell-based myocardial repair is currently being developed. In the near future, if new cell sources (e.g., iPS cells or cardiac progenitor cells) became easily available and new application methods (e.g., cell sheet application) increase in popularity, cell therapy could help patients suffering from otherwise untreatable ischemic heart disease.

References

1. Oettgen P, Boyle AJ, Schulman SP, Hare JM. Cardiac stem cell therapy. Need for optimization of efficacy and safety monitoring. Circulation. 2006;114:353–8.
2. Weir RA, McMurray JJ. Epidemiology of heart failure and left ventricular dysfunction after acute myocardial infarction. Curr Heart Fail Rep. 2006;3:175–80.
3. Taylor DO, Edwards LB, Aurora P, Christie JD, Dobbels F, Kirk R, Rahmel AO, Kucheryavaya AY, Hertz MI. Registry of the International Society for Heart and Lung Transplantation: twenty-fifth official adult heart transplant report-2008. J Heart Lung Transplant. 2008;27:943–56. doi:10.1016/j.healun.2008.06.017.
4. Mirhoseini M, Shelgikar S, Cayton MM. Transmyocardial laser revascularization: a review. J Clin Laser Med Surg. 1993;11:15–9.
5. Rosengart TK, Lee LY, Patel SR, Sanborn TA, Parikh M, Bergman GW, Hachamovitch R, Szulc M, Kligfield PD, Okin PM, Hahn RT, Devereux RB, Post MR, Hackett NR, Foster T, Grasso TM, Lesser ML, Isom OW, Crystal RG. Angiogenesis gene therapy: phase I assessment of direct intramyocardial administration of an adenovirus vector ex-pressing VEGF 121 cDNA to individuals with clinically significant severe coronary artery disease. Circulation. 1999;100:468–74.

6. Kobayashi T, Hamano K, Li T, Katoh T, Kobayashi S, Matsuzaki M, Esato K. Enhancement of angiogenesis by the implantation of self bone marrow cells in a rat ischemic heart model. J Surg Res. 2000;89:189–95.
7. Hamano K, Nishida M, Hirata K, Mikamo A, Li TS, Harada M, Miura T, Matsuzaki M, Esato K. Local implantation of autologous bone marrow cells for therapeutic angiogenesis in patients with ischemic heart disease:clinical trial and preliminary results. Jpn Circ J. 2001;65:845–7.
8. Fisher SA, Dorée C, Brunskill SJ, Mathur A, Martin-Rendon E. Bone marrow stem cell treatment for ischemic heart disease in patients with no option of revascularization: a systematic review and meta-analysis. PLoS One. 2013;8:e64669.
9. Martin-Rendon E, Brunskill SJ, Hyde CJ, Stanworth SJ, Mathur A, Watt SM. Autologous bone marrow stem cells to treat acute myocardial infarction: a systematic review. Eur Heart J. 2008;29:1807–18.
10. Afzal MR, Samanta A, Shah ZI, Jeevanantham V, Abdel-Latif A, Zuba-Surma EK, Dawn B. Adult bone marrow cell therapy for ischemic heart disease: evidence and insights from randomized controlled trials. Circ Res. 2015;117:558–75.
11. Strauer BE, Yousef M, Schannwell CM. The acute and long-term effects of intracoronary stem cell transplantation in 191 patients with chronic heARt failure: the STAR-heart study. Eur J Heart Fail. 2010;12:721–9.
12. Cheng K, Li TS, Malliaras K, Davis DR, Zhang Y, Marbán E. Magnetic targeting enhances engraftment and functional benefit of iron-labeled cardiosphere-derived cells in myocardial infarction. Circ Res. 2010;106:1570–81. doi:10.1161/CIRCRESAHA.109.212589.
13. Gnecchi M, Zhang Z, Ni A, Dzau VJ. Paracrine mechanisms in adult stem cell signaling and therapy. Circ Res. 2008;103:1204–19. doi:10.1161/CIRCRESAHA.108.176826.
14. Takahashi M, Li TS, Suzuki R, Kobayashi T, Ito H, Ikeda Y, Matsuzaki M, Hamano K. Cytokines produced by bone marrow cells can contribute to functional improvement of the infarcted heart by protecting cardiomyocytes from ischemic injury. Am J Physiol Heart Circ Physiol. 2006;291:H886–93.
15. Dimmeler S, Zeiher AM, Schneider MD. Unchain my heart: the scientific foundations of cardiac repair. J Clin Invest. 2005;115:572–83.
16. Amado LC, Saliaris AP, Schuleri KH, St John M, Xie JS, Cattaneo S, Durand DJ, Fitton T, Kuang JQ, Stewart G, Lehrke S, Baumgartner WW, Martin BJ, Heldman AW, Hare JM. Cardiac repair with intramyocardial injection of allogeneic mesenchymal stem cells after myocardial infarction. Proc Natl Acad Sci U S A. 2005;102:11474–9.
17. Orlic D, Kajstura J, Chimenti S, Jakoniuk I, Anderson SM, Li B, Pickel J, McKay R, Nadal-Ginard B, Bodine DM, Leri A, Anversa P. Bone marrow cells regenerate infarcted myocardium. Nature. 2001;410:701–5.
18. Smart N, Bollini S, Dubé KN, Vieira JM, Zhou B, Davidson S, Yellon D, Riegler J, Price AN, Lythgoe MF, Pu WT, Riley PR. De novo cardiomyocytes from within the activated adult heart after injury. Nature. 2011;474:640–4. doi:10.1038/nature10188.
19. Beltrami AP, Urbanek K, Kajstura J, Yan SM, Finato N, Bussani R, Nadal-Ginard B, Silvestri F, Leri A, Beltrami CA, Anversa P. Evidence that human cardiac myocytes divide after myocardial infarction. N Engl J Med. 2001;344:1750–7.
20. Bergmann O, Bhardwaj RD, Bernard S, Zdunek S, Barnabe-Heider F, Walsh S, Zupicich J, Alkass K, Buchholz BA, Druid H, Jovinge S, Frisen J. Evidence for cardiomyocyte renewal in humans. Science. 2009;324:98–102. doi:10.1126/science.1164680.
21. Eulalio A, Mano M, Dal Ferro M, Zentilin L, Sinagra G, Zacchigna S, Giacca M. Functional screening identifies miRNAs inducing cardiac regeneration. Nature. 2012;492:376–81. doi:10.1038/nature11739.
22. Wei K, Serpooshan V, Hurtado C, Diez-Cuñado M, Zhao M, Maruyama S, Zhu W, Fajardo G, Noseda M, Nakamura K, Tian X, Liu Q, Wang A, Matsuura Y, Bushway P, Cai W, Savchenko A, Mahmoudi M, Schneider MD, van den Hoff MJ, Butte MJ, Yang PC, Walsh K, Zhou B, Bernstein D, Mercola M, Ruiz-Lozano P. Epicardial FSTL1 reconstitution regenerates the adult mammalian heart. Nature. 2015;525:479–85. doi:10.1038/nature15372.

23. Strauer BE, Brehm M, Zeus T, Gattermann N, Hernandez A, Sorg RV, Kögler G, Wernet P. Intracoronary, human autologous stem cell transplantation for myocardial regeneration following myocardial infarction. Dtsch Med Wochenschr. 2001;126:932–8.
24. Strauer BE, Steinhoff G. 10 years of intracoronary and intramyocardial bone marrow stem cell therapy of the heart: from the methodological origin to clinical practice. J Am Coll Cardiol. 2011;58:1095–104.
25. Poh KK, Sperry E, Young RG, Freyman T, Barringhaus KG, Thompson CA. Repeated direct endomyocardial transplantation of allogeneic mesenchymal stem cells: safety of a high dose, "off-the-shelf", cellular cardiomyoplasty strategy. Int J Cardiol. 2007;117:360–4.
26. Hare JM, Traverse JH, Henry TD, Dib N, Strumpf RK, Schulman SP, Gerstenblith G, DeMaria AN, Denktas AE, Gammon RS, Hermiller Jr JB, Reisman MA, Schaer GL, Sherman W. A randomized, double-blind, placebo-controlled, dose-escalation study of intravenous adult human mesenchymal stem cells (prochymal) after acute myocardial infarction. J Am Coll Cardiol. 2009;54:2277–86. doi:10.1016/j.jacc.2009.06.055.
27. Price MJ, Chou CC, Frantzen M, Miyamoto T, Kar S, Lee S, Shah PK, Martin BJ, Lill M, Forrester JS, Chen PS, Makkar RR. Intravenous mesenchymal stem cell therapy early after reperfused acute myocardial infarction improves left ventricular function and alters electrophysiologic properties. Int J Cardiol. 2006;111:231–9.
28. Sekine H, Shimizu T, Dobashi I, Matsuura K, Hagiwara N, Takahashi M, Kobayashi E, Yamato M, Okano T. Cardiac cell sheet transplantation improves damaged heart function via superior cell survival in comparison with dissociated cell injection. Tissue Eng Part A. 2011;17:2973–80.
29. Yoon CH, Koyanagi M, Iekushi K, Seeger F, Urbich C, Zeiher AM, Dimmeler S. Mechanism of improved cardiac function after bone marrow mononuclear cell therapy: role of cardiovascular lineage commitment. Circulation. 2010;121:2001–11.
30. Schächinger V, Erbs S, Elsässer A, Haberbosch W, Hambrecht R, Hölschermann H, Yu J, Corti R, Mathey DG, Hamm CW, Süselbeck T, Assmus B, Tonn T, Dimmeler S, Zeiher AM, REPAIR-AMI Investigators. Intracoronary bone marrow-derived progenitor cells in acute myocardial infarction. N Engl J Med. 2006;355:1210–21.
31. Erbs S, Linke A, Schachinger V, Assmus B, Thiele H, Diederich KW, Hoffmann C, Dimmeler S, Tonn T, Hambrecht R, Zeiher AM, Schuler G. Restoration of microvascular function in the infarct-related artery by intracoronary transplantation of bone marrow progenitor cells in patients with acute myocardial infarction: the Doppler Substudy of the Reinfusion of Enriched Progenitor Cells and Infarct Remodeling in Acute Myocardial Infarction (REPAIR-AMI) trial. Circulation. 2007;116:366–74.
32. Lunde K, Solheim S, Aakhus S, Arnesen H, Abdelnoor M, Egeland T, Endresen K, Ilebekk A, Mangschau A, Fjeld JG, Smith HJ, Taraldsrud E, Grøgaard HK, Bjørnerheim R, Brekke M, Müller C, Hopp E, Ragnarsson A, Brinchmann JE, Forfang K. Intracoronary injection of mononuclear bone marrow cells in acute myocardial infarction. N Engl J Med. 2006;355(12):1199–209. PubMed PMID: 16990383.
33. Traverse JH, Henry TD, Ellis SG, Pepine CJ, Willerson JT, Zhao DX, Forder JR, Byrne BJ, Hatzopoulos AK, Penn MS, Perin EC, Baran KW, Chambers J, Lambert C, Raveendran G, Simon DI, Vaughan DE, Simpson LM, Gee AP, Taylor DA, Cogle CR, Thomas JD, Silva GV, Jorgenson BC, Olson RE, Bowman S, Francescon J, Geither C, Handberg E, Smith DX, Baraniuk S, Piller LB, Loghin C, Aguilar D, Richman S, Zierold C, Bettencourt J, Sayre SL, Vojvodic RW, Skarlatos SI, Gordon DJ, Ebert RF, Kwak M, Moyé LA, Simari RD, Cardiovascular Cell Therapy ResearchNetwork. Effect of intracoronary delivery of autologous bone marrow mononuclear cells 2 to 3 weeks following acute myocardial infarction on left ventricular function: the LateTIME randomized trial. JAMA. 2011;306:2110–9. doi:10.1001/jama.2011.1670.
34. Asahara T, Murohara T, Sullivan A, Silver M, van der Zee R, Li T, Witzenbichler B, Schatteman G, Isner JM. Isolation of putative progenitor endothelial cells for angiogenesis. Science. 1997;275:964–7.

35. Iwasaki H, Kawamoto A, Ishikawa M, Oyamada A, Nakamori S, Nishimura H, Sadamoto K, Horii M, Matsumoto T, Murasawa S, Shibata T, Suehiro S, Asahara T. Dose-dependent contribution of CD34-positive cell transplantation to concurrent vasculogenesis and cardiomyogenesis for functional regenerative recovery after myocardial infarction. Circulation. 2006;113:1311–25.
36. Losordo DW, Henry TD, Davidson C, Sup Lee J, Costa MA, Bass T, Mendelsohn F, Fortuin FD, Pepine CJ, Traverse JH, Amrani D, Ewenstein BM, Riedel N, Story K, Barker K, Povsic TJ, Harrington RA, Schatz RA, ACT34-CMI Investigators. Intramyocardial, autologous CD34+ cell therapy for refractory angina. Circ Res. 2011;109:428–36. doi:10.1161/CIRCRESAHA.111.245993.
37. Leobon B, Garcin I, Menasche P, Vilquin JT, Audinat E, Charpak S. Myoblasts transplanted into rat infarcted myocardium are functionally isolated from their host. Proc Natl Acad Sci U S A. 2003;100:7808–11.
38. Taylor DA, Atkins BZ, Hungspreugs P, Jones TR, Reedy MC, Hutcheson KA, Glower DD, Kraus WE. Regenerating functional myocardium: improved performance after skeletal myoblast transplantation. Nat Med. 1998;4:929–33.
39. Ghostine S, Carrion C, Souza LC, Richard P, Bruneval P, Vilquin JT, Pouzet B, Schwartz K, Menasché P, Hagège AA. Long-term efficacy of myoblast transplantation on regional structure and function after myocardial infarction. Circulation. 2002;106:I131–6.
40. Menasché P, Alfieri O, Janssens S, McKenna W, Reichenspurner H, Trinquart L, Vilquin JT, Marolleau JP, Seymour B, Larghero J, Lake S, Chatellier G, Solomon S, Desnos M, Hagège AA. The Myoblast Autologous Grafting in Ischemic Cardiomyopathy (MAGIC) trial: first randomized placebo-controlled study of myoblast transplantation. Circulation. 2008;117:1189–200. doi:10.1161/CIRCULATIONAHA.107.734103.
41. Sawa Y, Miyagawa S, Sakaguchi T, Fujita T, Matsuyama A, Saito A, Shimizu T, Okano T. Tissue engineered myoblast sheets improved cardiac function sufficiently to discontinue LVAS in a patient with DCM: report of a case. Surg Today. 2012;42:181–4. doi:10.1007/s00595-011-0106-4.
42. Toma C, Pittenger MF, Cahill KS, Byrne BJ, Kessler PD. Human mesenchymal stem cells differentiate to a cardiomyocyte phenotype in the adult murine heart. Circulation. 2002;105:93–8.
43. Chen SL, Fang WW, Ye F, Liu YH, Qian J, Shan SJ, Zhang JJ, Chunhua RZ, Liao LM, Lin S, Sun JP. Effect on left ventricular function of intracoronary transplantation of autologous bone marrow mesenchymal stem cell in patients with acute myocardial infarction. Am J Cardiol. 2004;94:92–5.
44. Ohnishi S, Ohgushi H, Kitamura S, Nagaya N. Mesenchymal stem cells for the treatment of heart failure. Int J Hematol. 2007;86:17–21.
45. Assmus B, Rolf A, Erbs S, Elsässer A, Haberbosch W, Hambrecht R, Tillmanns H, Yu J, Corti R, Mathey DG, Hamm CW, Süselbeck T, Tonn T, Dimmeler S, Dill T, Zeiher AM, Schächinger V, REPAIR-AMI Investigators. Clinical outcome 2 years after intracoronary administration of bone marrow-derived progenitor cells in acute myocardial infarction. Circ Heart Fail. 2010;3:89–96. doi:10.1161/CIRCHEARTFAILURE.108.843243.
46. Traverse JH, Henry TD, Pepine CJ, Willerson JT, Zhao DX, Ellis SG, Forder JR, Anderson RD, Hatzopoulos AK, Penn MS, Perin EC, Chambers J, Baran KW, Raveendran G, Lambert C, Lerman A, Simon DI, Vaughan DE, Lai D, Gee AP, Taylor DA, Cogle CR, Thomas JD, Olson RE, Bowman S, Francescon J, Geither C, Handberg E, Kappenman C, Westbrook L, Piller LB, Simpson LM, Baraniuk S, Loghin C, Aguilar D, Richman S, Zierold C, Spoon DB, Bettencourt J, Sayre SL, Vojvodic RW, Skarlatos SI, Gordon DJ, Ebert RF, Kwak M, Moyé LA, Simari RD, Cardiovascular Cell Therapy Research Network (CCTRN). Effect of the use and timing of bone marrow mononuclear cell delivery on left ventricular function after acute myocardial infarction: the TIME randomized trial. JAMA. 2012;308:2380–9.
47. Hare JM, Fishman JE, Gerstenblith G, DiFede Velazquez DL, Zambrano JP, Suncion VY, Tracy M, Ghersin E, Johnston PV, Brinker JA, Breton E, Davis-Sproul J, Schulman IH, Byrnes J, Mendizabal AM, Lowery MH, Rouy D, Altman P, Wong Po Foo C, Ruiz P, Amador A, Da Silva

J, IK MN, Heldman AW, George R, Lardo A. Comparison of allogeneic vs autologous bone marrow–derived mesenchymal stem cells delivered by transendocardial injection in patients with ischemic cardiomyopathy: the POSEIDON randomized trial. JAMA. 2012;308:2369–79.
48. Li TS, Cheng K, Malliaras K, Smith RR, Zhang Y, Sun B, Matsushita N, Blusztajn A, Terrovitis J, Kusuoka H, Marbán L, Marbán E. Direct comparison of different stem cell types and subpopulations reveals superior paracrine potency and myocardial repair efficacy with cardiosphere-derived cells. J Am Coll Cardiol. 2012;59:942–53.
49. Bolli R, Chugh AR, D'Amario D, Loughran JH, Stoddard MF, Ikram S, Beache GM, Wagner SG, Leri A, Hosoda T, Sanada F, Elmore JB, Goichberg P, Cappetta D, Solankhi NK, Fahşah I, Rokosh DG, Slaughter MS, Kajstura J, Anversa P. Cardiac stem cells in patients with ischaemic cardiomyopathy (SCIPIO): initial results of a randomised phase 1 trial. Lancet. 2011;378:1847–57.
50. Makkar RR, Smith RR, Cheng K, Malliaras K, Thomson LE, Berman D, Czer LS, Marbán L, Mendizabal A, Johnston PV, Russell SD, Schuleri KH, Lardo AC, Gerstenblith G, Marbán E. Intracoronary cardiosphere-derived cells for heart regeneration after myocardial infarction (CADUCEUS): a prospective, randomized phase 1 trial. Lancet. 2012;379:895–904. doi:10.1016/S0140-6736(12)60195-0.
51. Malliaras K, Makkar RR, Smith RR, Cheng K, Wu E, Bonow RO, Marbán L, Mendizabal A, Cingolani E, Johnston PV, Gerstenblith G, Schuleri KH, Lardo AC, Marbán E. Intracoronary cardiosphere-derived cells after myocardial infarction: evidence of therapeutic regeneration in the final 1-year results of the CADUCEUS trial (CArdiosphere-Derived aUtologous stem CElls to reverse ventricUlar dySfunction). J Am Coll Cardiol. 2014;63:110–22.
52. The Lancet Editors. Expression of concern: the SCIPIO trial. Lancet. 2014;383:1279. doi:10.1016/S0140-6736(14)60608-5.
53. Evans MJ, Kaufman MH. Establishment in culture of pluripotential cells from mouse embryos. Nature. 1981;292:154–6.
54. Martin GR. Isolation of a pluripotent cell line from early mouse embryos cultured in medium conditioned by teratocarcinoma stem cells. Proc Natl Acad Sci U S A. 1981;78:7634–8.
55. Thomson JA, Itskovitz-Eldor J, Shapiro SS, Waknitz MA, Swiergiel JJ, Marshall VS, Jones JM. Embryonic stem cell lines derived from human blastocysts. Science. 1998;282:1145–7.
56. Odorico JS, Kaufman DS, Thomson JA. Multilineage differentiation from human embryonic stem cell lines. Stem Cells. 2001;19:193–204.
57. Chong JJ, Yang X, Don CW, Minami E, Liu YW, Weyers JJ, Mahoney WM, Van Biber B, Cook SM, Palpant NJ, Gantz JA, Fugate JA, Muskheli V, Gough GM, Vogel KW, Astley CA, Hotchkiss CE, Baldessari A, Pabon L, Reinecke H, Gill EA, Nelson V, Kiem HP, Laflamme MA, Murry CE. Human embryonic-stem-cell-derived cardiomyocytes regenerate non-human primate hearts. Nature. 2014;510:273–7. doi:10.1038/nature13233.
58. Takahashi K, Yamanaka S. Induction of pluripotent stem cells from mouse embryonic and adult fibroblast cultures by defined factors. Cell. 2006;126:663–76.
59. Takahashi K, Tanabe K, Ohnuki M, Narita M, Ichisaka T, Tomoda K, Yamanaka S. Induction of pluripotent stem cells from adult human fibroblasts by defined factors. Cell. 2007;131:861–72.
60. Yu J, Vodyanik MA, Smuga-Otto K, Antosiewicz-Bourget J, Frane JL, Tian S, Nie J, Jonsdottir GA, Ruotti V, Stewart R, Slukvin II, Thomson JA. Induced pluripotent stem cell lines derived from human somatic cells. Science. 2007;318:1917–20.
61. Aasen T, Izpisúa Belmonte JC. Isolation and cultivation of human keratinocytes from skin or plucked hair for the generation of induced pluripotent stem cells. Nat Protoc. 2010;5:371–82. doi:10.1038/nprot.2009.241.
62. Mauritz C, Schwanke K, Reppel M, Neef S, Katsirntaki K, Maier LS, Nguemo F, Menke S, Haustein M, Hescheler J, Hasenfuss G, Martin U. Generation of functional murine cardiac myocytes from induced pluripotent stem cells. Circulation. 2008;118:507–17. doi:10.1161/CIRCULATIONAHA.108.778795.
63. Zhang J, Wilson GF, Soerens AG, Koonce CH, Yu J, Palecek SP, Thomson JA, Kamp TJ. Functional cardiomyocytes derived from human induced pluripotent stem cells. Circ Res. 2009;104:e30–41. doi:10.1161/CIRCRESAHA.108.192237.

64. Ye L, Chang YH, Xiong Q, Zhang P, Zhang L, Somasundaram P, Lepley M, Swingen C, Su L, Wendel JS, Guo J, Jang A, Rosenbush D, Greder L, Dutton JR, Zhang J, Kamp TJ, Kaufman DS, Ge Y, Zhang J. Cardiac repair in a porcine model of acute myocardial infarction with human induced pluripotent stem cell-derived cardiovascular cells. Cell Stem Cell. 2014;15:750–61. doi:10.1016/j.stem.2014.11.009.
65. Ma N, Stamm C, Kaminski A, Li W, Kleine HD, Müller-Hilke B, Zhang L, Ladilov Y, Egger D, Steinhoff G. Human cord blood cells induce angiogenesis following myocardial infarction in NOD/scid-mice. Cardiovasc Res. 2005;66:45–54.
66. Perin EC, Sanz-Ruiz R, Sánchez PL, Lasso J, Pérez-Cano R, Alonso-Farto JC, Pérez-David E, Fernández-Santos ME, Serruys PW, Duckers HJ, Kastrup J, Chamuleau S, Zheng Y, Silva GV, Willerson JT, Fernández-Avilés F. Adipose-derived regenerative cells in patients with ischemic cardiomyopathy: The PRECISE Trial. Am Heart J. 2014;168:88–95.e2. doi:10.1016/j.ahj.2014.03.022.
67. Kobayashi T, Hamano K, Li TS, Nishida M, Ikenaga S, Hirata K, Zempo N, Esato K. Angiogenesis induced by the injection of peripheral leukocytes and platelets. J Surg Res. 2002;103:279–86.
68. Menasche P. Cardiac cell therapy: lessons from clinical trials. J Mol Cell Cardiol. 2011;50:258–65.

Chapter 7
Angiogenic Therapy for Ischemic Cardiomyopathy with Cell Sheet Technology

Shigeru Miyagawa and Yoshiki Sawa

Abstract Although recent progress in medications, left ventricular assist device (LVAD), or heart transplantation did a favor for heart failure patients, additional treatments including cell therapy and tissue engineering are required because of some drawbacks such as infections or donor shortage in clinical situation.

Cell sheet technology has been developed, and angiogenic therapy by using skeletal cell sheet has been introduced to clinical with a large number of preclinical studies, aiming angiogenesis by cytokine paracrine effects. Moreover, to induce cardiomyogenesis in failed heart cardiomyocytes derived from induced pluripotent stem cells (iPS cells) has been reported to play a crucial part in myocardial regeneration in severely damaged myocardium. Regenerative technology with cell sheet has some potentials in the clinical treatment of heart failure which has little response to the internal medical or conventional surgical treatment, and this technology may open new era in the treatment of heart failure.

Keywords Heart failure • Angiogenesis • Cell sheet

7.1 Introduction

Despite of great advancement in medical treatment, heart failure mainly caused by ischemic or dilated cardiomyopathy (ICM or DCM) is a life-threatening disorder worldwide. In this situation cardiac surgeon and researchers introduced left ventricular assist device (LVAD) implantation [1] and heart transplantation [2] to overcome these problems. Owing to their untiring efforts, these permutation strategies have made great contributions to treatment for heart failure for the last several decades. But due to the durability of LVAD [3] and the donor shortage [4], this surgical treatment has some limitations such as infection in LVAD and donor shortage

S. Miyagawa • Y. Sawa (✉)
Department of Cardiovascular Surgery, Osaka University Graduate School of Medicine, 2-2 Yamadaoka, Suita, Osaka 565-0871, Japan
e-mail: sawa-p@surg1.med.osaka-u.ac.jp

Y. Higashi, T. Murohara (eds.), *Therapeutic Angiogenesis*,
DOI 10.1007/978-981-10-2744-4_7

in heart transplantation to treat end-stage heart failure. This clinical situation has led cardiologists or cardiac surgeon to consider alternative strategy of treating end-stage heart failure.

The recent remarkable progress in myocardial regeneration therapy is cellular cardiomyoplasty, and this technique has already introduced to the treatment of heart failure with skeletal myoblasts [5] or bone marrow mononuclear cells (BM-MNCs) [6]. Although these challenges demonstrated its feasibility and safety, its efficacy may be insufficient to completely damaged myocardium which surgeon targets. Considering these results in cellular cardiomyoplasty, next-generation strategy in myocardial regeneration therapy, tissue-engineered cardiomyoplasty by using cell sheet technique and cell-based scaffold implantation technique, was developed experimentally and clinically.

The aim of this section is to analyze recent advances about myocardial regeneration effectiveness induced by bioengineered cardiac tissue, especially cell sheet technology.

7.2 Recent Advancement of Bioengineered Myocardial Graft-Cell Sheet

Recently artificial cardiac tissues were created ex vivo and implanted to the distressed heart on behalf of scattered cells. This cardiac tissue will be implanted to failed heart only via surgical procedure, so only cardiac surgeons hold an unchallenged position as introducer to clinical situation. This method can provide viable muscle tissue to completely distressed myocardium which has little myocyte and massive fibrosis (myogenesis) and induce angiogenesis. This is a distinctive feature of bioengineered cardiac tissue and attracts many clinicians who enthuse to regenerate completely distressed myocardium.

In 1999 a three-dimensional cardiac-like tissue which has spontaneous contract ability by using the self-assembling properties of neonatal cardiomyocyte was developed [7]. Kelm et al. pioneered three-dimensional microtissue by using same approach [8], and Baar et al. obtained a cylindrical construct which spontaneously contracts and produces force [9]. Another encouraging approach to pioneer myocardial tissue without scaffold is cell sheet technique. Okano T et al. developed cell sheet technique [10], and this technology was applied to several diseased organs, such as the heart [11], eye [12], and kidney [13], experimentally and clinically. Cell sheets can be removed from special dishes which are coated by temperature-responsive polymer, poly(*N*-isopropylacrylamide) (PIPAAm), by using the ability of changes from hydrophobic to hydrophilic in lowering the temperature without destruction of cell-cell or cell-extracellular matrix (ECM) community in cell sheet. The strongest point of this technology is that cell sheet is made of cells and ECM produced by cells itself without artificial scaffolds [14]. Cell sheet has high ability of integration to native tissues because of the preserving adhesion molecules on the surface [15].

Shimizu et al. developed contractile chick cardiomyocyte sheet which were recognized to heart tissue-like structure and show electrical pulsatile amplitude by using this special dish without enzymatic or EDTA treatment [16]. They layered one cell sheet to make bilayer cell sheets (the electrically communicative three-dimensional cardiac construct) which show the spontaneous and synchronous pulsation and revealed that each cell sheet adhered together rapidly by detection of desmosomes and intercalated disks between cell sheets [17]. Four-layered neonatal rat cardiomyocyte sheet was developed, and they have electrical communication between sheets via connexin43. This pulsatile cardiac tissue could survive in the subcutaneous tissue up to 1 year and showed spontaneous beating, heart-like structure, neovascularization, and the improvement of its size, conduction velocity, and contractile force in proportion to the host growth [18, 19]. Cardiomyocyte sheet has a quality of flexibility and can change form easily. Myocardial tubes were created and this tube can produce the pressure and follow Frank-Starling mechanism [20]. Sekine et al. wrapped this tube to rat thoracic aorta and revealed that this tube can produce the pressure in vivo [21].

Interestingly, electrical coupling between two sheets start in approximately 34 min and completed in about 46 min by a multiple-electrode extracellular recording system, and histological examination reveals that connexin43 was detected within 30 min [22]. This data predict that cardiomyocyte sheet make an electrical coupling between cell sheet and host myocardium within at least 1 h after implantation to myocardium. Miyagawa et al. demonstrated that neonatal cardiomyocyte sheet survive in the infarct myocardium and communicate electrically with host myocardium by the suggestion of connexin43 existence and morphological changes of QRS wave and amplitude in action potential leading to the improvement of cardiac performance [11]. Similarly, another paper reports that there is electrical integration between neonatal myocyte sheets and host heart by electrophysiological study [23]. Moreover, functional gap junction and morphological integration via "bridging cardiomyocyte" between sheet and host myocardium is detected [24]. These in vitro and in vivo studies guarantee that there is electrical and morphological coupling between cell sheets and host myocardium exactly, and cell sheets may contractile synchronously with host heart beating and contribute to the improvement of regional systolic function.

To consider human application, thin-layered cardiomyocyte sheets which thickness of one cardiomyocyte sheet is approximately 45 μm [18] may not be enough to repair the end-stage heart failure, and the greatest disadvantage of this technique is the difficulty to obtain thick-layered cardiomyocyte sheets. The question is how many cardiomyocyte sheets can be layered at maximum in vivo? Problem is oxygen supply to cardiomyocyte sheet, that is, vascularization to cardiac tissue after implantation. When four- or six-layered neonatal cardiomyocyte sheet was implanted to the subcutaneous tissue of athymic rats, they are survived 1 year with angiogenesis which was rapidly organized within a few days, and its thickness was approximately 100 μm [19]. When over four sheets ware implanted, central necrosis was found because of loss of oxygen supply instead of rapid organization in microvasculature [25]. Polysurgery technique was developed to overcome poor vascularization and produce thick, vascularized cardiac tissue, and this multistep transplant technique can recreate approximately 1 mm thick myocardium with a well-organized vasculature network [25]. But the problem is how to implant to diseased myocardium after incubating thick-layered cardiomyocyte sheets in ectopic tissue, that is, how to connect microvasculature network in thick-layered cardiomyocyte sheets to host myocardium. This problem must be overcome for polysurgery technique to be widely used. About the vascularization processes after implantation, Sekiya et al. depicts that cardiomyocyte sheet has a potential angiogenic factors such as the expression of angiogenesis-related genes and endothelial cell network. Interestingly, vasculature in the layered cardiomyocyte sheets comes from the layered cardiomyocyte sheets itself, and vessels prolong from layered cardiomyocyte sheets to host myocardium to connect with host vasculature [26]. This is a very important paper to support the idea that additional angiogenetic factors such as endothelial cells and some angiogenic growth factors [27] enhance the angiogenesis for thick-layered cardiomyocyte sheets to survive in the damaged myocardium. Based on this paper, Sekine et al. reported that neonatal cardiomyocytes-endothelial cells cocultured sheet could enhance the improvement of cardiac performance in the original cardiomyocyte sheet with enhanced vascularization [28]. These techniques about the enhancement of angiogenesis may be a breakthrough for integration of thick-layered cardiomyocyte sheets incubated in ectopic tissue to damaged myocardium.

This cell sheet technique is applied to several cell sources such as skeletal myoblasts [29–31], mesenchymal stem cells (MSCs) [32], human smooth muscle cells [33], fibroblast cocultured with endothelial progenitor cells [34], and cardiomyocyte cocultured with endothelial cells [28]. In clinical setting, cellular cardiomyoplasty has been reported to have the potential of fundamental regenerative capability and has already been introduced in clinical trials with skeletal myoblast, and results suggest that it is a relatively feasible and safety therapy [35]. In this situation skeletal myoblast is a most close cell source to clinical application at this stage for tissue cardiomyoplasty. Memon et al. demonstrated that non-ligature implantation of skeletal myoblast sheet regenerated the damaged myocardium and improved global cardiac function by attenuating the cardiac remodeling in the rat ligation model by the mechanisms of the recruitment of hematopoietic stem cells and the release of some growth factors. Moreover, this cell delivery system by using cell sheets implantation showed better restoration of damaged myocardium compared with needle injection [30] with less arrhythmogenicity [36]. Skeletal myoblast sheet was applied to dilated cardiomyopathy hamster model, and this paper revealed that deteriorated myocardium was recovered accompanied with the reservation of alpha-sarcoglycan and beta-sarcoglycan expression on host myocyte and inhibition of fibrosis [29]. They implanted myoblast sheets to 27 weeks DCM hamster which is in a moderate heart failure stage (Fractional shortening 16%) and showed the preservation of distressed heart function and histology and prolongation of survival. Although these results depicts that skeletal myoblast sheet has a possibility to treat the moderate heart failure in DCM, efficacy to end-stage heart failure is unknown. Farther study is needed to elucidate the efficacy of cell sheet implantation for end-stage heart failure. Moreover, grafting of skeletal myoblast sheets attenuated cardiac remodeling and improved cardiac performance in pacing-induced canine heart failure model [31]. These papers demonstrate that skeletal myoblast sheet has a potential to regenerate the deteriorate myocardium induced by coronary artery diseases and DCM both in small and large animal model. But large animal study is only this paper, and there is no work to elucidate the long-term results after cell sheet implantation.

The mechanism of recovery in damaged myocardium has not been completely elucidated and is may be very complicated. Although the abovementioned paper revealed the possibility of cytokine release and hematopoietic stem cells recruitment about the mechanisms of regeneration, this is only one part of regenerative mechanisms. The restoration of structural proteins may be due to the relief of the myocyte stretch evidenced by the reduction of left ventricle dimension or growth factors. It has been already revealed that skeletal myoblasts cannot beat synchronously with host myocardium in vitro [37] and in vivo study [38]. After myoblast sheet implantation, the diastolic dysfunction in the distressed region of myocardium was significantly recovered (form our human and porcine studies, data not shown) leading to the improvement of systolic function in the same region without the contraction of implanted myoblasts. The improvement of diastolic function may depend on the elasticity of newly formed scar tissue by transplanted skeletal myoblast sheet and migrated cells. Histologically massive angiogenesis in the implanted region is one of the crucial characters, and we speculate that angiogenesis and recovery of diastolic function is a main part of regenerative mechanism in myoblast sheet implantation.

One of the main questions may be how long implanted cell sheet survive and whether implanted cell sheet can permanently work in ICM after implantation. We have already checked the survival of myoblast sheet by PCR analysis of Y chromosome in the experiments in which the male cell sheet was implanted to female rat infarction model and revealed that the number of implanted cells declined to about 20% compared with pre-value and implanted myoblast sheet disappear at least 3 months after implantation by histological analysis. Several hours after implantation, implanted myoblasts are exposed to ischemic condition, and they express HIF-1 gene which induce angiogenic cytokines such as HGF or VEGF, and stromal-derived factor-1 (SDF-1) is expressed in the implanted cell sheet which enhances the recruitment of mesenchymal stem cells from bone marrow which also play an important part in therapeutic angiogenesis. It has already been reported that main process in angiogenesis induced by cytokines may happen, and newly formed vasculature may be established only within 1 month after implantation, and the function in newly formed vasculature may be maintained by recruited mesenchymal stem cells in spite of survival in implanted myoblast sheet. So implanted myoblast sheet may be only an "opportunity" to induce angiogenesis in ischemic myocardium.

Although many paper reported that angiogenesis induced by paracrine effects in the cell sheet plays a crucial part in recovery in ICM, induced newly formed vasculature by cell sheet histologically seems to be immature vasculatures which possess only endothelial cells without smooth muscle cells. So how to induce mature vasculature in ischemic region may be an important issue in the next generation of myocardial regeneration therapy. Kainuma et al. reported that they implanted myoblast sheets to epicardium in ICM with omentum and revealed that this combination therapy induce therapeutical angiogenesis in ischemic myocardium with vascular maturation which showed both endothelial and smooth muscle cells, leading functional angiogenic recovery with improved coronary flow reserve [39].

The growth of MSCs sheet on the infarct myocardium contributes to the improvement of anterior wall thickness with new vessels, and some implanted cells differentiate to cardiomyocyte [32]. This paper is characterized by the self-incubation of cell sheet in vivo, and differentiated cardiomyocyte may not contribute to the improvement of systolic function because of low incidence of differentiation from MSCs to cardiac myocyte. Although maximum thickness of MSCs sheet, approximately 600 μm, is not enough to human end-stage heart failure, this method of self-incubation in vivo is very unique and may be one of the possible strategies to recreate thick-layered sheet in vivo.

Two-cell cocultured sheet was developed to enhance the angiogenesis [28]. Cultured cell sheet combined with fibroblast and endothelial progenitor cells enhanced blood vessel formation leading the functional improvement. This strategy may be good candidate for ischemic myocardium or peripheral arterial disease to target only angiogenesis. There is another paper which targets to angiogenesis for peripheral arterial disease. Cocultured cell sheet combined with fibroblast and human smooth muscle cell accelerates the secretion of angiogenic factor in vitro and increase the blood perfusion by the formation of new vessels [33]. Enhanced effectiveness induced by coculture of two cells is supported by a paper which reveals that BMC and myoblast co-implantation enhances the single-cell transplantation efficacy [40]. It attracts researchers' interest that cell sheet changes its character by coculture method and enhance the effectiveness in cell sheet derived from one kind of cell.

Based on physiological and pathological characters in diseases, clinician can decide which type cell sheet is appropriate to each disease such as ischemic myocardium (angiogenesis is needed), end-stage DCM (myogenesis is needed), acute myocardial infarction (angiogenesis is needed), and chronic ischemic cardiomyopathy (both angiogenesis and myogenesis are needed). Implantation of stem cell which can differentiate to cardiomyocyte in vivo can induce little myogenesis because of low incidence of differentiation to cardiomyocyte in vivo [41] and massive cell loss in needle injection method. Although cardiomyocyte sheet is only a candidate in myogenesis which can replace myocardial scar tissue with functioning contractile tissue because of no evidence of synchronized beating in myoblast in vivo. For the aim of cardiomyogenesis with angiogenesis in cardiomyopathy, cardiomyocyte sheet derived from pluripotent stem (iPS) cells has been already basically developed with the evidence of synchronous work with recipient myocardium [42] and with significant functional recovery [43], safety after in vivo implantation by immature cell elimination [44], and a large number of cell culture system [44, 45].

And also clinical application of myoblast sheet has been done for ICM patients and proved the safety in this therapy with possible efficacy [46], and myoblast sheet has already been approved by the government to clinical use as "heart sheet" [47]. But several clinical applications have only proved the possibility in efficacy, so further study may be needed to elucidate the efficacy in the myoblast sheet for ICM.

7.3 Conclusions

In this section, we surveyed exciting topics in myocardial generation therapy, especially in cell sheet. Cell-based treatment inclusive cellular cardiomyoplasty and tissue cardiomyoplasty having angiogenic ability makes remarkable progress in a very short period of time, and many researchers and clinicians have enthusiasm to develop new technology by using iPS cells to relieve diseased patients. Owing to these works, cell sheet was introduced to clinical application, but many things remain unknown or immature both in basic and clinical respect. However, we are just at the beginning of clinical myocardial regenerative therapy, and now cardiologists and cardiac surgeons may assimilate basic achievement to routine clinical therapy.

References

1. Akatsu T, Murai S, Kamiya S, Kojima K, Mizuhashi Y, Hasegawa H, Kitagawa Y. Perineal hernia as a rare complication after laparoscopic abdominoperineal resection: report of a case. Surg Today. 2009;39:340–3.
2. Barnard CN. The operation. A human cardiac transplant: an interim report of a successful operation performed at groote schuur hospital, cape town. S Afr Med J. 1967;41:1271–4.
3. El-Banayosy A, Korfer R, Arusoglu L, Kizner L, Morshuis M, Milting H, Tenderich G, Fey O, Minami K. Device and patient management in a bridge-to-transplant setting. Ann Thorac Surg. 2001;71:S98–102; discussion S114–105.
4. Piccione Jr W. Bridge to transplant with the heartmate device. J Card Surg. 2001;16:272–9.
5. Menasche P, Alfieri O, Janssens S, McKenna W, Reichenspurner H, Trinquart L, Vilquin JT, Marolleau JP, Seymour B, Larghero J, Lake S, Chatellier G, Solomon S, Desnos M, Hagege AA. The myoblast autologous grafting in ischemic cardiomyopathy (magic) trial: first randomized placebo-controlled study of myoblast transplantation. Circulation. 2008;117:1189–200.
6. Strauer BE, Brehm M, Zeus T, Kostering M, Hernandez A, Sorg RV, Kogler G, Wernet P. Repair of infarcted myocardium by autologous intracoronary mononuclear bone marrow cell transplantation in humans. Circulation. 2002;106:1913–8.
7. Akins RE, Boyce RA, Madonna ML, Schroedl NA, Gonda SR, McLaughlin TA, Hartzell CR. Cardiac organogenesis in vitro: reestablishment of three-dimensional tissue architecture by dissociated neonatal rat ventricular cells. Tissue Eng. 1999;5:103–18.
8. Kelm JM, Ehler E, Nielsen LK, Schlatter S, Perriard JC, Fussenegger M. Design of artificial myocardial microtissues. Tissue Eng. 2004;10:201–14.
9. Baar K, Birla R, Boluyt MO, Borschel GH, Arruda EM, Dennis RG. Self-organization of rat cardiac cells into contractile 3-d cardiac tissue. FASEB J. 2005;19:275–7.
10. Okano T, Yamada N, Sakai H, Sakurai Y. A novel recovery system for cultured cells using plasma-treated polystyrene dishes grafted with poly(n-isopropylacrylamide). J Biomed Mater Res. 1993;27:1243–51.

11. Miyagawa S, Sawa Y, Sakakida S, Taketani S, Kondoh H, Memon IA, Imanishi Y, Shimizu T, Okano T, Matsuda H. Tissue cardiomyoplasty using bioengineered contractile cardiomyocyte sheets to repair damaged myocardium: their integration with recipient myocardium. Transplantation. 2005;80:1586–95.
12. Nishida K, Yamato M, Hayashida Y, Watanabe K, Yamamoto K, Adachi E, Nagai S, Kikuchi A, Maeda N, Watanabe H, Okano T, Tano Y. Corneal reconstruction with tissue-engineered cell sheets composed of autologous oral mucosal epithelium. N Engl J Med. 2004;351:1187–96.
13. Kushida A, Yamato M, Isoi Y, Kikuchi A, Okano T. A noninvasive transfer system for polarized renal tubule epithelial cell sheets using temperature-responsive culture dishes. Eur Cell Mater. 2005;10:23–30; discussion 23–30.
14. Masuda S, Shimizu T, Yamato M, Okano T. Cell sheet engineering for heart tissue repair. Adv Drug Deliv Rev. 2008;60:277–85.
15. Kushida A, Yamato M, Konno C, Kikuchi A, Sakurai Y, Okano T. Decrease in culture temperature releases monolayer endothelial cell sheets together with deposited fibronectin matrix from temperature-responsive culture surfaces. J Biomed Mater Res. 1999;45:355–62.
16. Shimizu T, Yamato M, Kikuchi A, Okano T. Two-dimensional manipulation of cardiac myocyte sheets utilizing temperature-responsive culture dishes augments the pulsatile amplitude. Tissue Eng. 2001;7:141–51.
17. Shimizu T, Yamato M, Akutsu T, Shibata T, Isoi Y, Kikuchi A, Umezu M, Okano T. Electrically communicating three-dimensional cardiac tissue mimic fabricated by layered cultured cardiomyocyte sheets. J Biomed Mater Res. 2002;60:110–7.
18. Shimizu T, Yamato M, Isoi Y, Akutsu T, Setomaru T, Abe K, Kikuchi A, Umezu M, Okano T. Fabrication of pulsatile cardiac tissue grafts using a novel 3-dimensional cell sheet manipulation technique and temperature-responsive cell culture surfaces. Circ Res. 2002;90:e40.
19. Shimizu T, Sekine H, Isoi Y, Yamato M, Kikuchi A, Okano T. Long-term survival and growth of pulsatile myocardial tissue grafts engineered by the layering of cardiomyocyte sheets. Tissue Eng. 2006;12:499–507.
20. Kubo H, Shimizu T, Yamato M, Fujimoto T, Okano T. Creation of myocardial tubes using cardiomyocyte sheets and an in vitro cell sheet-wrapping device. Biomaterials. 2007;28:3508–16.
21. Sekine H, Shimizu T, Yang J, Kobayashi E, Okano T. Pulsatile myocardial tubes fabricated with cell sheet engineering. Circulation. 2006;114:I87–93.
22. Haraguchi Y, Shimizu T, Yamato M, Kikuchi A, Okano T. Electrical coupling of cardiomyocyte sheets occurs rapidly via functional gap junction formation. Biomaterials. 2006;27:4765–74.
23. Furuta A, Miyoshi S, Itabashi Y, Shimizu T, Kira S, Hayakawa K, Nishiyama N, Tanimoto K, Hagiwara Y, Satoh T, Fukuda K, Okano T, Ogawa S. Pulsatile cardiac tissue grafts using a novel three-dimensional cell sheet manipulation technique functionally integrates with the host heart, in vivo. Circ Res. 2006;98:705–12.
24. Sekine H, Shimizu T, Kosaka S, Kobayashi E, Okano T. Cardiomyocyte bridging between hearts and bioengineered myocardial tissues with mesenchymal transition of mesothelial cells. J Heart Lung Transplant. 2006;25:324–32.
25. Shimizu T, Sekine H, Yang J, Isoi Y, Yamato M, Kikuchi A, Kobayashi E, Okano T. Polysurgery of cell sheet grafts overcomes diffusion limits to produce thick, vascularized myocardial tissues. FASEB J. 2006;20:708–10.
26. Sekiya S, Shimizu T, Yamato M, Kikuchi A, Okano T. Bioengineered cardiac cell sheet grafts have intrinsic angiogenic potential. Biochem Biophys Res Commun. 2006;341:573–82.
27. Miyagawa S, Sawa Y, Taketani S, Kawaguchi N, Nakamura T, Matsuura N, Matsuda H. Myocardial regeneration therapy for heart failure: hepatocyte growth factor enhances the effect of cellular cardiomyoplasty. Circulation. 2002;105:2556–61.
28. Sekine H, Shimizu T, Hobo K, Sekiya S, Yang J, Yamato M, Kurosawa H, Kobayashi E, Okano T. Endothelial cell coculture within tissue-engineered cardiomyocyte sheets enhances neovascularization and improves cardiac function of ischemic hearts. Circulation. 2008;118: S145–52.

29. Kondoh H, Sawa Y, Miyagawa S, Sakakida-Kitagawa S, Memon IA, Kawaguchi N, Matsuura N, Shimizu T, Okano T, Matsuda H. Longer preservation of cardiac performance by sheet-shaped myoblast implantation in dilated cardiomyopathic hamsters. Cardiovasc Res. 2006;69:466–75.
30. Memon IA, Sawa Y, Fukushima N, Matsumiya G, Miyagawa S, Taketani S, Sakakida SK, Kondoh H, Aleshin AN, Shimizu T, Okano T, Matsuda H. Repair of impaired myocardium by means of implantation of engineered autologous myoblast sheets. J Thorac Cardiovasc Surg. 2005;130:1333–41.
31. Hata H, Matsumiya G, Miyagawa S, Kondoh H, Kawaguchi N, Matsuura N, Shimizu T, Okano T, Matsuda H, Sawa Y. Grafted skeletal myoblast sheets attenuate myocardial remodeling in pacing-induced canine heart failure model. J Thorac Cardiovasc Surg. 2006;132:918–24.
32. Miyahara Y, Nagaya N, Kataoka M, Yanagawa B, Tanaka K, Hao H, Ishino K, Ishida H, Shimizu T, Kangawa K, Sano S, Okano T, Kitamura S, Mori H. Monolayered mesenchymal stem cells repair scarred myocardium after myocardial infarction. Nat Med. 2006;12:459–65.
33. Hobo K, Shimizu T, Sekine H, Shin'oka T, Okano T, Kurosawa H. Therapeutic angiogenesis using tissue engineered human smooth muscle cell sheets. Arterioscler Thromb Vasc Biol. 2008;28:637–43.
34. Kobayashi H, Shimizu T, Yamato M, Tono K, Masuda H, Asahara T, Kasanuki H, Okano T. Fibroblast sheets co-cultured with endothelial progenitor cells improve cardiac function of infarcted hearts. J Artif Organs. 2008;11:141–7.
35. Dib N, Michler RE, Pagani FD, Wright S, Kereiakes DJ, Lengerich R, Binkley P, Buchele D, Anand I, Swingen C, Di Carli MF, Thomas JD, Jaber WA, Opie SR, Campbell A, McCarthy P, Yeager M, Dilsizian V, Griffith BP, Korn R, Kreuger SK, Ghazoul M, MacLellan WR, Fonarow G, Eisen HJ, Dinsmore J, Diethrich E. Safety and feasibility of autologous myoblast transplantation in patients with ischemic cardiomyopathy: four-year follow-up. Circulation. 2005;112:1748–55.
36. Patila T, Miyagawa S, Imanishi Y, Fukushima S, Siltanen A, Mervaala E, Kankuri E, Harjula A, Sawa Y. Comparison of arrhythmogenicity and proinflammatory activity induced by intramyocardial or epicardial myoblast sheet delivery in a rat model of ischemic heart failure. PLoS One. 2015;10:e0123963.
37. Itabashi Y, Miyoshi S, Yuasa S, Fujita J, Shimizu T, Okano T, Fukuda K, Ogawa S. Analysis of the electrophysiological properties and arrhythmias in directly contacted skeletal and cardiac muscle cell sheets. Cardiovasc Res. 2005;67:561–70.
38. Leobon B, Garcin I, Menasche P, Vilquin JT, Audinat E, Charpak S. Myoblasts transplanted into rat infarcted myocardium are functionally isolated from their host. Proc Natl Acad Sci U S A. 2003;100:7808–11.
39. Kainuma S, Miyagawa S, Fukushima S, Pearson J, Chen YC, Saito A, Harada A, Shiozaki M, Iseoka H, Watabe T, Watabe H, Horitsugi G, Ishibashi M, Ikeda H, Tsuchimochi H, Sonobe T, Fujii Y, Naito H, Umetani K, Shimizu T, Okano T, Kobayashi E, Daimon T, Ueno T, Kuratani T, Toda K, Takakura N, Hatazawa J, Shirai M, Sawa Y. Cell-sheet therapy with omentopexy promotes arteriogenesis and improves coronary circulation physiology in failing heart. Mol Ther. 2015;23:374–86.
40. Memon IA, Sawa Y, Miyagawa S, Taketani S, Matsuda H. Combined autologous cellular cardiomyoplasty with skeletal myoblasts and bone marrow cells in canine hearts for ischemic cardiomyopathy. J Thorac Cardiovasc Surg. 2005;130:646–53.
41. Laflamme MA, Murry CE. Regenerating the heart. Nat Biotechnol. 2005;23:845–56.
42. Higuchi T, Miyagawa S, Pearson JT, Fukushima S, Saito A, Tsuchimochi H, Sonobe T, Fujii Y, Yagi N, Astolfo A, Shirai M, Sawa Y. Functional and electrical integration of induced pluripotent stem cell-derived cardiomyocytes in a myocardial infarction rat heart. Cell Transplant. 2015;24:2479–89.
43. Kawamura M, Miyagawa S, Miki K, Saito A, Fukushima S, Higuchi T, Kawamura T, Kuratani T, Daimon T, Shimizu T, Okano T, Sawa Y. Feasibility, safety, and therapeutic efficacy of human induced pluripotent stem cell-derived cardiomyocyte sheets in a porcine ischemic cardiomyopathy model. Circulation. 2012;126:S29–37.

44. Matsuura K, Kodama F, Sugiyama K, Shimizu T, Hagiwara N, Okano T. Elimination of remaining undifferentiated induced pluripotent stem cells in the process of human cardiac cell sheet fabrication using a methionine-free culture condition. Tissue Eng Part C Methods. 2015;21:330–8.
45. Miyagawa S, Fukushima S, Imanishi Y, Kawamura T, Mochizuki-Oda N, Masuda S, Sawa Y. Building a new treatment for heart failure-transplantation of induced pluripotent stem cell-derived cells into the heart. Curr Gene Ther. 2016;16:5–13.
46. Imamura T, Kinugawa K, Sakata Y, Miyagawa S, Sawa Y, Yamazaki K, Ono M. Improved clinical course of autologous skeletal myoblast sheet (tcd-51073) transplantation when compared to a propensity score-matched cardiac resynchronization therapy population. J Artif Organs. 2016;19(1):80–6.
47. Sawa Y, Yoshikawa Y, Toda K, Fukushima S, Yamazaki K, Ono M, Sakata Y, Hagiwara N, Kinugawa K, Miyagawa S. Safety and efficacy of autologous skeletal myoblast sheets (tcd-51073) for the treatment of severe chronic heart failure due to ischemic heart disease. Circ J. 2015;79:991–9.

Part II
Gene Therapy

Chapter 8
The Role of VEGF in the Extremities

Brendan A.S. McIntyre, Takayuki Asahara, and Cantas Alev

Abstract Vascular endothelial growth factor (VEGF) is a pro-angiogenic cytokine that has a strong stimulatory effect on endothelial cell proliferation and migration. The use of VEGF in treating vascular conditions affecting the extremities, such as peripheral artery disease (PAD) and critical limb ischemia (CLI), has been studied in both basic research and clinical settings for over 30 years. In this chapter, we will discuss the animal models and applied treatments that have been attempted in this arena, with an emphasis on clinical translation. Novel combined treatments pairing cellular therapy with VEGF gene therapy will also be discussed.

Keywords VEGF • $VEGF_{165}$ • $VEGF_{121}$ • Ischemia • Peripheral artery disease • Critical limb ischemia

8.1 VEGF in Therapeutic Angiogenesis and VEGF Isoforms

Vascular endothelial growth factor (VEGF) was originally discovered more than 30 years ago and has been one of the most extensively studied cytokines known to the field of biomedicine [1]. VEGF is an angiogenic factor that has been associated with enhanced endothelial cell migration, proliferation, and vessel repair [2, 3]. The VEGF family of genes is comprised of five members, VEGF-A, VEGF-B, VEGF-C, VEGF-D, and PGF or placental growth factor, with signaling occurring through one of three receptors, VEGFR1 (FLT-1), VEGFR2 (FLK-1/KDR) or

B.A.S. McIntyre
Miltenyi Biotec Inc., 2303 Lindbergh St, Auburn, CA, USA

T. Asahara
Tokai University School of Medicine, 143 Shimokasuya, Isehara 259-1193, Kanagawa, Japan

C. Alev (✉)
Kyoto University, Center for iPS Cell Research and Application (CiRA),
53 Kawahara-cho, Shogoin, Sakyo-ku, 606-8507 Kyoto, Japan
e-mail: alev@cira.kyoto-u.ac.jp

Y. Higashi, T. Murohara (eds.), *Therapeutic Angiogenesis*,
DOI 10.1007/978-981-10-2744-4_8

VEGFR3 (FLT-4), and co-receptor neuropilin-1 (NRP1). Interestingly, two additional VEGF-A homologues not expressed in humans have been found to occur outside of the traditional VEGF family, known as VEGF-E and VEGF-F. VEGF-E occurs as a single exon-less cDNA found in the genomes of *Orf* viruses [4], and VEGF-F refers to venom-type VEGFs found in a number of venomous snake species [5]; both VEGF-E and VEGF-F preferentially bind to VEGFR2 and cause edema and vessel leakiness.

In the field of biomedicine, the VEGF family member most commonly associated with clinical intervention due to its strong pro-angiogenic activity in both normal development and pathological settings is VEGF-A [3]. Despite the existence of at least seven reported isoforms, there are four major isoforms of VEGF-A, VEGF-121, VEGF-165, VEGF-189, and VEGF-206, which are produced via alternative splicing and differ by the presence or absence of sequences in exons 6 and 7 [6]. VEGF-165 and VEGF-121 are the only VEGF isoforms to have been used in a clinical setting to treat CLI to date and will be the focus of this chapter. Of these, VEGF-165 (or VEGF-A_{165}) is the most extensively studied and has demonstrated the most potent and bioactive pro-angiogenic functions in vitro and in vivo [7].

8.2 Evidence of Efficacy of VEGF in Experimental Preclinical Models

8.2.1 Generation One: VEGF Plasmid DNA and Recombinant Protein Injection

The use of VEGF for the revascularization of ischemic tissue is a field of research and therapy built upon the pioneering work performed in the laboratory of Jeffrey Isner [8–10]. Rather than focusing on directly abrogating vascular smooth muscle (SMC) proliferation in the neointima, these early VEGF-based studies using VEGF-165 ($VEGF_{165}$) were based on the idea that re-endothelialization through VEGF-mediated enhancement of endothelial cell (EC) migration and proliferation would stabilize vessel integrity and thus attenuate hyperplasia [10, 11]. The mechanism of this remodeling had been shown to occur as a result of nitric oxide (NO) and prostacyclin production in the endothelium [12]. Expanding upon these and other discoveries, a rabbit model of hind limb ischemia [13] was employed to test the effect of intramuscular (IM) injection of recombinant VEGF protein [8] and later VEGF DNA, using the $phVEGF_{165}$ plasmid as a vector [14]. This latter study opened the way for intramuscular naked DNA injection (plasmid DNA in saline without liposome encapsulation) as a viable clinical alternative form of gene transfer in the treatment of limb ischemia. A subsequent study using a rat model of hind limb ischemia, where the animals were injected with a replication-deficient adenoviral vector expressing $VEGF_{165}$ (AdVEGF), further validated this work, with both models showing sustained transgene expression for greater than 1 week postinjection [15].

The authors in both studies argued that in comparison to recombinant protein, expression from plasmid DNA or adenoviral particles would provide a more sustained level of VEGF in the host. Presumably, the use of recombinant human VEGF (rhVEGF) would necessitate frequent and multiple injections, making it an impractical form of therapy. The half-life of VEGF protein under normoxic conditions is 15–20 min, with VEGF mRNA being only slightly better at 30 min. Both VEGF protein and mRNA half-life are further extended in hypoxic environments, as would be encountered in the ischemic limb. However, the mRNA half-life of VEGF under hypoxic conditions is much longer, remaining stable for 6–8 h [16–19]. Further, due to the leader sequence present in exon 1 of the VEGF gene, VEGF is packaged for extracellular export when expressed by host cells [20]. These studies exploring the basic functions of the VEGF gene, combined with the numerous preclinical models employing mice, rats, and rabbits, provided the groundwork that paved the way for the translation of VEGF-based therapies to the clinic. However, even during this translational phase, some mouse models had exhibited certain shortcomings, including instances where VEGF alone was not sufficient to restore circulation, such as cases of endothelial nitric oxide synthase (eNOS) insufficiency in the *Akt1* knockout mouse [21]. This study demonstrated that concomitant active eNOS is required for VEGF-induced revascularization [21, 22]. Furthermore, functional VEGF receptors are also required for proper signal transduction in order to see an observable effect of VEGF-mediated therapies. For instance, it has been shown that hind limb ischemia models using BALB/c mice are refractory to VEGF-mediated therapies, due to impaired expression of VEGF receptor 2/Flk-1/KDR [23]. This brings up the interesting perspective regarding the need for proper patient-based genetic screening when applying VEGF-based therapies in the clinic, a practice that is not currently performed.

In addition to studies of VEGF-165, the use of VEGF-121 has been investigated in preclinical models of ischemia in animals. VEGF-121 is the only form of VEGF that does not contain a basic heparin-binding region, due to the absence of 15 basic amino acids within exon 7, and is thus freely diffusible [24]. This isoform of VEGF was tested in both rats and rabbits using an adenoviral vector, AdCMV.VEGF121 [25]. An interesting variation in this trial is that the transgene was administered into healthy limbs prior to induction of hind limb ischemia. Results from this study showed significant improvement following a single dose four-point IM injection, with respect to tissue perfusion and vascularization. Not only did this pave the way for the potential clinical use of this therapy as a preventive measure against ischemia, but it also demonstrated for the first time that therapeutic angiogenesis could be used to induce vessel growth in normal, healthy skeletal muscle tissue [25].

8.2.2 Generation Two: Viral Vectors and Combined VEGF plus Cell-Based Therapies

One of the major concerns brought up regarding initial VEGF-based therapies was that vessels formed following non-dose-controlled injections are immature and leaky [26]. In order to overcome this obstacle, an elegant approach employing adeno-associated viruses was developed enabling the controlled application of inducible VEGF expression, allowing a more tightly controlled dosing of the transgene [27–29]. In doing so, dose rates of VEGF were controlled and importantly switched off, allowing for proper maturation of formed vessels. Although this approach has yet to be translated to the clinic, it provides a tantalizing model for dose-controlled VEGF gene therapy.

Another way proposed to make gene therapy more effective is to use a therapeutic cell type as a targeted vector. Endothelial progenitor cells or EPCs offer such a potential target cell type. EPCs were initially identified as circulating progenitors with enhanced angiogenic properties [30] and continue to hold promise in the treatment of ischemic limbs [31]. Early on in their discovery, it was noted that EPCs themselves could serve as vectors of VEGF delivery, combining not only the regenerative capacity of the growth factor but that of the cellular vector as well [32]. In order to test this, a mouse model of hind limb ischemia was used in conjunction with adenoviral gene transfer of VEGF. This approach resulted in significant improvement in the therapeutic effect of administered EPCs, as measured primarily by lower limb amputation rates [32]. But this study also noted that in order to translate these findings to the clinic, as much as 12 L of peripheral blood would be required to obtain a sufficient therapeutic dose of EPCs in humans—an impossible feat given the current limits of ex vivo cell expansion technology.

An alternative albeit still hypothetical approach could be the use of human embryonic stem cells (ESCs) or induced pluripotent stem cells (iPSCs) (collectively called human pluripotent stem cells or hPSCs), which have been shown to give rise to endothelial cells and progenitors of the cardiovascular system in vitro [33–35]. Human PSC-derived vascular progenitor cells could be used as cellular vectors for the targeted delivery of VEGF or other pro-angiogenic growth factors in vivo, thus freeing possible future clinical applications from the dependency on and limitations of primary blood or bone marrow cells (Table 8.1).

Table 8.1 Summary of preclinical models used for VEGF therapy

Method of delivery	Lead author	Year	Gene (isoform)	Animal model	Detail of therapeutic	Major findings, advances, or limitations
A. Single-factor studies						
Protein	Takeshita [8]	1994	$VEGF_{165}$	Rabbit (limb ischemia)	rhVEGF (IM)	Dose-dependent augmentation in limb perfusion and vascularization
Plasmid DNA	Asahara [10]	1996	$VEGF_{165}$	Rabbit (femoral artery balloon injury)	$phVEGF_{165}$ (IA via catheter)	Tested EC response after balloon injury, therapeutic delivered via hydrogel polymer-coated balloon catheter
Plasmid DNA	Tsurumi [14]	1996	$VEGF_{165}$	Rabbit (limb ischemia)	$phVEGF_{165}$ (IM)	Demonstrated use of IM injection of naked plasmid DNA as an effective option for PAD, CLI
Adenovirus	Mack [15]	1998	$VEGF_{165}$	Rat (limb ischemia)	AdVEGF (IM)	Shows adenovirus as a viable option for delivery of transgenes for PAD, CLI
Adenovirus	Gowdak [25]	2000	$VEGF_{121}$	Rat and rabbit (limb ischemia)	$AdCMV.VEGF_{121}$ (IM)	Demonstrated ability of $VEGF_{121}$ to stimulate angiogenesis in experimental model
Adeno-associated virus (AAV)	Tafuro [27]	2009	$VEGF_{165}$	Mouse (limb ischemia)	pAAV-TRE2/VEGF (IM)	Used a tetracycline inducible system to control dosage of VEGF

(continued)

Table 8.1 (continued)

Method of delivery	Lead author	Year	Gene (isoform)	Animal model	Detail of therapeutic	Major findings, advances, or limitations
Hydrogel-embedded VEGF protein	Phelps [62]	2010	$VEGF_{121}$	Mouse (limb ischemia)	$rhVEGF_{121}$	Showed that hydrogel-embedded VEGF121 is more effective, likely due to enhanced stability and retention
AAV—conditionally silenced VEGF	Boden [29]	2016	$VEGF_{165}$	Mouse (limb ischemia)	dsAAV1-CS-hVEGF (IM)	Modified AAV vector using conditional silencing; VEGF expressed only under hypoxia
B. Combination therapies						
Dual protein	Asahara [9]	1995	$VEGF_{165}$ + FGF2	Rabbit (limb ischemia)	rhVEGF and rhFGF2 (IA via catheter)	Synergistic effects of dual FGF2 (bFGF) and VEGF treatment on vascularization
EPC overexpressing VEGF	Iwaguro [32]	2002	$VEGF_{164}$ (mouse $VEGF_{165}$)	Mouse (limb ischemia)	EPC transduced with ad/VEGF (IV via tail vein)	30× reduction in EPC number administered to observe improvement, compared to previous studies
Plasmid DNA (VEGF + Ang1)	Shyu [67]	2003	$VEGF_{165}$ + Ang1	Rabbit (limb ischemia)	phVEGF165 + pAng1 (IM)	Ang1 potentiates response of VEGF in hind limb ischemia
AAV (VEGF + Ang1)	Chen [68]	2007	$VEGF_{165}$ + Ang1	Rabbit (limb ischemia)	AAV-VEGF/Ang1 (IM)	Encoding of two genes on an adeno-associated viral vector and improved vessel integrity when Ang1 was added to VEGF

VEGF mimetic peptide	Santulli [73]	2009	VEGF/KDR binding interface	Rat (limb ischemia)	QK (IM)	Performed as well as $VEGF_{165}$ in revascularization
VEGF mimetic peptide amphiphiles (nanofibers)	Webber [75]	2011	VEGF nanofiber	Mouse (limb ischemia)	VEGF PA (IM)	Improved activity over $VEGF_{165}$ due to the ability of nanofiber to enhance display of epitope
Chimeric protein (VEGF + Ang1)	Anisimov [71]	2013	Chimeric protein, receptor-binding region	Mouse (limb ischemia)	rhVA1 (IM)	Receptor-binding parts of VEGF and Ang1 used to make chimeric protein VA1; more mature vessels observed with VA1 than VEGF only

Table lists the type of therapeutic used, lead author on the publication reporting its use, year of publication, the isoform of VEGF used in the study, the animal model used to test efficacy, detail of the therapeutic and how it was administered, and the major findings, advances, and limitations of the study. Table is divided into single-factor (A) and combined-factor (B) studies. *IA* intra-arterial, *IM* intramuscular, *IV* intravenous, *PA* peptide amphiphile, *rh* recombinant human

8.3 Clinical Trials Using VEGF for Neoangiogenesis

8.3.1 Plasmid-Based Clinical Trials Using VEGF

Clinical translation of the initial experimental models employing VEGF in the treatment of PAD was proposed in 1996 [36], and case studies began to surface already in that same year with balloon angioplasty administration of phVEGF$_{165}$ in one patient [37]. This is a logical progression, given the limited treatment options available to physicians, resulting in high levels of limb amputation, juxtaposed to the minimally invasive nature of the proposed therapy. In fact, CLI, an exacerbation of PAD, is responsible for the vast majority of limb amputations in the developed world according to a 17-year retrospective study conducted on Finnish medical data up to the year 2000 [38]. Recent molecular observations made from clinical cases have also pointed out that patients with CLI may benefit from supplementing VEGF levels in affected limbs based on measurements of local VEGF expression levels [39, 40]. In the initial clinical gene transfer studies, a plasmid encoding VEGF-A$_{165}$ under the cytomegalovirus (CMV) promoter was used. FDA approval had been granted for a phase I clinical trial, and within a couple of years, trial results were being reported [41]. The treatment regimen of this first ischemic limb trial consisted of two doses of 2 mg IM injections of plasmid DNA into four sites in the affected limbs of nine patients. Results from this study looking primarily at safety and feasibility of the gene transfer treatment were extremely encouraging, with improved blood flow in the limbs of eight patients, healing of ulcers occurring in four of seven affected limbs, and successful limb salvage occurring in three patients, all of which prompted and encouraged further clinical work in this area [41].

The same treatment course was applied to six patients affected by Buerger's disease, or thromboangiitis obliterans (TBO), a vascular occlusive disease seen in young smokers, characterized by inflammation and blockage of small- to medium-sized arteries in the extremities. In this small trial, which included two patients from the trial of Baumgartner et al., all patients demonstrated increased blood flow and formation of new vessels. Additionally, three of five limbs afflicted with ulcers healed [42]. As with the previous study, this trial did note that successful limb salvage could only occur if gene therapy is instituted before the establishment of necrotic lesions.

After gaining global recognition, trials were initiated in Korea, Taiwan, the Netherlands, and Russia. The Korean trial used VEGF$_{165}$ plasmid DNA injected IM in nine patients, seven of whom had Buerger's disease and reported good tolerance without any significant side effects related to gene transfer [43]. With its primary goals being to establish safety and dose-escalation effects, the patient pool was not large enough to draw conclusions on the efficacy of the applied treatment. However, dose escalation of up to 8 mg of injected plasmid did not result in improved vascular conductivity, as assessed by ABI (also used by the Isner-led studies), but did note that serum VEGF levels were not increased, even at high doses [43]. One of the major unique aspects of the Kim et al. study is that they used a different vector, pCK-$VEGF_{165}$, to drive VEGF expression. The difference between pCK-$VEGF_{165}$ and ph$VEGF_{165}$ vectors is that the promoter used in pCK has been shown to drive higher expression levels in the skeletal muscles in animal models [44]. This was achieved by virtue of retaining a longer upstream sequence of the CMV promoter [44].

At around the same time, a Taiwanese group reported a larger clinical trial of 27 CLI patients using ph$VEGF_{165}$ [45]. Both of these non-US-based studies reported transient edema as the major patient complication observed, which is a consistent side effect observed in multiple studies [46]. But intriguingly, even at much lower doses, 2 × 0.4 to 2 mg for the Taiwanese trial versus 2 × 2 to 8 mg for the Korean trial, which used the modified pCK plasmid, with reported higher expression of VEGF IM [44], the lower-dose Taiwanese study showed significantly elevated levels of circulating VEGF, which were not seen in the Korean study. In an attempt to determine the minimal effective dose, the Taiwanese study also determined that no less than 2 × 1.2 mg was needed to see a positive effect on vascularization as measured by ABI. By showing patient tolerability, the Taiwanese study also addressed concerns about applying pro-angiogenic VEGF treatments outside of Western patient pools.

Following these two studies, a larger clinical trial conducted in the Netherlands used ph$VEGF_{165}$ with the same 2 × 2 mg IM injection regimen in a larger cohort of diabetic patients suffering from CLI [47]. This larger trial included 27 placebo/control and 27 VEGF-treated individuals, with patients injected in the thigh (two aliquots) and calf muscle (two aliquots) of the most ischemic limb and repeated once after a span of 4 weeks. Again, results failed to demonstrate significant differences in amputation rates, although secondary measures such as improvement in skin ulcers and circulation were better in the VEGF-treated group in addition to 14 subjects in the treatment group reporting clinical improvement versus only three in the control group [47].

More recently, a report detailing a Russian phase 2b/3 multicenter trial with 100 enrolled patients was published [48]. In this randomized, controlled clinical trial, the largest study conducted to date using $VEGF_{165}$ plasmid DNA under the CMV promoter, the authors concluded that VEGF gene transfer was an effective way to treat CLI based on pain-free walking at 2 years post-treatment. Yet again, amputation rates between placebo and treated patients were not significantly different, but showed a trend toward improvement in the treated patients, another theme consistent with all previous trials. Intriguingly, quality of life (QoL) scores were not found to be significantly different in the two groups despite improved mobility, blood flow, and healing rates.

Cumulatively, the results of this trial were sufficient to register the drug *Neovasculgen* for the treatment of patients with moderate-to-severe claudication due to stages 2a to 3 atherosclerotic lower limb ischemia in Russia [48]. The authors went on to recommend further studies with larger patient cohorts to evaluate pCMV-$VEGF_{165}$ in patients with resting pain from peripheral atherosclerosis, ischemia caused by diabetes mellitus or autoimmune disorders, and those who undergo peripheral artery revascularization in order to expand the number of conditions to apply this therapeutic. Despite the inherent caveats of this and previous studies, the evidence for the clinical benefits of IM injection of naked or plasmid DNA is becoming apparent, with multicenter trials now conducted in five countries over the past three decades [49]. The current holder of drug rights to pCMV-$VEGF_{165}$, the Human Stem Cells Institute (HSCI), currently has approvals for marketing authorization of *Neovasculgen* in Russia and the Ukraine, with approval processes initiated in 2016 for the FDA in the USA and in China (http://eng.hsci.ru/products/neovasculgen). The HSCI also plans to explore the use of *Neovasculgen* for ischemic heart disease, diabetic foot syndrome, and treatment of trauma-induced peripheral nerve damage.

8.3.2 Non-plasmid-Based Clinical Trials Using VEGF

The first adenoviral trial using VEGF was reported in 2002 using a replication-deficient adenovirus as a vector to deliver cDNA encoding VEGF-121, $Ad_{GV}VEGF_{121.10}$ [50]. With only 11 of the 15 VEGF-121-treated and 1 of the 3 placebo-treated patients completing the 1-year follow-up, conclusions with respect to efficacy could not be made at that time [50, 51]. However, this initial study led to a larger cohort of patients being recruited for the Regional Angiogenesis with VEGF (RAVE) trial, which was a phase 2, randomized, double-blind, placebo-controlled, dose-finding study [52]. Dosing in this study was between 4 billion viral particles for low dose and 40 billion viral particles for high dose administered at one time in the form of 20 IM injections. At present, final results of this trial have not been reported in the form of a peer-reviewed publication, but subsequent reports regarding the RAVE findings have discussed the lack of improvement with respect to blood flow measured by ankle- and toe-brachial index (ABI and TBI), QoL, and wound healing [47]. This may likely be due to the high diffusibility of $VEGF_{121}$ in the bloodstream compared with the more localized effects attributed to $VEGF_{165}$, which binds to heparan sulfate-bearing proteins (proteoglycans) and is thus much more prone to remain sequestered in the extracellular matrix [53]. Questions have also been raised regarding the half-life of $VEGF_{121}$, particularly that a higher turnover of this protein does not permit prolonged and complete angiogenesis as does $VEGF_{165}$ [54].

A second trial directly comparing adenovirus to liposome-complexed plasmid DNA was also reported in 2002 [55]. Statistical significance was not reached with any of the endpoints in this study. Unlike other preceding studies, administration of vector adenoviral particles or plasmid DNA was conducted intra-arterially, as opposed to with IM injections. It was argued that this method of delivery was chosen over IM injection due to the hypothesis that it would lead to a more widespread distribution of the transgene [55]. The authors concluded there was no improvement in VEGF gene transfer or even activity levels with adenoviral administration, contrary to their hypothesis. However, they did conclude that VEGF treatment resulted in improved vascularity in treated limbs, in agreement with previous trials in which naked DNA was injected IM [55] (Table 8.2).

Table 8.2 Chronological listing of VEGF-based clinical therapies according to publication date of findings

Year	Lead author	No. of patients	Trial type	Dosage	Detail of therapeutic	Major findings and limitations
1996	Isner [37]	1	Case study	2 mg via hydrogel-coated angioplasty balloon to the popliteal artery	$phVEGF_{165}$	Increased angiogenesis and circulation. Limitations: single-patient case study
1998	Baumgartner [41]	9	Phase I	2 × 2 mg injections of plasmid DNA injected IM	$phVEGF_{165}$	Improved blood flow in 8/9 patients; healing of 4/7 affected limb ulcers
1998	Isner [42]	6	Phase I	2 or 4 mg of plasmid DNA injected IM	$phVEGF_{165}$	Improved blood flow in 7/7 limbs, ulcers healing in 3/5 affected limbs. Limitations: two patients used from the above study
2002	Makinen [55]	54	Phase II	2×10^{10} pfu AV particles or 2 mg of liposome-complexed plasmid DNA; arterial infusion following percutaneous transluminal angioplasty	pCMV-$VEGF_{165}$ or $VEGF_{165}$-Ad	Placebo-controlled trial showing increased vascularity in treated limbs
2003	Rajagopalan [52]	105	Phase II	4 or 40×10^9 pfu AV particles injected IM	$Ad_{GV}VEGF_{121.10}$	Lack of statistical improvement for all measurements
2003	Shyu [45]	27	Phase I	2 × 0.4 to 2 mg of plasmid DNA injected IM	$phVEGF_{165}$	Improved blood flow, ulcer healing, and rest pain

(continued)

Table 8.2 (continued)

Year	Lead author	No. of patients	Trial type	Dosage	Detail of therapeutic	Major findings and limitations
2004	Kim [43]	9	Phase I	2 × 2 to 8 mg of plasmid DNA injected IM	pCK-$VEGF_{165}$	Improved blood flow
2006	Kusumanto [47]	54	Phase I	2 × 2 mg of plasmid DNA injected IM	ph$VEGF_{165}$	Placebo-controlled trial showing increased blood flow and wound healing in treated patients
2015	Skora [49]	32	Phase I	2 mg of plasmid DNA injected IM with 0.8 to 3.9 × 10^9 viral particles injected	ph$VEGF_{165}$ + BM MNCs	Autologous BM-MNCs administered with VEGF plus trial were compared to control using vasodilator pentoxifylline
2015	Deev [48]	100	Phase IIb/III	2 × 1.2 mg of plasmid DNA injected IM + BM-MNCs	pCMV-$VEGF_{165}$	Improvement in pain-free walking in treatment group

Only VEGF treatments dealing with PAD or CLI are included. Table also reports major findings of each study. Despite showing measures of patient improvement, all listed studies have failed to demonstrate significant reduction in amputation rates. *AV* adenoviral, *BM-MNC* bone marrow mononuclear cells, *IM* intramuscular

8.4 Perspectives and Future of VEGF as a Treatment for Ischemic Limb Diseases

Therapeutic intramuscular gene therapy holds much promise for the treatment of CLI and PAD. Evidence from animal models has shown that this relatively simple, minimally invasive procedure can promote vessel development resulting in increased circulation and healing of ulcers in treated limbs. However, there have been significant challenges in translating the successes of animal models to the clinic in this arena. The limitations of preclinical models are raised in the context of taking an otherwise healthy animal and inducing ischemia through a sudden trauma such as ligation and incision, as compared to the human condition which develops gradually and where adjacent tissue is not otherwise healthy but potentially diseased and atherosclerotic, leaving questions as to whether cytokine responses will be equivalent [56].

Yet, despite their limitations, animal models are able to inform on the effectiveness of angiogenic induction and, importantly, offer a setting with which to study more complex combinations of multiplexed therapies such as combined cellular and gene, multigene, or gene and drug treatments.

With respect to the clinical findings of the completed VEGF studies, a common and somewhat surprising result is the lack of improvement of quality of life (QoL) scores associated with these treatments. In an attempt to understand this finding, despite the improvement of circulation, wound healing, and rest pain, it has been speculated that there may be underlying psychological depression associated with individuals experiencing PAD and CLI, making quantitative measures of improvement difficult [48]. However, in terms of patient safety, all trials conducted to date have reported that plasmid or "naked DNA" injection of VEGF is well tolerated. The most common side effects encountered to date have been transient edema, which is resolved without the use of diuretics [43, 45]. Incidences of cancer, or retinopathy, originally hypothesized to potentially result from high serum levels of VEGF, have not materialized.

While all studies report improvement in vascularization and wound healing, the major clinically important benchmark, reduction in limb amputation, has failed to show significant improvement, despite data trends indicating otherwise. The evaluation of VEGF in treating neuropathy is another factor to consider when analyzing clinical efficacy. Neuropathy, causing numbness in the limbs affected by CLI, can be associated with increased morbidity potentially leading to amputation as a result of loss of pain sensation. As the focus of this chapter has been on vascularization and neoangiogenesis resulting from VEGF treatment for PAD, this therapeutic aspect has not been explored in detail. However, reports of clinical improvement of chronic neuropathy associated with VEGF gene transfer in CLI have been described in at least one trial [57] and were supported by work using animal models [58]. A positive effect for VEGF leads to cautious speculation that ulceration and eventual limb amputation rates should decrease as a result of restored sensation [59]. Conversely, studies have also reported that a placebo effect resulting from improved patient care over the course of clinical trials may be occurring as well, which would lead to lower amputation rates in both placebo- and VEGF-treated patients [47]. This fact may provide clinicians with a relatively simple, yet potentially costly, guideline to keep amputation rates down. Put succinctly, improved patient support and care would appear to translate to lower amputation rates for CLI.

One possible improvement that could be incorporated into future trials using single-factor VEGF gene therapy alone would be to complex liposomes with the $phVEGF_{165}$ plasmid. Liposome complexes have been shown to improve efficacy of gene transfer in animal models [60] and have been used in clinical trials of VEGF gene therapy [55]. However, they have not been used with the most commonly adopted form of clinical VEGF gene transfer, IM injection, and in the past were only delivered with arterial infusion following percutaneous transluminal angioplasty. Giving further hope for this improvement is the fact that it has been known for some time that liposome-mediated gene transfer is effective IM [61].

Along the same lines as the improvement of gene transfer via liposomes, testing the use of hydrogels such as polyethylene glycol (PEG) matrices, which can be used to tether biomolecules such as growth factors, cytokines, peptides, DNA vectors, and drugs in place, offers potential for improvement as well. In the arena of pro-angiogenic therapy, PEG hydrogels have been shown to significantly improve the rate of retention and activity of recombinant $VEGF_{121}$ when complexed with these materials [62]. This model also demonstrated functional differences in the ability of hydrogel-embedded VEGF to reperfuse ischemic mouse limbs at a greater rate than recombinant protein alone [62]. Fine-tuning such models by combining fibrin hydrogels with a fusion VEGF protein that allows enzymatic cleavage and release from these gels provides a further level of control over cytokine delivery in vivo [63]. These methods could be used for adenoviral gene delivery as well, ensuring more controlled and local release of the therapeutic, specifically to its target location. This approach has been employed for lentiviral gene delivery of VEGF, which was held in place via porous PEG hydrogels with heparin-chitosan nanoparticles used to retain vectors in situ [64].

The combination of growth factors with cellular therapy, as both a vector for delivery and adjuvant to treatment, is an avenue of research that is showing much promise and will also find its translation to the clinic for CLI and PAD. Future clinical studies will surely aim to combine cellular plus gene therapy approaches as demonstrated in animal models combining VEGF gene transfer with EPCs [32]. In fact clinical progress in this arena is already underway with a 3.7 million USD study set up to fund the phase I trial of lentiviral-transduced mesenchymal stromal cells (MSCs) engineered to overexpress VEGF. Initial results in four patients have already demonstrated a therapeutic effect in terms of improved blood flow in affected ischemic limbs. It remains to be seen if these combination therapies are able to demonstrate reduction in limb amputation rates, which has remained elusive to VEGF-based gene therapies thus far.

The clinical trials examined here using VEGF gene therapy for the treatment of PAD and CLI have shown that there are some clear advantages to the use of the $VEGF_{165}$ isoform. It has consistently led to improvements in vascular flow, wound healing, and in some cases lower pain at rest. Nevertheless, important concerns with respect to the rigor of these clinical trials have been raised, such as the limits of using ankle-brachial index (ABI) to measure increased perfusion, which is a common measure in all trials [65]. This is due to the fact that the ABI is based on the blood pressure in the distal posterior or anterior tibial artery, which is contrary to the smaller arteries branching off these main trunks where increased vascularity from VEGF-based gene therapy is predominantly observed. The use of more sophisticated measures such as MRI-based arterial peak flow (APF) in future studies will surely help to elucidate the true effects of VEGF-based gene therapy for CLI and PAD [66].

Moving away from simpler single-factor trials, combination therapies have been evaluated experimentally for more than a decade and are progressing into the clinic as well [67]. They bring with them the obvious complication of dosing and delivering one or more factors which may exhibit biological interference and may not be uniformly expressed, distributed, or retained within the host tissue, thus making success with rigorously tested experimental models an important benchmark before translation to the clinic. Despite these hurdles, initial studies have demonstrated that combination therapies with $VEGF_{165}$ in addition to angiopoietin-1 (Ang1), which has been shown to inhibit VEGF leakiness and stabilize newly formed blood vessels, are more effective than either factor alone in rabbit models of hind limb ischemia when administered either by IM injection of plasmid DNA [67] or with adeno-associated viral (AAV) transduction using a vector encoding both genes [68]. Complex combination therapies will surely become more and more prevalent as the field evolves. As evidence of this, a recent phase I trial was reported in which cellular therapy and dual gene transfer were combined. In this study, a venous endothelial and smooth muscle cell product, MultiGeneAngio (MGA), was engineered to express $VEGF_{165}$ and Ang1 [69]. More recently, results from the phase 1b dose-escalation MGA study were published showing patient tolerability and significant amputation-free survival and healing rates [70].

Taking the idea of combined gene therapies one step further, using chimeric molecules made by fusing active domains of two biomolecules, is another evolution in this avenue of research. Although not yet tested clinically, chimeric molecules, which can be engineered to lack the unwanted side effects of native proteins while still giving clinical benefit, have already been created. For example, VEGF-Ang1 chimeric proteins such as VA1 have been evaluated experimentally, showing increased perfusion in animal models of limb ischemia, with less vessel leakage and inflammation than VEGF alone [71]. VEGF mimetics are also emerging as a new tool in the arsenal of treating CLI and PAD. The compound QK was designed to mimic a region of the VEGF binding interface, which binds the VEGF receptor KDR in solution and stimulates EC proliferation and signaling [72]. QK was put to the test in a rat hind limb ischemia model where it performed with $VEGF_{165}$ in inducing revascularization [73]. More recently, this helical VEGF mimetic peptide was shown to induce vessel sprouting both in vivo and in vitro and, importantly, showed that it could be used as a complimentary therapy alongside VEGF, as putting both together resulted in augmented effectiveness [74]. In addition to QK, another mimetic using self-assembling nanofiber technology to display VEGF peptides on their surface was found to give functional recovery in a mouse hind limb ischemia model [75]. It remains to be seen how VEGF mimetics will perform in combination with multigene therapies or cellular plus gene and drug therapy approaches, especially in a clinical setting. An overview of the therapeutic funnel, which includes treatments tested in animal models and those also validated in the clinical setting, has been compiled in Fig. 8.1.

As an additional emerging technology, endonuclease-based genome-editing tools offer new approaches for the endogenous activation of VEGF in situ. While proof-of-concept studies utilizing, e.g., clustered regularly interspaced short palindromic repeat (CRISPR)/Cas9-based genome editing have demonstrated that local activation of transcript levels is achievable [76], it remains to be seen how this will emerge as a treatment for limb ischemia [77]. In this context it was already shown that expression of VEGF-A can be activated in mammalian cells in vitro using either

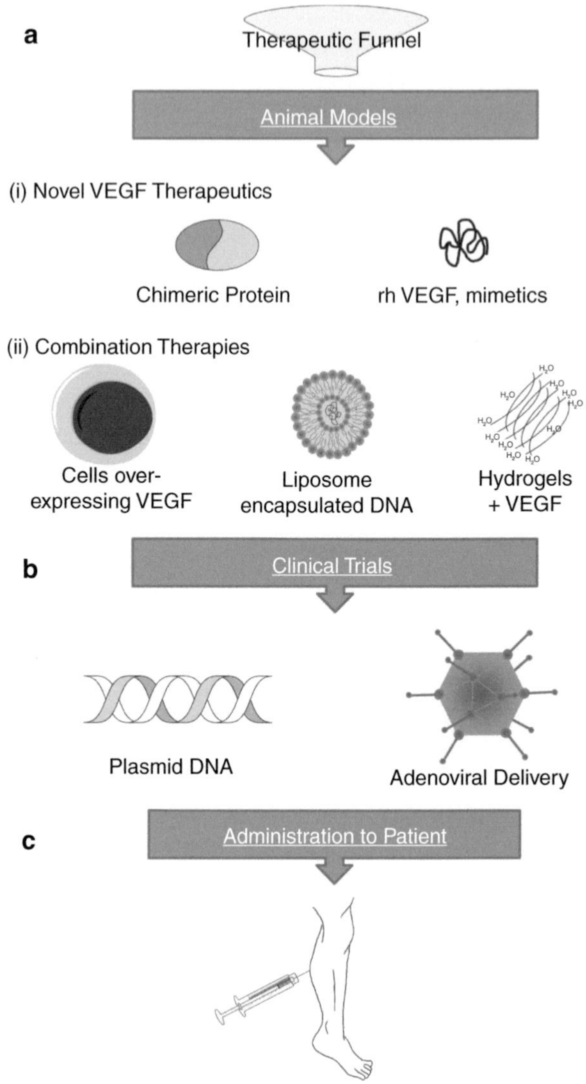

Fig. 8.1 Experimental and clinical therapies involving VEGF for patients suffering from ischemic limb diseases. (**a**) The therapeutic funnel with (i) novel examples of VEGF therapies such as chimeric proteins and mimetic peptides and (ii) combination therapies where cells are used a vectors, or VEGF is incorporated into liposomes or hydrogels. (**b**) Therapies tested in the clinical setting, which include plasmid DNA and adenoviral particles. (**c**) Administration of therapeutics to the patient. Intramuscular (IM) delivery, the most common method of VEGF administration-based therapy, is shown (images not to scale)

transcription activator-like effector (TALE) or CRISPR/Cas9-based genome-editing technology [78, 79]. Advanced genome-engineering tools will likely contribute to the establishment of improved cell-based therapies, utilizing either adult tissue or pluripotent stem cell (PSC)-derived stem or somatic cells, modified to enhance VEGF expression without the use of inserted transgenes. As such, we can envision autologous cellular therapies harnessing, for example, human-induced pluripotent stem cells (hiPSCs), differentiated to venous or arterial ECs, or autologous in vitro-induced MSCs or EPCs and modified ex vivo for enhanced VEGF expression prior to reintroduction into an ischemic limb. Re-expressing wild-type forms of mutant proteins has already been achieved for β-thalassemia [80] and cystic fibrosis [81], and expanding this strategy to overexpress VEGF and/or other pro-angiogenic factors for the treatment of CLI and PAD could take advantage of these advances in cellular therapy and gene editing.

In summary, there are a number of studies examining the efficacy of VEGF in treating PAD and CLI. Animal models have tended to give results that were more favorable than clinical trials, a discrepancy resulting from a variety of factors, including health status prior to injury, lack of preexisting ulcers or lesions, and genetic and lifestyle homogeneity. Future therapeutics will aim to bring new levels of sophistication to these treatments, yet with new treatments come new challenges. Thus, VEGF, and in particular $VEGF_{165}$, remains an important line of therapy and experimentation for the clinical treatment of CLI and PAD.

References

1. Senger DR, Galli SJ, Dvorak AM, Perruzzi CA, Harvey VS, Dvorak HF. Tumor cells secrete a vascular permeability factor that promotes accumulation of ascites fluid. Science. 1983;219(4587):983–5.
2. Dvorak HF, Brown LF, Detmar M, Dvorak AM. Vascular permeability factor/vascular endothelial growth factor, microvascular hyperpermeability, and angiogenesis. Am J Pathol. 1995;146(5):1029–39.
3. Hoeben A, Landuyt B, Highley MS, Wildiers H, Van Oosterom AT, De Bruijn EA. Vascular endothelial growth factor and angiogenesis. Pharmacol Rev. 2004;56(4):549–80. doi:10.1124/pr.56.4.3.
4. Ogawa S, Oku A, Sawano A, Yamaguchi S, Yazaki Y, Shibuya M. A novel type of vascular endothelial growth factor, VEGF-E (NZ-7 VEGF), preferentially utilizes KDR/Flk-1 receptor and carries a potent mitotic activity without heparin-binding domain. J Biol Chem. 1998;273(47):31273–82.
5. Yamazaki Y, Matsunaga Y, Tokunaga Y, Obayashi S, Saito M, Morita T. Snake venom vascular endothelial growth factors (VEGF-Fs) exclusively vary their structures and functions among species. J Biol Chem. 2009;284(15):9885–91. doi:10.1074/jbc.M809071200.
6. Ferrara N, Davis-Smyth T. The biology of vascular endothelial growth factor. Endocr Rev. 1997;18(1):4–25. doi:10.1210/edrv.18.1.0287.
7. Byrne AM, Bouchier-Hayes DJ, Harmey JH. Angiogenic and cell survival functions of vascular endothelial growth factor (VEGF). J Cell Mol Med. 2005;9(4):777–94.
8. Takeshita S, Pu LQ, Stein LA, Sniderman AD, Bunting S, Ferrara N, et al. Intramuscular administration of vascular endothelial growth factor induces dose-dependent collateral artery augmentation in a rabbit model of chronic limb ischemia. Circulation. 1994;90 (5 Pt 2):II228–34.
9. Asahara T, Bauters C, Zheng LP, Takeshita S, Bunting S, Ferrara N, et al. Synergistic effect of vascular endothelial growth factor and basic fibroblast growth factor on angiogenesis in vivo. Circulation. 1995;92(9 Suppl):II365–71.
10. Asahara T, Chen D, Tsurumi Y, Kearney M, Rossow S, Passeri J, et al. Accelerated restitution of endothelial integrity and endothelium-dependent function after phVEGF165 gene transfer. Circulation. 1996;94(12):3291–302.
11. Asahara T, Bauters C, Pastore C, Kearney M, Rossow S, Bunting S, et al. Local delivery of vascular endothelial growth factor accelerates reendothelialization and attenuates intimal hyperplasia in balloon-injured rat carotid artery. Circulation. 1995;91(11):2793–801.

12. Laitinen M, Zachary I, Breier G, Pakkanen T, Hakkinen T, Luoma J, et al. VEGF gene transfer reduces intimal thickening via increased production of nitric oxide in carotid arteries. Hum Gene Ther. 1997;8(15):1737–44. doi:10.1089/hum.1997.8.15-1737.
13. Takeshita S, Gal D, Leclerc G, Pickering JG, Riessen R, Weir L, et al. Increased gene expression after liposome-mediated arterial gene transfer associated with intimal smooth muscle cell proliferation. In vitro and in vivo findings in a rabbit model of vascular injury. J Clin Invest. 1994;93(2):652–61. doi:10.1172/JCI117017.
14. Tsurumi Y, Takeshita S, Chen D, Kearney M, Rossow ST, Passeri J, et al. Direct intramuscular gene transfer of naked DNA encoding vascular endothelial growth factor augments collateral development and tissue perfusion. Circulation. 1996;94(12):3281–90.
15. Mack CA, Magovern CJ, Budenbender KT, Patel SR, Schwarz EA, Zanzonico P, et al. Salvage angiogenesis induced by adenovirus-mediated gene transfer of vascular endothelial growth factor protects against ischemic vascular occlusion. J Vasc Surg. 1998;27(4):699–709.
16. Shima DT, Deutsch U, D'Amore PA. Hypoxic induction of vascular endothelial growth factor (VEGF) in human epithelial cells is mediated by increases in mRNA stability. FEBS Lett. 1995;370(3):203–8.
17. Levy AP, Levy NS, Wegner S, Goldberg MA. Transcriptional regulation of the rat vascular endothelial growth factor gene by hypoxia. J Biol Chem. 1995;270(22):13333–40.
18. Levy NS, Chung S, Furneaux H, Levy AP. Hypoxic stabilization of vascular endothelial growth factor mRNA by the RNA-binding protein HuR. J Biol Chem. 1998;273(11):6417–23.
19. Stein I, Neeman M, Shweiki D, Itin A, Keshet E. Stabilization of vascular endothelial growth factor mRNA by hypoxia and hypoglycemia and coregulation with other ischemia-induced genes. Mol Cell Biol. 1995;15(10):5363–8.
20. Leung DW, Cachianes G, Kuang WJ, Goeddel DV, Ferrara N. Vascular endothelial growth factor is a secreted angiogenic mitogen. Science. 1989;246(4935):1306–9.
21. Murohara T, Asahara T, Silver M, Bauters C, Masuda H, Kalka C, et al. Nitric oxide synthase modulates angiogenesis in response to tissue ischemia. J Clin Invest. 1998;101(11):2567–78. doi:10.1172/JCI1560.
22. Qian HS, Liu P, Huw LY, Orme A, Halks-Miller M, Hill SM, et al. Effective treatment of vascular endothelial growth factor refractory hindlimb ischemia by a mutant endothelial nitric oxide synthase gene. Gene Ther. 2006;13(18):1342–50. doi:10.1038/sj.gt.3302781.
23. Fukino K, Sata M, Seko Y, Hirata Y, Nagai R. Genetic background influences therapeutic effectiveness of VEGF. Biochem Biophys Res Commun. 2003;310(1):143–7.
24. Robinson CJ, Stringer SE. The splice variants of vascular endothelial growth factor (VEGF) and their receptors. J Cell Sci. 2001;114(Pt 5):853–65.
25. Gowdak LH, Poliakova L, Wang X, Kovesdi I, Fishbein KW, Zacheo A, et al. Adenovirus-mediated VEGF(121) gene transfer stimulates angiogenesis in normoperfused skeletal muscle and preserves tissue perfusion after induction of ischemia. Circulation. 2000;102(5):565–71.
26. Ozawa CR, Banfi A, Glazer NL, Thurston G, Springer ML, Kraft PE, et al. Microenvironmental VEGF concentration, not total dose, determines a threshold between normal and aberrant angiogenesis. J Clin Invest. 2004;113(4):516–27. doi:10.1172/JCI18420.
27. Tafuro S, Ayuso E, Zacchigna S, Zentilin L, Moimas S, Dore F, et al. Inducible adeno-associated virus vectors promote functional angiogenesis in adult organisms via regulated vascular endothelial growth factor expression. Cardiovasc Res. 2009;83(4):663–71. doi:10.1093/cvr/cvp152.
28. Alfranca A. VEGF therapy: a timely retreat. Cardiovasc Res. 2009;83(4):611–2. doi:10.1093/cvr/cvp228.
29. Boden J, Lassance-Soares RM, Wang H, Wei Y, Spiga MG, Adi J, et al. Vascular regeneration in ischemic hindlimb by adeno-associated virus expressing conditionally silenced vascular endothelial growth factor. J Am Heart Assoc. 2016;5(6):e001815. doi:10.1161/JAHA.115.001815.
30. Asahara T, Murohara T, Sullivan A, Silver M, van der Zee R, Li T, et al. Isolation of putative progenitor endothelial cells for angiogenesis. Science. 1997;275(5302):964–7.

31. Kalka C, Masuda H, Takahashi T, Kalka-Moll WM, Silver M, Kearney M, et al. Transplantation of ex vivo expanded endothelial progenitor cells for therapeutic neovascularization. Proc Natl Acad Sci U S A. 2000;97(7):3422–7. doi:10.1073/pnas.070046397.
32. Iwaguro H, Yamaguchi J, Kalka C, Murasawa S, Masuda H, Hayashi S, et al. Endothelial progenitor cell vascular endothelial growth factor gene transfer for vascular regeneration. Circulation. 2002;105(6):732–8.
33. Adams WJ, Zhang Y, Cloutier J, Kuchimanchi P, Newton G, Sehrawat S, et al. Functional vascular endothelium derived from human induced pluripotent stem cells. Stem Cell Rep. 2013;1(2):105–13. doi:10.1016/j.stemcr.2013.06.007.
34. Fujiwara M, Yan P, Otsuji TG, Narazaki G, Uosaki H, Fukushima H, et al. Induction and enhancement of cardiac cell differentiation from mouse and human induced pluripotent stem cells with cyclosporin-A. PLoS One. 2011;6(2):e16734. doi:10.1371/journal.pone.0016734.
35. Yoder MC. Differentiation of pluripotent stem cells into endothelial cells. Curr Opin Hematol. 2015;22(3):252–7. doi:10.1097/MOH.0000000000000140.
36. Isner JM, Walsh K, Symes J, Pieczek A, Takeshita S, Lowry J, et al. Arterial gene transfer for therapeutic angiogenesis in patients with peripheral artery disease. Hum Gene Ther. 1996;7(8):959–88. doi:10.1089/hum.1996.7.8-959.
37. Isner JM, Pieczek A, Schainfeld R, Blair R, Haley L, Asahara T, et al. Clinical evidence of angiogenesis after arterial gene transfer of phVEGF165 in patient with ischaemic limb. Lancet. 1996;348(9024):370–4.
38. Eskelinen E, Lepantalo M, Hietala EM, Sell H, Kauppila L, Maenpaa I, et al. Lower limb amputations in Southern Finland in 2000 and trends up to 2001. Eur J Vasc Endovasc Surg. 2004;27(2):193–200. doi:10.1016/j.ejvs.2003.10.011.
39. Brandao D, Costa C, Canedo A, Vaz G, Pignatelli D. Endogenous vascular endothelial growth factor and angiopoietin-2 expression in critical limb ischemia. Int Angiol. 2011;30(1):25–34.
40. Bleda S, de Haro J, Acin F, Varela C, Esparza L. Enhanced vascular endothelial growth factor gene expression in ischaemic skin of critical limb ischaemia patients. Int J Vasc Med. 2012;2012:691528. doi:10.1155/2012/691528.
41. Baumgartner I, Pieczek A, Manor O, Blair R, Kearney M, Walsh K, et al. Constitutive expression of phVEGF165 after intramuscular gene transfer promotes collateral vessel development in patients with critical limb ischemia. Circulation. 1998;97(12):1114–23.
42. Isner JM, Baumgartner I, Rauh G, Schainfeld R, Blair R, Manor O, et al. Treatment of thromboangiitis obliterans (Buerger's disease) by intramuscular gene transfer of vascular endothelial growth factor: preliminary clinical results. J Vasc Surg. 1998;28(6):964–73; discussion 73–5.
43. Kim HJ, Jang SY, Park JI, Byun J, Kim DI, Do YS, et al. Vascular endothelial growth factor-induced angiogenic gene therapy in patients with peripheral artery disease. Exp Mol Med. 2004;36(4):336–44. doi:10.1038/emm.2004.44.
44. Lee Y, Park EJ, Yu SS, Kim DK, Kim S. Improved expression of vascular endothelial growth factor by naked DNA in mouse skeletal muscles: implication for gene therapy of ischemic diseases. Biochem Biophys Res Commun. 2000;272(1):230–5. doi:10.1006/bbrc.2000.2758.
45. Shyu KG, Chang H, Wang BW, Kuan P. Intramuscular vascular endothelial growth factor gene therapy in patients with chronic critical leg ischemia. Am J Med. 2003;114(2):85–92.
46. Baumgartner I, Rauh G, Pieczek A, Wuensch D, Magner M, Kearney M, et al. Lower-extremity edema associated with gene transfer of naked DNA encoding vascular endothelial growth factor. Ann Intern Med. 2000;132(11):880–4.
47. Kusumanto YH, van Weel V, Mulder NH, Smit AJ, van den Dungen JJ, Hooymans JM, et al. Treatment with intramuscular vascular endothelial growth factor gene compared with placebo for patients with diabetes mellitus and critical limb ischemia: a double-blind randomized trial. Hum Gene Ther. 2006;17(6):683–91. doi:10.1089/hum.2006.17.683.
48. Deev RV, Bozo IY, Mzhavanadze ND, Voronov DA, Gavrilenko AV, Chervyakov YV, et al. pCMV-vegf165 intramuscular gene transfer is an effective method of treatment for patients with chronic lower limb ischemia. J Cardiovasc Pharmacol Ther. 2015;20(5):473–82. doi:10.1177/1074248415574336.

49. Skora J, Pupka A, Janczak D, Barc P, Dawiskiba T, Korta K, et al. Combined autologous bone marrow mononuclear cell and gene therapy as the last resort for patients with critical limb ischemia. Arch Med Sci. 2015;11(2):325–31. doi:10.5114/aoms.2013.39935.
50. Rajagopalan S, Trachtenberg J, Mohler E, Olin J, McBride S, Pak R, et al. Phase I study of direct administration of a replication deficient adenovirus vector containing the vascular endothelial growth factor cDNA (CI-1023) to patients with claudication. Am J Cardiol. 2002;90(5):512–6.
51. Mohler 3rd ER, Rajagopalan S, Olin JW, Trachtenberg JD, Rasmussen H, Pak R, et al. Adenoviral-mediated gene transfer of vascular endothelial growth factor in critical limb ischemia: safety results from a phase I trial. Vasc Med. 2003;8(1):9–13.
52. Rajagopalan S, Mohler 3rd E, Lederman RJ, Saucedo J, Mendelsohn FO, Olin J, et al. Regional angiogenesis with vascular endothelial growth factor (VEGF) in peripheral arterial disease: design of the RAVE trial. Am Heart J. 2003;145(6):1114–8. doi:10.1016/S0002-8703(03)00102-9.
53. Xu D, Fuster MM, Lawrence R, Esko JD. Heparan sulfate regulates VEGF165- and VEGF121-mediated vascular hyperpermeability. J Biol Chem. 2011;286(1):737–45. doi:10.1074/jbc.M110.177006.
54. Berger JS, Hiatt WR. Medical therapy in peripheral artery disease. Circulation. 2012;126(4):491–500. doi:10.1161/CIRCULATIONAHA.111.033886.
55. Makinen K, Manninen H, Hedman M, Matsi P, Mussalo H, Alhava E, et al. Increased vascularity detected by digital subtraction angiography after VEGF gene transfer to human lower limb artery: a randomized, placebo-controlled, double-blinded phase II study. Mol Ther. 2002;6(1):127–33. doi:10.1006/mthe.2002.0638.
56. Waters RE, Terjung RL, Peters KG, Annex BH. Preclinical models of human peripheral arterial occlusive disease: implications for investigation of therapeutic agents. J Appl Physiol (1985). 2004;97(2):773–80. doi:10.1152/japplphysiol.00107.2004.
57. Simovic D, Isner JM, Ropper AH, Pieczek A, Weinberg DH. Improvement in chronic ischemic neuropathy after intramuscular phVEGF165 gene transfer in patients with critical limb ischemia. Arch Neurol. 2001;58(5):761–8.
58. Schratzberger P, Walter DH, Rittig K, Bahlmann FH, Pola R, Curry C, et al. Reversal of experimental diabetic neuropathy by VEGF gene transfer. J Clin Invest. 2001;107(9):1083–92. doi:10.1172/JCI12188.
59. Veves A, King GL. Can VEGF reverse diabetic neuropathy in human subjects? J Clin Invest. 2001;107(10):1215–8. doi:10.1172/JCI13038.
60. Sakurai F, Nishioka T, Saito H, Baba T, Okuda A, Matsumoto O, et al. Interaction between DNA-cationic liposome complexes and erythrocytes is an important factor in systemic gene transfer via the intravenous route in mice: the role of the neutral helper lipid. Gene Ther. 2001;8(9):677–86. doi:10.1038/sj.gt.3301460.
61. Trivedi RA, Dickson G. Liposome-mediated gene transfer into normal and dystrophin-deficient mouse myoblasts. J Neurochem. 1995;64(5):2230–8.
62. Phelps EA, Landazuri N, Thule PM, Taylor WR, Garcia AJ. Bioartificial matrices for therapeutic vascularization. Proc Natl Acad Sci U S A. 2010;107(8):3323–8. doi:10.1073/pnas.0905447107.
63. Sacchi V, Mittermayr R, Hartinger J, Martino MM, Lorentz KM, Wolbank S, et al. Long-lasting fibrin matrices ensure stable and functional angiogenesis by highly tunable, sustained delivery of recombinant VEGF164. Proc Natl Acad Sci U S A. 2014;111(19):6952–7. doi:10.1073/pnas.1404605111.
64. Thomas AM, Gomez AJ, Palma JL, Yap WT, Shea LD. Heparin-chitosan nanoparticle functionalization of porous poly(ethylene glycol) hydrogels for localized lentivirus delivery of angiogenic factors. Biomaterials. 2014;35(30):8687–93. doi:10.1016/j.biomaterials.2014.06.027.
65. Baumgartner I. Intramuscular vascular endothelial growth factor gene therapy: fact or fiction? Am J Med. 2003;114(2):156–7.

66. Brinkhuizen T, Weijzen CA, Eben J, Thissen MR, van Marion AM, Lohman BG, et al. Immunohistochemical analysis of the mechanistic target of rapamycin and hypoxia signalling pathways in basal cell carcinoma and trichoepithelioma. PLoS One. 2014;9(9):e106427. doi:10.1371/journal.pone.0106427.
67. Shyu KG, Chang H, Isner JM. Synergistic effect of angiopoietin-1 and vascular endothelial growth factor on neoangiogenesis in hypercholesterolemic rabbit model with acute hindlimb ischemia. Life Sci. 2003;73(5):563–79.
68. Chen F, Tan Z, Dong CY, Chen X, Guo SF. Adeno-associated virus vectors simultaneously encoding VEGF and angiopoietin-1 enhances neovascularization in ischemic rabbit hind-limbs. Acta Pharmacol Sin. 2007;28(4):493–502. doi:10.1111/j.1745-7254.2007.00527.x.
69. Grossman PM, Mohler 3rd ER, Roessler BJ, Wilensky RL, Levine BL, Woo EY, et al. Phase I study of multi-gene cell therapy in patients with peripheral artery disease. Vasc Med. 2016;21(1):21–32. doi:10.1177/1358863X15612148.
70. Flugelman MY, Halak M, Yoffe B, Schneiderman J, Rubinstein C, Bloom AI, et al. Phase Ib safety, two-dose study of multigeneangio in patients with chronic critical limb ischemia. Mol Ther. 2017;25(3):816–25. doi:10.1016/j.ymthe.2016.12.019.
71. Anisimov A, Tvorogov D, Alitalo A, Leppanen VM, An Y, Han EC, et al. Vascular endothelial growth factor-angiopoietin chimera with improved properties for therapeutic angiogenesis. Circulation. 2013;127(4):424–34. doi:10.1161/CIRCULATIONAHA.112.127472.
72. D'Andrea LD, Iaccarino G, Fattorusso R, Sorriento D, Carannante C, Capasso D, et al. Targeting angiogenesis: structural characterization and biological properties of a de novo engineered VEGF mimicking peptide. Proc Natl Acad Sci U S A. 2005;102(40):14215–20. doi:10.1073/pnas.0505047102.
73. Santulli G, Ciccarelli M, Palumbo G, Campanile A, Galasso G, Ziaco B, et al. In vivo properties of the proangiogenic peptide QK. J Transl Med. 2009;7:41. doi:10.1186/1479-5876-7-41.
74. Finetti F, Basile A, Capasso D, Di Gaetano S, Di Stasi R, Pascale M, et al. Functional and pharmacological characterization of a VEGF mimetic peptide on reparative angiogenesis. Biochem Pharmacol. 2012;84(3):303–11. doi:10.1016/j.bcp.2012.04.011.
75. Webber MJ, Tongers J, Newcomb CJ, Marquardt KT, Bauersachs J, Losordo DW, et al. Supramolecular nanostructures that mimic VEGF as a strategy for ischemic tissue repair. Proc Natl Acad Sci U S A. 2011;108(33):13438–43. doi:10.1073/pnas.1016546108.
76. Ousterout DG, Kabadi AM, Thakore PI, Majoros WH, Reddy TE, Gersbach CA. Multiplex CRISPR/Cas9-based genome editing for correction of dystrophin mutations that cause Duchenne muscular dystrophy. Nat Commun. 2015;6:6244. doi:10.1038/ncomms7244.
77. Kabadi AM, Ousterout DG, Hilton IB, Gersbach CA. Multiplex CRISPR/Cas9-based genome engineering from a single lentiviral vector. Nucleic Acids Res. 2014;42(19):e147. doi:10.1093/nar/gku749.
78. Maeder ML, Linder SJ, Cascio VM, Fu Y, Ho QH, Joung JK. CRISPR RNA-guided activation of endogenous human genes. Nat Methods. 2013;10(10):977–9. doi:10.1038/nmeth.2598.
79. Maeder ML, Linder SJ, Reyon D, Angstman JF, Fu Y, Sander JD, et al. Robust, synergistic regulation of human gene expression using TALE activators. Nat Methods. 2013;10(3):243–5. doi:10.1038/nmeth.2366.
80. Yang Y, Zhang X, Yi L, Hou Z, Chen J, Kou X, et al. Naive induced pluripotent stem cells generated from beta-thalassemia fibroblasts allow efficient gene correction with CRISPR/Cas9. Stem Cells Transl Med. 2016;5(1):8–19. doi:10.5966/sctm.2015-0157.
81. Firth AL, Menon T, Parker GS, Qualls SJ, Lewis BM, Ke E, et al. Functional gene correction for cystic fibrosis in lung epithelial cells generated from patient iPSCs. Cell Rep. 2015;12(9):1385–90. doi:10.1016/j.celrep.2015.07.062.

Chapter 9
HGF Gene Therapy for Therapeutic Angiogenesis in Peripheral Artery Disease

Jun Muratsu, Fumihiro Sanada, Yoshiaki Taniyama, Yuka Ikeda-Iwabu, Rei Otsu, Kana Shibata, May Kanako Brulé, Hiromi Rakugi, and Ryuichi Morishita

Abstract Peripheral arterial disease (PAD) is a chronic arterial occlusive disease in the lower and upper extremities due primarily to atherosclerosis. The clinical consequences of severe PAD and critical limb ischemia include pain while walking (intermittent claudication), pain at rest, and the loss of tissue integrity in the distal ischemic limbs. Moreover, PAD is strongly associated with elevated morbidity and mortality with cardiovascular disease. In general, the best therapeutic option for chronic arterial occlusion is revascularization with percutaneous transluminal angioplasty or surgical bypass. Unfortunately, some patients are poor candidates for revascularization due to the anatomical distribution of severe stenosis, calcification of the artery, poor outflow in the distal lower limb, and medical comorbidities. Thus, specific strategies are needed to deliver sufficient blood flow to eliminate the symptoms in PAD patients. Many studies have been performed using cell-based therapy (e.g., endothelial progenitor cells, bone marrow mononuclear cells, and mesenchymal stem cells) and gene therapy (e.g., vascular endothelial growth factor, fibroblast growth factor, and hepatocyte growth factor (HGF)) for the induction of angiogenesis. Among them, our group has been working on the delivery of the HGF gene into ischemic limbs. It can stimulate angiogenesis without the induction of vascular inflammation and vascular permeability. In this review article, we focus on the

---

J. Muratsu • Y. Taniyama
Department of Clinical Gene Therapy, Osaka University, Graduate School of Medicine, 2-2 Yamada-oka, Suita, Osaka 565-0871, Japan

Department of Geriatric Medicine and Nephrology, Osaka University, Graduate School of Medicine, 2-2 Yamada-oka, Suita, Osaka 565-0871, Japan

F. Sanada • Y. Ikeda-Iwabu • R. Otsu • K. Shibata • M.K. Brulé
R. Morishita, M.D., Ph.D. (✉)
Department of Clinical Gene Therapy, Osaka University, Graduate School of Medicine, 2-2 Yamada-oka, Suita, Osaka 565-0871, Japan
e-mail: morishit@cgt.med.osaka-u.ac.jp

H. Rakugi
Department of Geriatric Medicine and Nephrology, Osaka University, Graduate School of Medicine, 2-2 Yamada-oka, Suita, Osaka 565-0871, Japan

Y. Higashi, T. Murohara (eds.), *Therapeutic Angiogenesis*,
DOI 10.1007/978-981-10-2744-4_9

results of clinical trials using HGF gene transfer and discuss the molecular differences between HGF and other growth factors.

Keywords Cell therapy • Gene therapy • Hepatocyte growth factor

9.1 Introduction

Peripheral artery disease (PAD) is a disease of the extremities caused by atherosclerosis and thrombosis [1]. It causes pain in the extremities that is triggered by activities such as walking, leading to less motility and an impaired quality of life (QOL). PAD is estimated to develop in 500–1000 per one million individuals per year and affects 20% of all people aged over 70 years and 4–12% of the population aged 55–70 years [2, 3]. As atherosclerosis is the principal pathological disorder responsible for both PAD and cardiovascular disease (CVD), PAD is strongly associated with systemic atherosclerotic CVD [4–6]. Of note, the presence of PAD gives a three- to sixfold increase in CVD mortality, and more than half of PAD patients have coronary heart disease and cerebrovascular disease [7–9]. Because of the high risk in CVD, the 2011 American College of Cardiology Foundation/American Heart Association (ACCF/AHA) guidelines recommend a secondary prevention strategy for PAD patients that includes risk factor modification [10–14]. For that reason, systemic treatments, such as antiplatelet drugs, cholesterol-lowering drugs, and inhibitors of the renin-angiotensin system, have been widely used for the treatment of PAD. However, none of these agents currently significantly improve perfusion to the lower extremities in patients with PAD. Other options for treating PAD patients are surgical or endovascular interventional revascularization. However, many PAD patients are poor candidates for revascularization due to the anatomical distributions of multiple severe stenoses, calcification of the arteries, poor outflow in the distal lower limbs, and medical comorbidities. Therefore, specific strategies are needed to deliver sufficient blood flow to eliminate symptoms in such patients. Gene therapies and cell-based therapies have been developed to initiate therapeutic angiogenesis in severe PAD and critical limb ischemia (CLI) [15, 16]. Initially, gene therapy using proangiogenic factors, such as vascular endothelial growth factor (VEGF) [17], fibroblast growth factor (FGF) [18], and hepatocyte growth factors (HGF) [19], was intensively investigated in basic and clinical studies. Cell therapies using bone marrow mononuclear cells (BMMNCs), mesenchymal stem cells (BMMSCs), and endothelial progenitor cells (EPCs) have also been carried out in animal models and in early stage of clinical studies [20]. Based on the favorable results of basic research studies, clinical phase I/II trials using proangiogenic gene delivery have been performed, and the results demonstrate the safety of these approaches and their potential for symptomatic improvement in CLI [15, 16]. However, only HGF provides benefit in phase III clinical trials thus far. In this review, we discuss the advantages and the features of HGF on angiogenesis and inflammation over other angiogenic growth factors. Additionally, we discuss our recent findings regarding cilostazol, the only FDA-approved drug for the treatment of intermittent claudication (IC), on angiogenesis through HGF production in the vasculature.

9.2 Pathology of Advanced PAD

Atherosclerosis is not simply a disorder of lipid deposition but is considered an advancing pathophysiological process arising from a combination of inflammation and endothelial cell dysfunction [21]. These changes are induced by cardiovascular risk factors such as hyperlipidemia, diabetes mellitus, smoking, and hypertension. At sites of occlusion, angiogenic sprouting and capillary lumen formation induces collateral circulation to alleviate the effect of blood flow deprivation. However, these mechanisms are not operative in CLI patients. Endothelial cell (EC) dysfunction [22] and impaired endothelial progenitor cell (EPC) function [23] in PAD patients has been demonstrated previously. Inflammatory cytokines, such as angiotensin II, interleukin-6, and monocyte chemoattractant protein-1, secreted from monocyte/macrophage or vascular cells in atherosclerotic plaques, accelerate the premature senescence of ECs, EPCs, and other stem/progenitor cells in a telomere-independent manner [24, 25] and tissue fibrosis, which prevents oxygen diffusion and stem/progenitor cell migration toward the tissue regions where tissue repair is required. These changes can induce rest pain in the ischemic limb, chronic nonhealing wounds, and gangrene. Successful angiogenesis using gene therapy for CLI patients should offer relief from ischemic pain and ulcers and reduce mortality and amputation risk. Additionally, it should not promote the process of atherosclerosis. Considering the several potential complications that may arise in CLI patients, precise examinations from preclinical studies are required to develop successful clinical trials based on angiogenic gene transfer.

9.3 HGF Gene Therapy for PAD

Although PAD patients receive aggressive treatment, such treatments often do not promote sufficient blood flow to eliminate their symptoms. Thus, many studies have been performed using angiogenic gene therapy. The administration of plasmid DNA-encoding growth factors into ischemic tissue increases the local concentration of angiogenic factors and aims to induce EC proliferation and migration and the formation of new blood vessels in the ischemic leg. The goal of therapeutic angiogenesis is to induce the development of new arterial vessels to improve perfusion of the ischemic tissue. Here, we summarize HGF gene therapy for the treatment of PAD patients.

HGF was originally discovered as a potent mitogen for adult rat hepatocytes [26]. Later, its angiogenic properties on ECs via tyrosine phosphorylation of its receptor,

Table 9.1 Clinical trials of human HGF plasmid for PAD

Trials (trial name)	Strategy	Delivery route	Patient's number	Reference
Morishita et al. (AMG 0001-JN-100)	Naked pDNA	IM	22	[37]
Makino et al. (AMG 0001-JN-100, long-term follow up)	Naked pDNA	IM	22	[38]
Powell et al. (AG-CLI-0202)	Naked pDNA	IM	93	[39]
Powell et al. (AG-CLI-0205)	Naked pDNA	IM	27	[40]
Shigematsu et al. (AMG-JN-101)	Naked pDNA	IM	44	[41]
Shigematsu et al. (AMG-JN-102)	Naked pDNA	IM	10	[42]

IM intramuscular injection

c-Met, were identified. The receptor c-Met is expressed on ECs, smooth muscle cells (SMCs), and also bone marrow-derived EPCs [27], and its expression is induced by tissue hypoxia [28, 29]. HGF is now appreciated as not only a promising angiogenic growth factor but also as an anti-fibrotic and anti-inflammatory factor in chronic inflammatory diseases, such as atherosclerosis, chronic renal disease, and heart failure. Indeed, several animal experiments have documented that HGF plasmid DNA transfer can induce significant increases in blood flow and capillary density in the ischemic hind limbs compared to control plasmid without increasing vascular permeability and inflammation [30–35]. Additionally, recent findings have documented that a combination of HGF and VEGF can facilitate neovascularization by enhancing intracellular signaling and increasing the proliferation, migration, and survival of ECs while decreasing VEGF-induced vascular permeability [36]. Based on these preclinical results, human clinical trials using intramuscular injections of naked human HGF plasmid were conducted (Table 9.1). To date, five clinical trials have been completed.

9.3.1 Phase I/II Study (AMG0001-JN-100)

A phase I/II study (AMG0001-JN-100) was the first clinical trial of human HGF plasmid gene transfer for PAD. Twenty-two patients with CLI due to arteriosclerosis obliterans (ASO) and Buerger's disease were included. HGF gene transfer was carried out by two intramuscular injections of naked HGF plasmid at either 2 or 4 mg. This clinical trial comprised two stages; stage I was an open-label study, and stage II was a randomized, dose–response, open-label study. Stage I included patients with rest pain (Fontaine stage 3) and ischemic ulcers or necrosis (Fontaine stage 4) for an assessment of the acute allergic reactions, safety, and therapeutic effect. Stage II included patients with severe claudication (Fontaine stage 2b) for the evaluation of the safety and therapeutic effect of HGF plasmid gene transfer. The improvement rates for each end point obtained from the combined results of Stages I and II were as follows: 61.5% (8/13) for rest pain, 63.6% (7/11) for ischemic ulcers, and 64.7% (11/17) for ankle brachial index (ABI). No evidence of leg edema or serious adverse side effects were found. This initial clinical trial showed that intramuscular injection of naked human HGF plasmid is safe and feasible and can improve ischemic limb perfusion [37]. Subsequently, a long-term follow-up observation also documented its sustaining effects [38]. An ABI > 0.1 (78.6%, 11 of 14 patients), reduction in rest pain (100%, 9 of 9 patients), and decrease in the size of ulcers (90%, 9 of 10 patients) were observed at the 2-year follow-up without severe complications or adverse effects.

9.3.2 Phase II Studies (AG-CLI-0202 and AG-CLI-0205)

These phase II studies were double-blind, placebo-controlled trials in the USA and evaluated the safety and efficacy of the human HGF plasmid in ASO patients with CLI. AG-CLI-0202 was designed to compare the dose-responses of angiogenesis to HGF gene transfer in four categories: a low-dose group, 0.05 mg per site at eight sites in the affected leg, three times (0.4 mg at days 0, 14, and 28, $n = 23$); a middle-dose group, 0.5 mg per site at eight sites in the affected leg, two times (4 mg at days 0 and 28, $n = 18$); a high-dose group, 0.5 mg per site at eight sites in the ischemic leg, three times (4 mg at days 0, 14, and 28, $n = 15$); and a placebo control group ($n = 17$). The measurements of transcutaneous oxygen pressure (TcPO2) were increased at 6 months in the high-dose group when compared with the placebo, low-dose, and middle-dose groups. There was no difference between the groups at the secondary end points, including ABI, toe-brachial index (TBI), pain relief, wound healing, or major amputation [39]. AG-CLI-0205 was a trial at a dose of 4 mg three times at 2-week intervals. A total of 27 subjects (26 patients in the HGF plasmid group and 6 patients in the placebo control group) were enrolled. Changes in TBI and the visual analogue scale (VAS) improved significantly from baseline at 6 months in the HGF-treated group compared with the placebo (0.05 ± 0.05 vs. -0.17 ± 0.04; P value 0.047). Complete ulcer healing at 12 months occurred in 31% of the HGF-treated group and 0% of the placebo control group. There was no difference in the major amputation rate of the treated limb (HGF 29% vs. placebo 33%) or mortality at 12 months (HGF 19% vs. placebo 17%) between the two groups. Overall, hemodynamic measurements increased in the HGF-treated group, suggesting better blood flow in the lower extremities [40].

9.3.3 Phase III Studies (AMG-JN-101 and AMG-JN-102)

AMG-JN-101 was a phase III, randomized, double-blind, placebo-controlled trial. Forty-four patients with CLI were recruited and received intramuscular injections of human HGF plasmid at a dose of 0.5 mg per site at eight sites in the affected leg, two times, 4 weeks apart. A significant improvement in the primary end point (rest pain and ischemic ulcer size) and QOL in the HGF-treated group was observed. Although this trial failed to demonstrate an augmentation of ABI or amputation rate, there were no major safety problems [41]. AMG-JN-102 was an open-label trial evaluating the efficacy and safety of the HGF plasmid in patients with Buerger's disease. The treatment regimen was at a dose of 0.5 mg per site at eight sites in the ischemic leg, two times, 4 weeks apart. A total of 66.7% (6/9) of the patients who received the human HGF plasmid showed improvements in ischemic ulcer size at week 12. Again, there were no major safety problems [42]. Following these favorable outcomes in the HGF-treated patients, a global multicenter Phase III clinical trial aimed at recruiting >500 CLI patients was initiated. In spite of the relatively small phase III clinical trials using HGF gene transfer, so far, only HGF has shown beneficial effects in the phase III clinical trial for the treatment of PAD.

9.4 The Differences Between HGF and Other Growth Factors

Initially, the HGF protein was isolated from the plasma of patients with fulminate hepatic failure [26]. Thereafter, several functions of the HGF/c-Met system have been uncovered in pathophysiological conditions, which include cell survival, cell proliferation, anti-inflammation, and anti-fibrosis [27–29]. Thus, HGF is now appreciated as a key growth factor for the attenuation of both acute and chronic disease progression in the heart [43], kidney [44], liver [45], and vasculature [24, 32]. Among the beneficial functions of HGF, anti-inflammatory and anti-fibrotic actions are unique. Tissue inflammation initiates the progression of a wide range of chronic diseases, including PAD [46–49]. Inflammatory cytokines such as IL-6 and IL-8 have been associated with cellular senescence [50–53], and in turn, senescent cells secrete multiple inflammatory cytokines, which are considered to contribute to sustained low-grade inflammation. Apparently, a tissue inflammation—cellular senescence—cytokine secretion loop exists. Nonetheless, the majority of preclinical animal studies of PAD have been performed in the absence of tissue inflammation caused by vascular risk factors present in the clinical setting. To explore the possible explanations for the successes and failures of HGF and VEGF clinical trials, we compared the angiogenic potential of HGF and VEGF under Ang II administration in a mouse hind limb ischemia model. Interestingly, HGF, but not VEGF, attenuated Ang II-induced senescence of ECs and EPCs by reducing oxidative stress (ROS) [24]. When HGF stimulates its receptor c-Met, the SH2-domain-containing inositol 5-phosphatase (SHIP)-2 binds to c-Met directly and initiates downregulation of the epidermal growth factor receptor (EGFR) in a ligand-dependent manner. The degradation of EGFR by the HGF/c-Met system occurs through the ubiquitin proteasome system [32]. This system similarly operates when cells are stimulated by lipopolysaccharide, ET-1, and TGF-β1. Because these cytokines produce ROS through transactivation of EGFR, the HGF/c-Met system can reduce ROS production, leading to anti-senescence and anti-inflammatory actions. These results imply that in the clinic, the mechanism of ligand-dependent EGFR downregulation by the HGF/c-Met system contributes to the anti-inflammatory and antioxidant actions in chronic inflammatory diseases such as atherosclerosis [32, 54]. Interestingly, it has been shown that the expression level of VEGF increases after vascular injury and recruits monocyte-lineage cells; in contrast, HGF decreases the recruitment of cells after injury [55, 56]. Importantly, HGF and VEGF synergistically induce angiogenesis [57, 58], and HGF can reduce VEGF-induced leukocyte adhesion and adhesion molecule expression by suppressing VEGF-induced NFkB signaling [59]. Thus, HGF exerts its angiogenic properties while inhibiting inflammation, edema, and cellular senescence. Another unique function of HGF is its anti-fibrotic action, whereas VEGF and FGF induce tissue fibrosis [60, 61]. Our group has demonstrated that the upregulation of HGF significantly decreases fibrotic tissue area following acute myocardial infarction [43] and acute kidney injury. HGF significantly attenuates the transition of epithelial cells into mesenchymal cells (EMT), which is considered to be involved in perivascular fibrosis of the heart [62] and kidney [44]. The diminution of EMT and subsequent tissue fibrosis may serve to minimize impediments to tissue regeneration. It is conceivable that the engraftment of circulating stem/progenitor cells and oxygen diffusion may be limited by perivascular fibrosis and result in an impairment of tissue regeneration and oxygenation. Local levels of HGF significantly decrease in an ischemic tissue and subsequently cause intramuscular and perivascular fibrosis disease progression [63]. Therefore, HGF administration by gene therapy can significantly influence organ regeneration [64]. Thus, under pathological conditions, HGF, VEGF, and FGF may differentially influence angiogenesis. In the future, we should acknowledge the limitations of preclinical studies and the complexities of clinical settings.

Recently, we have demonstrated that cilostazol, the only FDA-approved drug for the treatment of IC, significantly improved ischemic limb perfusion through HGF production. In this study, the effects of cilostazol on ischemic leg were investigated in a mouse hind limb ischemia model. The administration of cilostazol significantly increased the expression of HGF, VEGF, angiopoietin-1, and peroxisome proliferator-activated receptor-γ (PPARγ), especially in SMCs and ECs. The capillary density in the ischemic leg was significantly increased in the cilostazol treatment group when compared to control and aspirin treatment groups. Interestingly, an increase in capillary density and growth factor expression was almost completely abolished by coadministration of HGF-neutralizing antibodies. These data suggest that cilostazol enhances angiogenesis primarily through HGF. In vitro experiments revealed that cilostazol treatment increased HGF production in vascular smooth muscle cells via two major pathways: the PPARγ and cAMP pathways [65]. Similar to our report, other groups also indicate that PPARγ agonists exert their anti-fibrotic actions through HGF [66–68]. The distribution of PPARγ and cAMP might limit the benefits of these drugs.

9.5 Conclusion

HGF, VEGF, and FGF gene therapies for PAD have remarkable benefits in preclinical studies; however, in large clinical trials, such benefits were found only for HGF. HGF has angiogenic potential with unique anti-inflammatory and anti-fibrotic functions. Thus, it may be used to treat the pathology of CLI and to induce angiogenesis. A comprehensive understanding of the biology of angiogenesis under pathological conditions might provide better information for successful future trials of therapeutic angiogenesis.

Acknowledgments We thank the members of the Department of Clinical Gene Therapy at the Osaka University Graduate School of Medicine for their helpful discussions and technical support.

COI Statement: R. Morishita received honoraria, consulting fees, and funds from Novartis, Takeda, Shionogi, Astellas, Boehringer Ingelheim, Daiichi-Sankyo, and Pfizer.

References

1. Ouma GO, Jonas RA, Usman MH, Mohler 3rd ER. Targets and delivery methods for therapeutic angiogenesis in peripheral artery disease. Vasc Med. 2012;17:174–92.
2. Fowkes FG, Rudan D, Rudan I, Aboyans V, Denenberg JO, McDermott MM, Norman PE, Sampson UK, Williams LJ, Mensah GA, Criqui MH. Comparison of global estimates of prevalence and risk factors for peripheral artery disease in 2000 and 2010: a systematic review and analysis. Lancet. 2013;382:1329–40.
3. Leng GC, Lee AJ, Fowkes FG, Whiteman M, Dunbar J, Housley E, Ruckley CV. Incidence, natural history and cardiovascular events in symptomatic and asymptomatic peripheral arterial disease in the general population. Int J Epidemiol. 1996;25:1172–81.
4. Howell MA, Colgan MP, Seeger RW, Ramsey DE, Sumner DS. Relationship of severity of lower limb peripheral vascular disease to mortality and morbidity: a six-year follow-up study. J Vasc Surg. 1989;9:691–6.
5. Criqui MH, Langer RD, Fronek A, Feigelson HS, Klauber MR, McCann TJ, Browner D. Mortality over a period of 10 years in patients with peripheral arterial disease. N Engl J Med. 1992;326:381–6.
6. McKenna M, Wolfson S, Kuller L. The ratio of ankle and arm arterial pressure as an independent predictor of mortality. Atherosclerosis. 1991;87:119–28.
7. Klop RB, Eikelboom BC, Taks AC. Screening of the internal carotid arteries in patients with peripheral vascular disease by colour-flow duplex scanning. Eur J Vasc Surg. 1991;5:41–5.
8. Valentine RJ, Grayburn PA, Eichhorn EJ, Myers SI, Clagett GP. Coronary artery disease is highly prevalent among patients with premature peripheral vascular disease. J Vasc Surg. 1994;19:668–74.
9. McFalls EO, Ward HB, Moritz TE, Goldman S, Krupski WC, Littooy F, Pierpont G, Santilli S, Rapp J, Hattler B, Shunk K, Jaenicke C, Thottapurathu L, Ellis N, Reda DJ, Henderson WG. Coronary-artery revascularization before elective major vascular surgery. N Engl J Med. 2004;351:2795–804.
10. McDermott MM, Mehta S, Ahn H, Greenland P. Atherosclerotic risk factors are less intensively treated in patients with peripheral arterial disease than in patients with coronary artery disease. J Gen Intern Med. 1997;12:209–15.
11. Clark AL, Byrne JC, Nasser A, McGroarty E, Kennedy JA. Cholesterol in peripheral vascular disease—a suitable case for treatment? QJM. 1999;92:219–22.

12. Hiatt WR. Medical treatment of peripheral arterial disease and claudication. N Engl J Med. 2001;344:1608–21.
13. Mangiafico RA, Mangiafico M. Medical treatment of critical limb ischemia: current state and future directions. Curr Vasc Pharmacol. 2011;9:658–76.
14. Anderson JL, Halperin JL, Albert NM, Bozkurt B, Brindis RG, Curtis LH, DeMets D, Guyton RA, Hochman JS, Kovacs RJ, Ohman EM, Pressler SJ, Sellke FW, Shen WK. Management of patients with peripheral artery disease (compilation of 2005 and 2011 ACCF/AHA guideline recommendations): a report of the American College of Cardiology Foundation/American Heart Association Task Force on Practice Guidelines. Circulation. 2013;127:1425–43.
15. Sanada F, Taniyama Y, Azuma J, Yuka II, Kanbara Y, Iwabayashi M, Rakugi H, Morishita R. Therapeutic angiogenesis by gene therapy for critical limb ischemia: choice of biological agent. Immunol Endocr Metab Agents Med Chem. 2014;14(1):32–9.
16. Sanada F, Taniyama Y, Kanbara Y, Otsu R, Ikeda-Iwabu Y, Carracedo M, Rakugi H, Morishita R. Gene therapy in peripheral artery disease. Expert Opin Biol Ther. 2015;15(3):381–90.
17. Isner JM, Walsh K, Symes J, Pieczek A, Takeshita S, Lowry J, Rossow S, Rosenfield K, Weir L, Brogi E, et al. Arterial gene therapy for therapeutic angiogenesis in patients with peripheral artery disease. Circulation. 1995;91(11):2687–92.
18. Comerota AJ, Throm RC, Miller KA, Henry T, Chronos N, Laird J, Sequeira R, Kent CK, Bacchetta M, Goldman C, Salenius JP, Schmieder FA, Pilsudski R. Naked plasmid DNA-encoding fibroblast growth factor type 1 for the treatment of end-stage unreconstructible lower extremity ischemia: preliminary results of a phase I trial. J Vasc Surg. 2002;35(5):930–6.
19. Morishita R, Aoki M, Hashiya N, Yamasaki K, Kurinami H, Shimizu S, Makino H, Takesya Y, Azuma J, Ogihara T. Therapeutic angiogenesis using hepatocyte growth factor (HGF). Curr Gene Ther. 2004;4(2):199–206. Review.
20. Peeters Weem SM, Teraa M, de Borst GJ, Verhaar MC, Moll FL. Bone marrow derived cell therapy in critical limb ischemia: a meta-analysis of randomized placebo controlled trials. Eur J Vasc Endovasc Surg. 2015;50(6):775–83.
21. Libby P. Inflammation in atherosclerosis. Arterioscler Thromb Vasc Biol. 2012;32(9):2045–51.
22. Böger RH, Bode-Böger SM. Endothelial dysfunction in peripheral arterial occlusive disease: from basic research to clinical use. Vasa. 1997;26(3):180–4. Review.
23. Li TS, Kubo M, Ueda K, Murakami M, Ohshima M, Kobayashi T, Tanaka T, Shirasawa B, Mikamo A, Hamano K. Identification of risk factors related to poor angiogenic potency of bone marrow cells from different patients. Circulation. 2009;120(11 Suppl):S255–61.
24. Sanada F, Taniyama Y, Azuma J, et al. Hepatocyte growth factor, but not vascular endothelial growth factor, attenuates angiotensin II-induced endothelial progenitor cell senescence. Hypertension. 2009;53(1):77–82.
25. Minamino T, Komuro I. Vascular cell senescence: contribution to atherosclerosis. Circ Res. 2007;100(1):15–26.
26. Nakamura T, Teramoto H, Ichihara A. Purification and characterization of a growth factor from rat platelets for mature parenchymal hepatocytes in primary cultures. Proc Natl Acad Sci U S A. 1986;83(17):6489–93.
27. Wojakowski W, Tendera M, Michałowska A, Majka M, Kucia M, Maślankiewicz K, Wyderka R, Ochała A, Ratajczak MZ. Mobilization of CD34/CXCR4+, CD34/CD117+, c-met+ stem cells, and mononuclear cells expressing early cardiac, muscle, and endothelial markers into peripheral blood in patients with acute myocardial infarction. Circulation. 2004;110(20):3213–20.
28. Vasir B, Reitz P, Xu G, Sharma A, Bonner-Weir S, Weir GC. Effects of diabetes and hypoxia on gene markers of angiogenesis (HGF, cMET, uPA and uPAR, TGF-alpha, TGF-beta, bFGF and Vimentin) in cultured and transplanted rat islets. Diabetologia. 2000;43(6):763–72.
29. Rosová I, Dao M, Capoccia B, Link D, Nolta JA. Hypoxic preconditioning results in increased motility and improved therapeutic potential of human mesenchymal stem cells. Stem Cells. 2008;26(8):2173–82.
30. Morishita R, Nakamura S, Hayashi S, Taniyama Y, Moriguchi A, Nagano T, Taiji M, Noguchi H, Takeshita S, Matsumoto K, Nakamura T, Higaki J, Ogihara T. Therapeutic angiogenesis

induced by human recombinant hepatocyte growth factor in rabbit hind limb ischemia model as cytokine supplement therapy. Hypertension. 1999;33:1379–84.
31. Hayashi S, Morishita R, Nakamura S, Yamamoto K, Moriguchi A, Nagano T, Taiji M, Noguchi H, Matsumoto K, Nakamura T, Higaki J, Ogihara T. Potential role of hepatocyte growth factor, a novel angiogenic growth factor, in peripheral arterial disease: downregulation of HGF in response to hypoxia in vascular cells. Circulation. 1999;100:II301–8.
32. Sanada F, Taniyama Y, Iekushi K, et al. Negative action of hepatocyte growth factor/c-met system on angiotensin II signaling via ligand-dependent epithelial growth factor receptor degradation mechanism in vascular smooth muscle cells. Circ Res. 2009;105(7):667–75.
33. Koike H, Morishita R, Iguchi S, Aoki M, Matsumoto K, Nakamura T, Yokoyama C, Tanabe T, Ogihara T, Kaneda Y. Enhanced angiogenesis and improvement of neuropathy by cotransfection of human hepatocyte growth factor and prostacyclin synthase gene. FASEB J. 2003;17:779–81.
34. Taniyama Y, Morishita R, Hiraoka K, Aoki M, Nakagami H, Yamasaki K, Matsumoto K, Nakamura T, Kaneda Y, Ogihara T. Therapeutic angiogenesis induced by human hepatocyte growth factor gene in rat diabetic hind limb ischemia model: molecular mechanisms of delayed angiogenesis in diabetes. Circulation. 2001;104:2344–50.
35. Taniyama Y, Morishita R, Aoki M, Nakagami H, Yamamoto K, Yamazaki K, Matsumoto K, Nakamura T, Kaneda Y, Ogihara T. Therapeutic angiogenesis induced by human hepatocyte growth factor gene in rat and rabbit hindlimb ischemia models: preclinical study for treatment of peripheral arterial disease. Gene Ther. 2001;8:181–9.
36. Kaga T, Kawano H, Sakaguchi M, Nakazawa T, Taniyama Y, Morishita R. Hepatocyte growth factor stimulated angiogenesis without inflammation: differential actions between hepatocyte growth factor, vascular endothelial growth factor and basic fibroblast growth factor. Vasc Pharmacol. 2012;57:3–9.
37. Morishita R, Makino H, Aoki M, et al. Phase I/IIa clinical trial of therapeutic angiogenesis using hepatocyte growth factor gene transfer to treat critical limb ischemia. Arterioscler Thromb Vasc Biol. 2011;31(3):713–20.
38. Makino H, Aoki M, Hashiya N, et al. Long-term follow-up evaluation of results from clinical trial using hepatocyte growth factor gene to treat severe peripheral arterial disease. Arterioscler Thromb Vasc Biol. 2012;32(10):2503–9.
39. Powell RJ, Simons M, Mendelsohn FO, et al. Results of a double-blind, placebo-controlled study to assess the safety of intramuscular injection of hepatocyte growth factor plasmid to improve limb perfusion in patients with critical limb ischemia. Circulation. 2008;118(1):58–65.
40. Powell RJ, Goodney P, Mendelsohn FO, Moen EK, Annex BH, HGF-0205 Trial Investigators. Safety and efficacy of patient specific intramuscular injection of HGF plasmid gene therapy on limb perfusion and wound healing in patients with ischemic lower extremity ulceration: results of the HGF-0205 trial. J Vasc Surg. 2010;52:1525–30.
41. Shigematsu H, Yasuda K, Iwai T, et al. Randomized, double-blind, placebo-controlled clinical trial of hepatocyte growth factor plasmid for critical limb ischemia. Gene Ther. 2010;17(9):1152–61.
42. Shigematsu H, Yasuda K, Sasajima T, Takano T, Miyata T, Ohta T, Tanemoto K, Obitsu Y, Iwai T, Ozaki S, Ogihara T, Morishita R, HGF Study Group. Transfection of human HGF plasmid DNA improves limb salvage in Buerger's disease patient with critical limb ischemia. Int Angiol. 2011;30:140–9.
43. Taniyama Y, Morishita R, Nakagami H, et al. Potential contribution of a novel antifibrotic factor, hepatocyte growth factor, to prevention of myocardial fibrosis by angiotensin II blockade in cardiomyopathic hamsters. Circulation. 2000;102(2):246–52.
44. Iekushi K, Taniyama Y, Kusunoki H, et al. Hepatocyte growth factor attenuates transforming growth factor-beta-angiotensin II crosstalk through inhibition of the pten/akt pathway. Hypertension. 2011;58(2):190–6.

45. Ueki T, Kaneda Y, Tsutsui H, et al. Hepatocyte growth factor gene therapy of liver cirrhosis in rats. Nat Med. 1999;5(2):226–30.
46. Vlassara H, Cai W, Chen X, et al. Managing chronic inflammation in the aging diabetic patient with ckd by diet or sevelamer carbonate: a modern paradigm shift. J Gerontol A Biol Sci Med Sci. 2012;67(12):1410–6.
47. Hansson GK. Inflammation, atherosclerosis, and coronary artery disease. N Engl J Med. 2005;352(16):1685–95.
48. Mantzouranis EC, Rosen FS, Colten HR. Reticuloendothelial clearance in cystic fibrosis and other inflammatory lung diseases. N Engl J Med. 1988;319(6):338–43.
49. Brevetti G, Giugliano G, Brevetti L, Hiatt WR. Inflammation in peripheral artery disease. Circulation. 2010;122(18):1862–75.
50. Kuilman T, Michaloglou C, Vredeveld LC, et al. Oncogene-induced senescence relayed by an interleukin-dependent inflammatory network. Cell. 2008;133(6):1019–31.
51. Chandeck C, Mooi WJ. Oncogene-induced cellular senescence. Adv Anat Pathol. 2010;17(1):42–8.
52. Mooi WJ. Oncogene-induced cellular senescence: causal factor in the growth arrest of pituitary microadenomas? Horm Res. 2009;71(Suppl 2):78–81.
53. Imanishi T, Tsujioka H, Akasaka T. Endothelial progenitor cell senescence—is there a role for estrogen? Ther Adv Cardiovasc Dis. 2010;4(1):55–69.
54. Shimizu K, Taniyama Y, Sanada F, et al. Hepatocyte growth factor inhibits lipopolysaccharide-induced oxidative stress via epithelial growth factor receptor degradation. Arterioscler Thromb Vasc Biol. 2012;32(11):2687–93.
55. Ohtani K, Egashira K, Hiasa K, et al. Blockade of vascular endothelial growth factor suppresses experimental restenosis after intraluminal injury by inhibiting recruitment of monocyte lineage cells. Circulation. 2004;110(16):2444–52.
56. Matsumoto K, Morishita R, Moriguchi A, et al. Inhibition of neointima by angiotensin-converting enzyme inhibitor in porcine coronary artery balloon-injury model. Hypertension. 2001;37(2):270–4.
57. Van Belle E, Witzenbichler B, Chen D, et al. Potentiated angiogenic effect of scatter factor/hepatocyte growth factor via induction of vascular endothelial growth factor: the case for paracrine amplification of angiogenesis. Circulation. 1998;97(4):381–90.
58. Xin X, Yang S, Ingle G, et al. Hepatocyte growth factor enhances vascular endothelial growth factor-induced angiogenesis in vitro and in vivo. Am J Pathol. 2001;158(3):1111–20.
59. Min JK, Lee YM, Kim JH, et al. Hepatocyte growth factor suppresses vascular endothelial growth factor-induced expression of endothelial icam-1 and vcam-1 by inhibiting the nuclear factor-kappa B pathway. Circ Res. 2005;96(3):300–7.
60. Hamada N, Kuwano K, Yamada M, et al. Anti-vascular endothelial growth factor gene therapy attenuates lung injury and fibrosis in mice. J Immunol. 2005;175(2):1224–31.
61. Chaudhary NI, Roth GJ, Hilberg F, et al. Inhibition of PDGF, VEGF and FGF signaling attenuates fibrosis. Eur Respir J. 2007;29(5):976–85.
62. Okayama K, Azuma J, Dosaka N, et al. Hepatocyte growth factor reduces cardiac fibrosis by inhibiting endothelial-mesenchymal transition. Hypertension. 2012;59(5):958–65.
63. Nakano N, Morishita R, Moriguchi A, et al. Negative regulation of local hepatocyte growth factor expression by angiotensin ii and transforming growth factor-beta in blood vessels: potential role of hgf in cardiovascular disease. Hypertension. 1998;32(3):444–51.
64. Urbanek K, Rota M, Cascapera S, et al. Cardiac stem cells possess growth factor-receptor systems that after activation regenerate the infarcted myocardium, improving ventricular function and long-term survival. Circ Res. 2005;97(7):663–73.
65. Sanada F, Kanbara Y, Taniyama Y, Otsu R, Carracedo M, Ikeda-Iwabu Y, Muratsu J, Sugimoto K, Yamamoto K, Rakugi H, Morishita R. Induction of angiogenesis by a type III phosphodiesterase inhibitor, cilostazol, through activation of peroxisome proliferator-activated receptor-γ and cAMP pathways in vascular cells. Arterioscler Thromb Vasc Biol. 2016;36(3):545–52.

66. Li Y, Wen X, Spataro BC, Hu K, Dai C, Liu Y. hepatocyte growth factor is a downstream effector that mediates the antifibrotic action of peroxisome proliferator-activated receptor-gamma agonists. J Am Soc Nephrol. 2006;17(1):54–65.
67. ZP H, Fang XL, Qian HY, Fang N, Wang BN, Wang Y. Telmisartan prevents angiotensin II-induced endothelial dysfunction in rabbit aorta via activating HGF/Met system and PPARγ pathway. Fundam Clin Pharmacol. 2014;28(5):501–11.
68. Kusunoki H, Taniyama Y, Rakugi H, Morishita R. Cardiac and renal protective effects of irbesartan via peroxisome proliferator-activated receptorγ-hepatocyte growth factor pathway independent of angiotensin II Type 1a receptor blockade in mouse model of salt-sensitive hypertension. J Am Heart Assoc. 2013;2(2):e000103.

Chapter 10
Fibroblast Growth Factor in Extremities

Michiko Tanaka and Yoshikazu Yonemitsu

Abstract Fibroblast growth factor (FGF) is recognized as one of the major angiogenic factors. Among the FGF family members, FGF-1 (acidic fibroblast growth factor, aFGF) and FGF-2 (basic fibroblast growth factor, bFGF) have been extensively studied at both preclinical and clinical stages for applications in therapeutic angiogenesis. In vivo studies have reported various synergistic effects of FGF-2, and gene therapy has achieved promising results in early preclinical studies. However, few clinical trials have been conducted. In this chapter, we provide an overview and future perspectives of the progression of research into angiogenic gene therapy with FGF for peripheral artery disease. The focus of a new gene transfer vector based on a non-transmissible recombinant Sendai virus that expresses the human FGF-2 gene is also introduced as a frontier of therapeutic angiogenesis in this field.

Keywords Angiogenesis • FGF-2 • Sendai viral vector • DVC1-0101

10.1 Introduction

Fibroblast growth factors (FGFs) belong to a family of more than 20 proteins. They are angiogenic factors based on their induction of endothelial cell proliferation, migration, and morphogenesis, extracellular matrix degradation, and vessel maturation. The activities of FGFs in angiogenesis are not caused by a single factor but are modulated by synergic effects with various extracellular matrix-associated molecules such as vascular endothelial growth factor (VEGF) [1]. The crosstalk of

M. Tanaka • Y. Yonemitsu (✉)
R&D Laboratory for Innovative Biotherapeutics, Graduate School of Pharmaceutical Sciences, Kyushu University, Room 601, Collaborative Station I, 3-1-1 Maidashi, Higashi-ku, Fukuoka 812-8582, Japan
e-mail: yonemitu@med.kyushu-u.ac.jp

Y. Higashi, T. Murohara (eds.), *Therapeutic Angiogenesis*,
DOI 10.1007/978-981-10-2744-4_10

synergic effects among FGFs, VEGFs, and inflammatory cytokines/chemokines modulates blood vessel growth in various pathological conditions [2]. The bulk of experimental data related to the prototypic FGFs have been obtained in vitro and in vivo to establish their potential for therapeutic angiogenesis. The FGF family, especially FGF-1, FGF-2, FGF-4, and FGF-5, has been widely investigated with a particular emphasis on FGF-2 [2, 3].

10.2 Preclinical Studies of Animal Models Using FGF-2 in Extremities

Various synergistic effects of FGF-2 induce pronounced angiogenesis in animal ischemic limbs. FGF-2 targets non-endothelial mesenchymal cells independently to stimulate arteriogenic (monocyte chemoattractant protein-1, MCP-1) pathways. Thus, MCP-1/C-C chemokine receptor-2 is essential for the adaptive and FGF-2-mediated recovery of blood flow in murine ischemic limbs [4]. The platelet-derived growth factor receptor (PDGFR)-α-p70S6K pathway is also an essential regulator of FGF-2-mediated therapeutic neovascularization [5]. Transient delivery of both FGF-2 and PDGF-BB naked DNA may increase angiogenesis compared with single gene delivery [6]. In addition, co-delivery of FGF-2 and human recombinant granulocyte colony-stimulating factor leads to superior reperfusion and mature vessel formation in a murine hindlimb ischemic model [7]. Moreover, several approaches have been reported for the optimization of FGF-2 carriers [8, 9]. The use of a recombinant Sendai virus (SeV) as a vector appears to be efficient [10–12].

10.2.1 SeV Vectors

SeV is a negative-sense, single-stranded RNA virus. It belongs to the *Paramyxoviridae* family including measles, mumps, and human respiratory syncytial virus and is known to cause respiratory tract infections in mice and rats [13]. However, it is not reported to be pathogenic to humans. The structure of SeV is shown in Fig. 10.1.

SeV-based vectors have a markedly higher gene expression levels and several theoretical and practical advantages over other viral vectors and plasmids [10, 14, 15]. The SeV genome does not integrate into the host genome, reducing the risk of insertional mutagenesis [16, 17]. Unlike other viral vectors, SeV is a cytoplasmic negative-stranded RNA virus that replicates entirely in the cytoplasm of cells and does not have a DNA intermediate. These aspects are very important in terms of

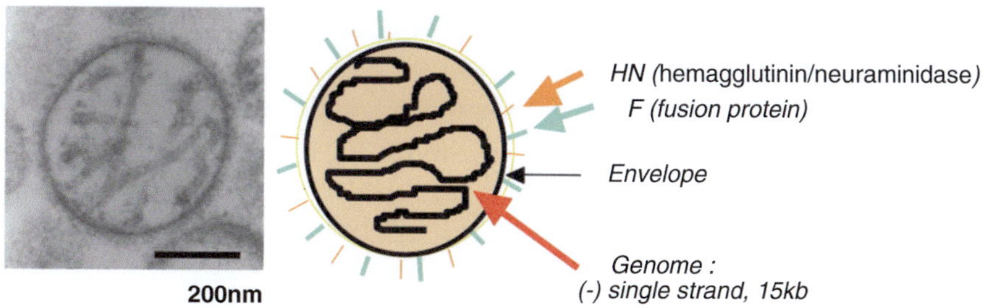

Fig. 10.1 Structure of Sendai virus (SeV)

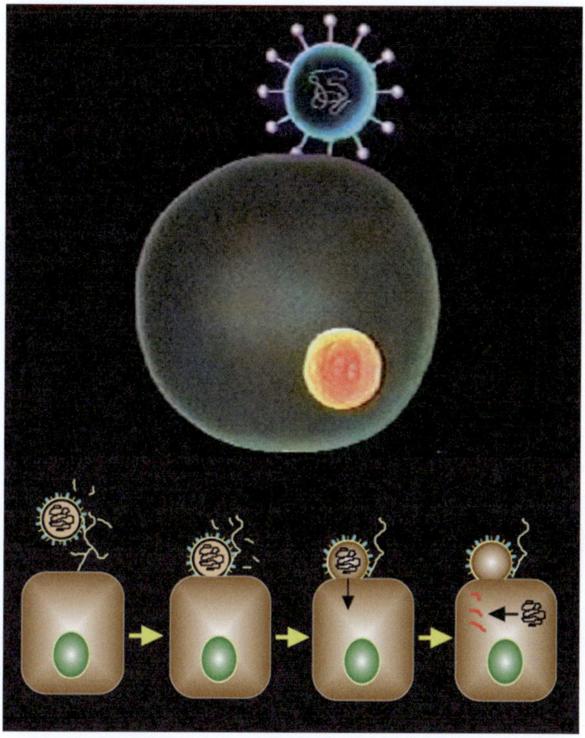

Fig. 10.2 Infection manner of SeV. First, adsorption to and degradation of sialic acid occur, followed by contact with the surface of the cell membrane, fusion with the cell membrane, Sendai virus entry, and finally replication and transcription in the cytoplasm

safety because SeV vectors do not cause genetic changes that are passed onto subsequent generations. Furthermore, SeV is not dependent on cell division for infection of target cells and requires only brief contact with cells for cellular uptake (Fig. 10.2).

10.2.2 Therapeutic Angiogenesis Using a SeV Vector Carrying FGF-2

Significant inhibition of auto-amputation by SeV/FGF-2 has been reported, although expression of VEGF from SeV/VEGF165 accelerates limb loss [11, 18]. Furthermore, the vector dose dependency of FGF-2 expression was examined in local muscles. As a result, net FGF-2 in the muscles increased proportionally to the titer of the input vector (Fig. 10.3a). However, the therapeutic outcomes demonstrated no such correlation. The highest therapeutic efficacy was observed at doses of 3×10^6 to 1×10^7 plaque-forming units, although all doses exhibited therapeutic potential (Fig. 10.3b). Taking the preclinical results together, the new gene transfer vector based on a non-transmissible recombinant SeV expressing the human FGF-2 gene (rSeV/df-hFGF-2) might be an effective tool for treatment of peripheral arterial disease (PAD) (Fig. 10.4).

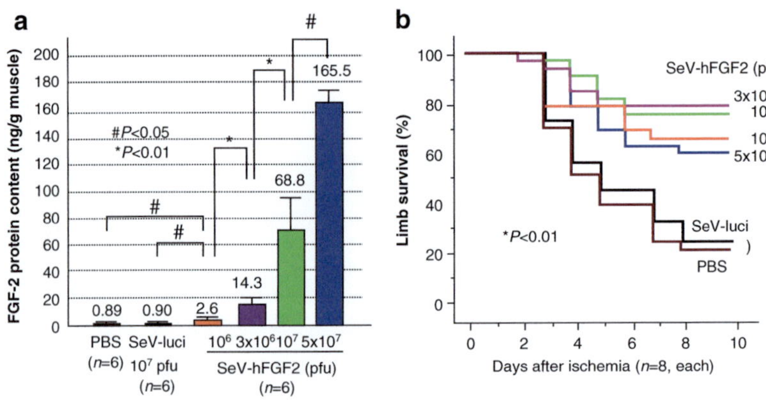

Fig. 10.3 FGF-2 expression level and corresponding limb salvaging effects of rSeV/dF (murine auto-amputation model). Expression of fibroblast growth factor (FGF)-2 (**a**) and limb prognosis curve (**b**) in ischemic thigh muscles of the murine auto-amputation model after SeV-mediated FGF-2 gene transfer, indicating the possible relationship between the levels of transgene expression and therapeutic effects. (**a**) Dose-dependent increase of net expression of FGF-2 including endogenous and exogenous FGF-2. Two days after the operation and vector injection, a posterior portion of the thigh muscles was subjected to an ELISA that detected both murine and human FGF-2 (hFGF-2). Each group included six animals. Data are expressed as means ± standard error. (**b**) Limb prognosis curve following intramuscular injection of PBS, control vector (SeV luciferase), or SeV-hFGF2. Each group included eight mice with an auto-amputation model. Limb survival curves were constructed using the Kaplan-Mayer method. Data were analyzed using the log-rank test. *PBS* phosphate-buffered saline, *pfu* plaque-forming unit. $\#P \leq 0.05$; $*P \leq 0.01$. Adapted from Shoji et al. [11]

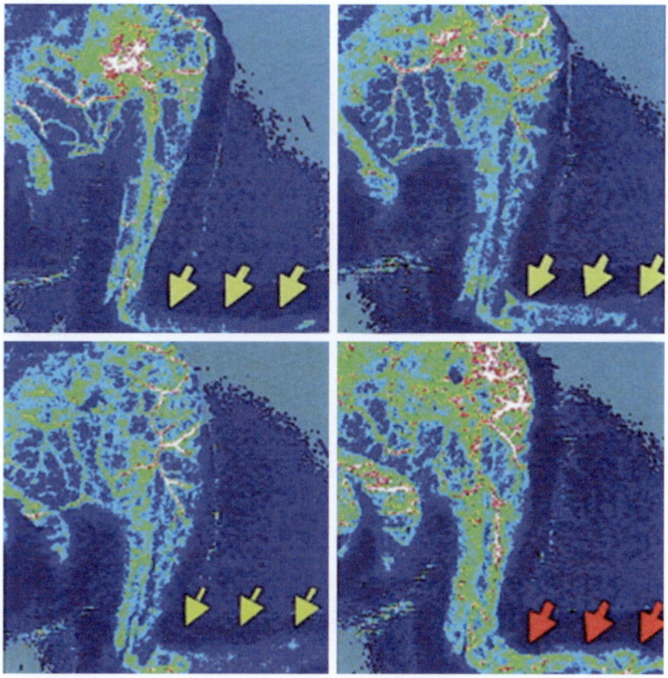

Fig. 10.4 Blood flow imaging by the laser Doppler method. Blood flow of the hind leg in SeV-hFGF2-administered animals apparently increased (*red arrows*) in comparison with that of the control (SeV-luci administered animals) at week 6. Adapted from Shoji et al. [11]

10.3 Clinical Studies

Among the published angiogenic trials using FGF for PAD patients, a plasmid-encoding FGF-1 has failed proof of concept, but a SeV vector carrying FGF-2 appears to be promising.

10.3.1 Clinical Trials Using Human Naked Plasmid DNA Encoding FGF-1

Intramuscular administration of human naked plasmid DNA encoding FGF-1 (NV1FGF) to a hamster model of ischemic limbs fed a cholesterol-rich diet promotes vascular growth in ischemic muscles [19]. This finding indicates a potential method to overcome perfusion defects in patients with PAD. To achieve preclinical success, a phase I clinical trial was performed using NV1FGF administered intramuscularly into the ischemic thigh and calf of patients with unreconstructable PAD. Increasing single and repeated doses of NV1FGF were injected into patients. As a result, NV1FGF was well tolerated and perfusion was improved, although there was no evidence of a dose-dependent effect [20]. Subsequently, the local NV1FGF plasmid distribution and transgene expression were tested to support the concept of multiple site injection for therapeutic use [21].

A randomized, double-blinded, placebo-controlled phase II called the TALISMAN 201 trial was conducted to administer eight intramuscular injections of a placebo or 16 mg NV1FGF to "non-optional" critical limb ischemia (CLI) patients [22]. There was a significant reduction in amputation risk at 12 months after administration of NV1FGF, while no difference was observed in ulcer healing between placebo and NV1FGF groups. A long-term evaluation of patients with CLI treated with NV1FGF or the placebo was reported, but the study was too small to evaluate safety and efficacy [23].

The later phase III, randomized, double-blinded, placebo-controlled TAMARIS study was conducted in 30 countries at 171 sites for 525 CLI patients [24]. The patients received 1.6 mg/mL NV1FGF or placebo on days 1, 15, 29, and 43 by intramuscular injections into the index leg. The primary endpoint of the trial was the time to major amputation of the treated leg or death during the study period of 12 months. The results were negative with no evidence of a reduction in amputation or death in the NV1FGF group [25].

10.3.2 Clinical Trials Using a SeV Vector Encoding FGF-2

The first-in-men clinical trial using a new gene transfer vector, rSeV/df-hFGF-2 (named DVC1-0101), was completed in March 2011 in Japan. The study was a phase I/IIa, open-label, four-dose-escalation clinical trial conducted in 12 patients with non-optional CLI to administer DVC1-0101 intramuscularly. The study cohort included ten cases of arteriosclerosis obliterans and two cases of thromboangiitis obliterans with a mean age of 65 years (range: 48–82 years). DVC1-0101 was administered at 30 sites with varying ischemic conditions as 5×10^7, 2×10^8, 1×10^9, and 5×10^9 cell infectious units (ciu)/60 kg body weight in a dose-escalating manner (Fig. 10.5). The study results showed that DVC1-0101 was safe and well tolerated and significantly improved walking functions [26].

A total of 120 adverse events were observed in the 6-month follow-up after gene transfer, and 12 serious advance events (SAEs) were reported and reviewed by the Safety Monitoring Board Committee at the facility and by the Health Sciences Council of the Japan Ministry of Health, Labour and Welfare at 5 years after completing the protocol (Table 10.1). No dose-response relationship was observed between DVC1-0101 and adverse events.

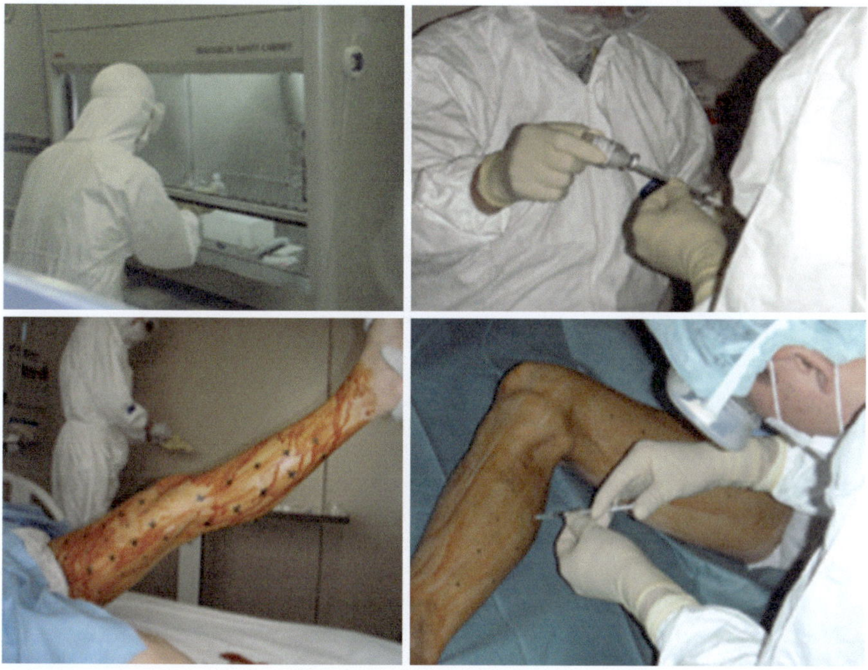

Fig. 10.5 Administration of a non-transmissible recombinant SeV expressing the hFGF-2 gene (*rSeV/df-hFGF-2*) for peripheral arterial disease patients

Some efficacy-related parameters showed positive responses. The improvement rate of clinical staging (Rutherford classification) was 58.3% ($n = 7/12$, $P = 0.7744$) and the improvement of stages 2–4 was 77.8% ($n = 7/9$, $P = 0.1797$). Pain reduction, excluding two amputation-provided patients, was seen for over 6 months ($P < 0.01$ or $P < 0.05$). No significant change was observed in ankle-brachial index or toe-brachial index at any time point for up to 6 months after treatment. Seven of the ten patients showed negative pulse-volume recording (PVR) values at pretreatment, and six subjects occasionally exhibited significant appearance of PVR values after treatment. The blood flow ratio according to the laser Doppler perfusion index showed no significant change after treatment. However, the foot skin temperature showed a significant decrease in differences of footpad temperatures (degrees Celsius = treated limb − untreated limb) at 1 month after treatment ($P < 0.05$, paired *t*-test). No ulcer healing was observed in the two stage 1 patients, but complete healing was observed in the stage 4 patient. Walking ability assessed by a treadmill showed significant improvement. Changes in absolute claudication distance (ΔACD; %, $n = 5$) were observed up to 321.5% at 8 months, and significance was shown at 2, 3, 5, 7, and 12 months after treatment ($P < 0.05$). Overall, the previous study concluded that DVC1-0101 may contribute to the improvement of walking performance in patients with PAD (Fig. 10.6) [26].

Table 10.1 5-year follow-up for serious adverse events (SAEs) in the phase I/IIa clinical trial using a SeV vector encoding FGF-2

Case #	SAE	Onset after treatment	Prognosis of event	Review result
102	Major amputation of treated limb (below knee)	3 years and 11 months	Healed stump	Not related
103	Major amputation of treated limb (below knee)	15 days	Healed stump	Cannot be denied
105	Major amputation of treated limb (3–5 toes) + salvaging bypass	3 months	Healed stump	Not related
	Myelodysplastic syndrome	1 year and 8 months	No change	Not related
	Acute-on-chronic progression of preexisting interstitial pneumonitis	2 years and 1 month	Dead	Not related
201	Compression fracture of lumbar bone	3 years	Recovered	Not related (incidental)
204	Coronary bypass for preexisting coronary aneurysm	2 years and 10 months	Recovered	Not related (incidental)
303	Heart failure	4 years and 8 months	Dead	Not related
403	Bleeding of gastric ulcer	1 year and 1 month	Recovered	Not related
	Minor amputation of treated limb (first toe)	1 year and 4 months	Healed stump	Not related
405	Acute myocardial infarction	2 years and 10 months	Recovered	Not related
	Cholelithiasis	2 years and 11 months	Recovered	Not related

10.3.2.1 Case Report

The findings for a 65-year-old Japanese woman at stage 2 (case 201) are presented in Fig. 10.6. She was diagnosed as arteriosclerosis obliterans with the symptom of 200 m of intermittent claudication (IC). She had been treated by pharmacotherapy and was stable, but then developed rest pain in her right foot. Her angiogram showed complete obstruction of whole superficial femoral to popliteal arteries and all main arteries at the ankle level in both legs (Fig. 10.7a, left panel). A soft X-ray (Fig. 10.7a, right panel) suggested difficulty in surgical management for her because of severe calcification throughout superficial femoral to popliteal arteries and an insufficient size of both saphenous veins.

DVC1-0101 was administrated at 30 sites in her right leg (10 sites at the upper and lower thigh and calf, respectively) for a total of 1.63×10^8 ciu. She had no remarkable adverse events during the 6-month observation period. Efficacy was demonstrated by an increase in foot temperature (Fig. 10.7b) and the appearance of

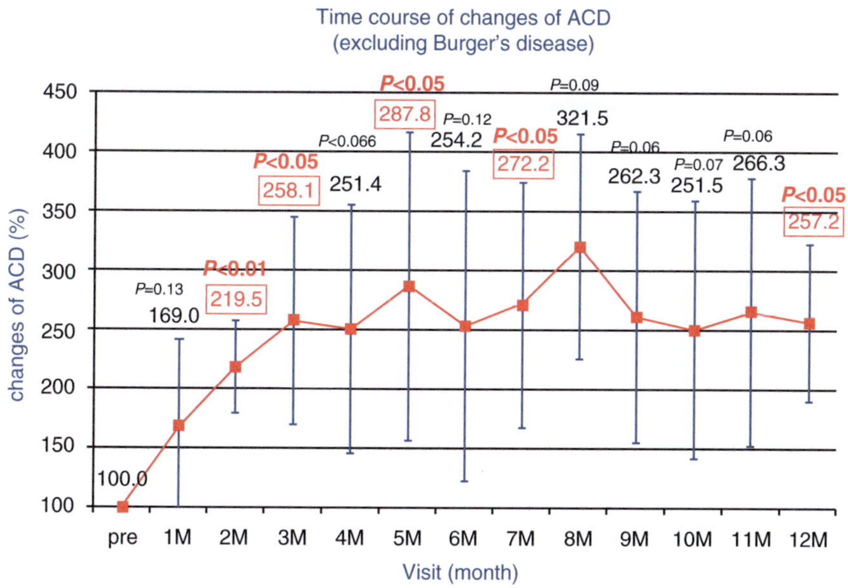

Fig. 10.6 Time course of walking activity examined by a treadmill. Redrawn with permission from [26]. Change in ACD (ΔACD; %, $n = 5$). Data were excluded from two patients with Burger's disease. *ACD* absolute claudication distance, *M* month

toe-specific pulsation (Fig. 10.7c) at her right leg. The most noteworthy results were complete disappearance of rest pain within 1 week after gene transfer and a linear increase in claudication onset distance and absolute claudication distance after gene transfer (Fig. 10.7d). These improvements in clinical symptoms have been observed for 5 years posttreatment.

10.3.2.2 Ongoing Trials

Subsequently, a phase IIb, randomized, double-blinded clinical trial with IC patients began in August 2014 and is ongoing to enroll subjects. The primary objective of this trial is to investigate the safety and clinical efficacy of DVC1-0101 (1 and 5×10^9 ciu/leg) in patients with IC. It also aims to examine the dose-response relationship using the rate of improvement in walking as a primary indicator. DVC1-0101 is administrated at 30 sites (15 sites in each upper and lower leg) intramuscularly. If the ACD is less than 200 m measured using a treadmill load test at two or more visits at 4, 5, or 6 months after the first administration, the high dose (5×10^9 ciu/leg) of DVC1-0101 will be administrated as the second administration [27]. The outcomes of this trial will provide valuable proof-of-concept data for DVC1-0101 in PAD patients and will lead to a better understanding of the gene therapy.

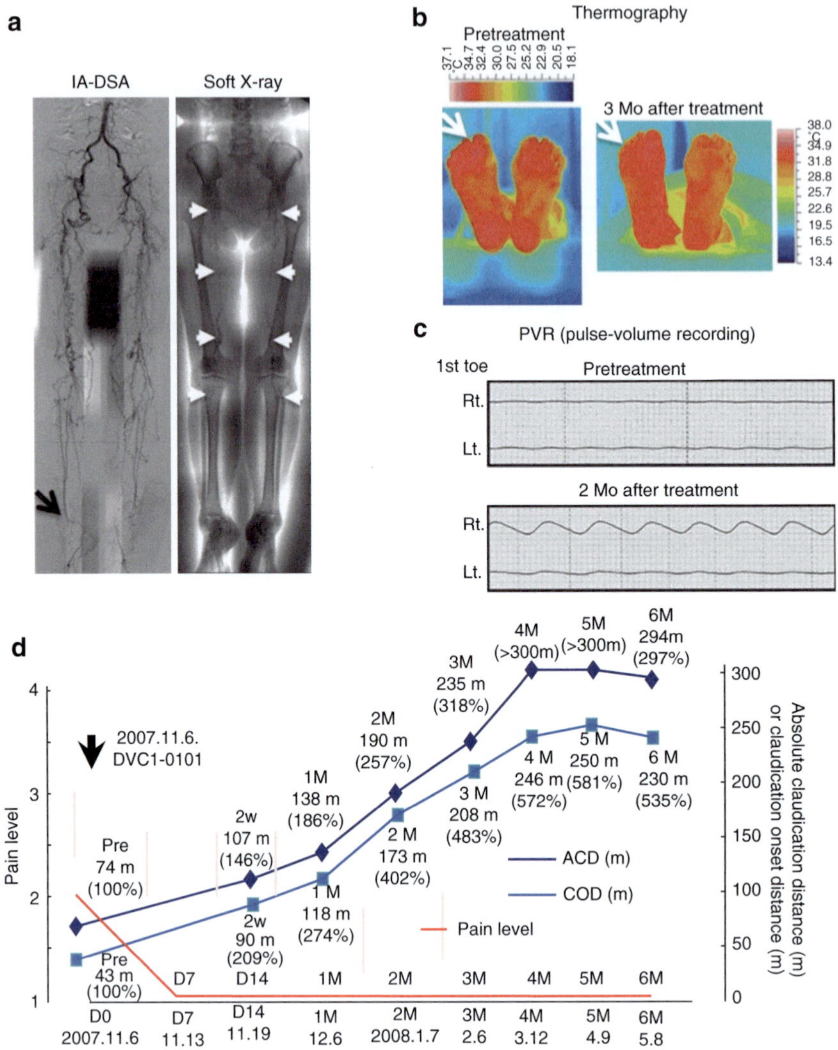

Fig. 10.7 Findings in a 65-year-old Japanese woman who received DVC1-0101 at stage 2 (case 201). Both limbs were affected with approximately 200 m of intermittent claudication caused by peripheral artery disease (PAD). Her symptoms had been stable for more than 10 years, but she developed rest pain in the right foot. (**a**) Her angiogram (*left panel*) showed complete obstruction of whole superficial femoral to popliteal arteries and all main arteries at the ankle level (*black arrow*) in both limbs. A soft X-ray (*right panel*, *white arrows*) suggested difficulty in surgical management because of severe calcification throughout superficial femoral to popliteal arteries and an insufficient size of both saphenous veins. (**b**) The result of thermography examination showed a right limb-specific increase in foot temperature (*white arrows*) at 3 months after gene transfer. (**c**) A pulse volume recording (PVR) showed toe-specific pulsation on the right after gene transfer. (**d**) Clinical course. Her rest pain completely disappeared within 1 week after the gene transfer, and her treadmill examination displayed a linear increase of absolute claudication distance (ACD) and claudication onset distance (COD) at 4 months after the gene transfer. These improvements in her clinical symptoms have been maintained for 5 years after the gene transfer. *IA-DSA* intra-arterial-digital subtraction angiography, *Mo* months. Adapted from Yonemitsu et al. [26]

10.3.3 Other Clinical Trials Using FGF

Although it is not a gene therapy, the safety and efficacy of intra-arterial injections of FGF-2 have been tested in patients with mild IC in a phase I trial. The findings revealed that FGF-2 was safe and well tolerated and improved lower extremity blood flow [28]. In addition, the TRAFFIC study, a well-controlled clinical trial of FGF-2, provided the proof of concept. Recombinant FGF-2 was infused intra-arterially in the phase II, randomized, double-blinded, placebo-controlled study of 190 patients with moderate-to-severe IC caused by infrainguinal atherosclerosis, resulting in a significant increase in peak walking time at 90 days [29]. However, the TRAFFIC study was not powered to detect differences at 180 days with type II errors [3]. The limited clinical outcomes and the mild-to-moderate adverse events owing to the systemic leakage of FGF-2 caused the TRAFFIC study to be terminated. It probably needed a higher and sustained concentration of FGF-2 in local target muscles without systemic leakage [30].

The safety and potential clinical efficacy of adenovirus-delivered FGF-4 (Ad5FGF-4) were evaluated in patients with unreconstructable CLI. The study concluded that intramuscular injection of Ad5FGF-4 appeared to be safe but failed to induce clinical efficacy because of limited assessed doses and a small patient cohort [31]. FGF-4 gene therapy was tested in patients with coronary artery disease, but the trials were halted early because an interim analysis indicated that a significant difference of efficacy in the primary endpoint was unlikely [32].

10.4 Future Directions

Experimental studies and clinical trials have established that gene therapy can be an alternative method of treatment for PAD [33]. Despite the success of proangiogenic gene therapy using FGF in preclinical in vitro and in vivo studies, the medical needs still do not fulfill expectations. There are many boundaries in gene therapy approaches that are not only using FGF but also other angiogenic factors. The issues are safety and trial design including the administration route and the dose and type of vector [34].

Specific theoretical concerns related to angiogenesis are hemorrhages, tumor growth, retinal neovascularization, enhanced hypertension, inflammatory responses, and atherogenesis [35]. All gene therapies for PAD appear to be safe and well tolerated, except for peripheral edema in treatments using recombinant VEGF. However, at this point, one patient who died from adenoviral infection has heightened awareness of the potential dangers [36]. Long-term follow-up will be continuously required to establish the safety and efficacy of gene therapy.

Although the data from pivotal studies are not yet available, a 3-year follow-up of gene transfer using non-viral FGF-1 for PAD has shown no difference in the number of strokes, myocardial infarctions, or deaths in the NV1FGF groups of five phase I and II trials compared with the placebo group [23]. Long-term follow-up of other angiogenic factor treatments, such as using naked plasmid DNA encoding hepatocyte growth factor [37] and VEGF gene transfer for coronary artery disease patients [38], has reported that gene therapy is relatively safe. Safety monitoring of long-term follow-ups for FGF treatments is still required for large trials [39].

References

1. Onimaru M, Yonemitsu Y, Fujii T, Tanii M, Nakano T, Nakagawa K, et al. VEGF-C regulates lymphangiogenesis and capillary stability by regulation of PDGF-B. Am J Physiol Heart Circ Physiol. 2009;297(5):H1685–96.
2. Presta M, Dell'Era P, Mitola S, Moroni E, Ronca R, Rusnati M. Fibroblast growth factor/fibroblast growth factor receptor system in angiogenesis. Cytokine Growth Factor Rev. 2005;16(2):159–78.
3. Aviles RJ, Annex BH, Lederman RJ. Testing clinical therapeutic angiogenesis using basic fibroblast growth factor (FGF-2). Br J Pharmacol. 2003;140(4):637–46.
4. Fujii T, Yonemitsu Y, Onimaru M, Tanii M, Nakano T, Egashira K, et al. Nonendothelial mesenchymal cell-derived MCP-1 is required for FGF-2-mediated therapeutic neovascularization: critical role of the inflammatory/arteriogenic pathway. Arterioscler Thromb Vasc Biol. 2006;26(11):2483–9.
5. Tsutsumi N, Yonemitsu Y, Shikada Y, Onimaru M, Tanii M, Okano S, et al. Essential role of PDGFRalpha-p70S6K signaling in mesenchymal cells during therapeutic and tumor angiogenesis in vivo: role of PDGFRalpha during angiogenesis. Circ Res. 2004;94(9):1186–94.
6. Li J, Wei Y, Liu K, Yuan C, Tang Y, Quan Q, et al. Synergic effects of FGF-2 and PDGF-BB on angiogenesis and muscle regeneration in rabbit hindlimb ischemia model. Microvasc Res. 2010;80:10–7.
7. Layman H, Sacasa M, Murphy AE, Murphy AM, Pham SM, Andreopoulos FM. Co-delivery of FGF-2 and G-CSF from gelatin-based hydrogels as angiogenic therapy in a murine critical limb ischemic model. Acta Biomater. 2009;5(1):230–9.
8. Safi Jr J, DiPaula Jr AF, Riccioni T, Kajstura J, Ambrosio G, Becker LC, et al. Adenovirus-mediated acidic fibroblast growth factor gene transfer induces angiogenesis in the nonischemic rabbit heart. Microvasc Res. 1999;58(3):238–49.
9. Layman H, Spiga MG, Brooks T, Pham SM, Webster KA, Andreopoulos FM. The effect of the controlled release of basic fibroblast growth factor from ionic gelatin-based hydrogels on angiogenesis in a murine critical limb ischemic model. Biomaterials. 2007;28(16):2646–54.
10. Yonemitsu Y, Kitson C, Ferrari S, Griesenbach U, Judd D, Steel R, et al. Efficient gene transfer to airway epithelium using recombinant Sendai virus. Nat Biotechnol. 2000;18(9):970–3.
11. Shoji T, Yonemitsu Y, Komori K, Tanii M, Itoh H, Sata S, et al. Intramuscular gene transfer of FGF-2 attenuates endothelial dysfunction and inhibits intimal hyperplasia of vein grafts in poor-runoff limbs of rabbit. Am J Physiol Heart Circ Physiol. 2003;285(1):H173–82.
12. Onimaru M, Yonemitsu Y, Tanii M, Nakagawara K, Masaki I, Okano S, et al. Fibroblast growth factor-2 gene transfer can stimulate hepatocyte growth factor expression irrespective of hypoxia-mediated downregulation in ischemic limbs. Circ Res. 2002;91(10):923–30.
13. Hirata T, Iida A, Shiraki-Iida T, Kitazono K, Kato A, Nagai Y, et al. An improved method for recovery of F-defective Sendai virus expressing foreign genes from cloned cDNA. J Virol Methods. 2002;104:125–33.
14. Iida A, Hasegawa M. Cytoplasmic RNA vector derived from nontransmissible Sendai virus: product and use. Gene Ther Protocols. 2002;69:361–70.
15. Inoue M, Tokusumi Y, Ban H, Shirakawa M, Kanaya T, Yoshizaki M, et al. Recombinant Sendai virus vectors deleted in both the matrix and the fusion genes: efficient gene transfer with preferable properties. J Gene Med. 2004;6:1069–81.

16. Baumgartner I. Lessons learned from human gene therapy in patients with chronic critical limb ischemia. J Invasive Cardiol. 2001;13:330–2.
17. Bitzer M, Ungerechts G, Bossow S, Graepler F, Sedlmeier R, Armeanu S, et al. Negative-strand RNA viral vectors: intravenous application of Sendai virus vectors for the systemic delivery of therapeutic genes. Mol Ther. 2003;7:210–7.
18. Masaki I, Yonemitsu Y, Yamashita A, Sata S, Tanii M, Komori K, et al. Angiogenic gene therapy for experimental critical limb ischemia: acceleration of limb loss by overexpression of vascular endothelial growth factor 165 but not of fibroblast growth factor-2. Circ Res. 2002;90(9):966–73.
19. Caron A, Michelet S, Caron A, Sordello S, Ivanov MA, Delaere P, et al. Human FGF-1 gene transfer promotes the formation of collateral vessels and arterioles in ischemic muscles of hypercholesterolemic hamsters. J Gene Med. 2004;6(9):1033–45.
20. Comerota AJ, Throm RC, Miller KA, Henry T, Chronos N, Laird J, et al. Naked plasmid DNA encoding fibroblast growth factor type 1 for the treatment of end-stage unreconstructible lower extremity ischemia: preliminary results of a phase I trial. J Vasc Surg. 2002;35(5):930–6.
21. Baumgartner I, Chronos N, Comerota A, Henry T, Pasquet JP, Finiels F, et al. Local gene transfer and expression following intramuscular administration of FGF-1 plasmid DNA in patients with critical limb ischemia. Mol Ther. 2009;17(5):914–21.
22. Nikol S, Baumgartner I, Belle EV, Diehm C, Visona A, Capogrossi MC, et al. Therapeutic angiogenesis with intramuscular NV1FGF improves amputation-free survival in patients with critical limb ischemia. Mol Ther. 2008;16(5):972–8.
23. Niebuhr A, Henry T, Goldman J, Baumgartner I, Belle EV, Gerss J, et al. Long-term safety of intramuscular gene transfer of non-viral FGF1 for peripheral artery disease. Gene Ther. 2012;19(3):264–70.
24. Adis R&D profile. Riferminogene pecaplasmide. Am J Cardiovasc Drugs. 2010;10(5):343–6.
25. Belch J, Hiatt WR, Baumgartner I, Driver IV, Nikol S, Norgren L, et al. Effect of fibroblast growth factor NV1FGF on amputation and death: a randomised placebo-controlled trial of gene therapy in critical limb ischaemia. Lancet. 2011;377(9781):1929–37.
26. Yonemitsu Y, Matsumoto T, Itoh H, Okazaki J, Uchiyama M, Yoshida K, et al. DVC1-0101 to treat peripheral arterial disease: a phase I/IIa open-label dose-escalation clinical trial. Mol Ther. 2013;21(3):707–14.
27. Tanaka M, Matsumoto T, Morisaki K, Kyuragi R, Fujino Y, Yoshida K, et al. Efficacy and safety of DVC1-0101 for intermittent claudication secondary to peripheral artery disease: study protocol of a randomized phase IIb trial. J Clin Trials. 2012;3:138.
28. Lazarous DF, Unger EF, Epstein SE, Stine A, Arevalo JL, Chew EY, et al. Basic fibroblast growth factor in patients with intermittent claudication: results of a phase I trial. J Am Coll Cardiol. 2000;36(4):1239–44.
29. Lederman RJ, Mendelsohn FO, Anderson RD, Saucedo JF, Tenaglia AN, Hermiller JB, et al. Therapeutic angiogenesis with recombinant fibroblast growth factor-2 for intermittent claudication (the TRAFFIC study): a randomised trial. Lancet. 2002;359(9323):2053–8.
30. Yonemitsu Y, Matsumoto T, Maehara Y. Gene therapy for peripheral arterial disease using Sendai virus vector: from preclinical studies to the phase I/IIa clinical trial. In: Nagai Y, editor. Sendai virus vector: advantages and applications. Tokyo: Springer; 2013. p. 185–99.
31. Matyas L, Schulte KL, Dormandy JA, Norgren L, Sowade O, Grotzbach G, et al. Arteriogenic gene therapy in patients with unreconstructable critical limb ischemia: a randomized, placebo-controlled clinical trial of adenovirus 5-delivered fibroblast growth factor-4. Hum Gene Ther. 2005;16(10):1202–11.
32. Henry TD, Grines CL, Watkins MW, Dib N, Barbeau G, Moreadith R, et al. Effects of Ad5FGF-4 in patients with angina: an analysis of pooled data from the AGENT-3 and AGENT-4 trials. J Am Coll Cardiol. 2007;50:1038–46.
33. Grochot-Przeczek A, Dulak J, Jozkowicz A. Therapeutic angiogenesis for revascularization in peripheral artery disease. Gene. 2013;525(2):220–8.
34. Gupta R, Tongers J, Losordo DW. Human studies of angiogenic gene therapy. Circ Res. 2009;105(8):724–36.

35. Yla-Herttuala S, Alitalo K. Gene transfer as a tool to induce therapeutic vascular growth. Nat Med. 2003;9(6):694–701.
36. Freedman SB. Clinical trials of gene therapy for atherosclerotic cardiovascular disease. Curr Opin Lipidol. 2002;13(6):653–61.
37. Makino H, Aoki M, Hashiya N, Yamasaki K, Azuma J, Sawa Y, et al. Long-term follow-up evaluation of results from clinical trial using hepatocyte growth factor gene to treat severe peripheral arterial disease. Arterioscler Thromb Vasc Biol. 2012;32(10):2503–9.
38. Hedman M, Muona K, Hedman A, Kivelä A, Syvänne M, Eränen J, et al. Eight-year safety follow-up of coronary artery disease patients after local intracoronary VEGF gene transfer. Gene Ther. 2009;16(5):629–34.
39. Tongers J, Roncalli JG, Losordo DW. Therapeutic angiogenesis for critical limb ischemia: microvascular therapies coming of age. Circulation. 2008;118(1):9–16.

Part III
Other Therapy

Chapter 11
Low-Intensity Pulsed Ultrasound

Akimichi Iwamoto, Takayuki Hidaka, Yasuki Kihara, Hiroshi Kubo, and Yukihito Higashi

Abstract Cell and gene therapies are invasive and may have complexity of handling and some safety concerns for patients with critical limb ischemia. Low-intensity pulsed ultrasound (LIPUS) irradiation has been widely used to promote the healing of bone fractures in humans. It is postulated that the initial step of bone fracture healing by LIPUS is angiogenesis, and subsequent neovascularization leads to osteoblasts being conveyed to the bone fracture site. Indeed, a few studies have shown that LIPUS induces angiogenesis in experimental hind-limb ischemic models. Unfortunately, there is no information on the effects of LIPUS on angiogenesis in humans. However, LIPUS irradiation should have beneficial effects on ischemia of extremities by inducing angiogenesis in patients with peripheral arterial disease (PAD). The use of LIPUS would be a noninvasive, safe, and effective method for improving clinical symptoms in patients with PAD. It is expected that the use of LIPUS would be a novel strategy for therapeutic angiogenesis in PAD patients with a wide range of limb ischemia severities. In this chapter, we focus on the effect of LIPUS on angiogenesis, the mechanism of LIPUS-induced angiogenesis, and the possibility of improvement of clinical symptoms by LIPUS irradiation in patients with PAD.

A. Iwamoto, M.D., Ph.D. • T. Hidaka, M.D., Ph.D. • Y. Kihara, M.D., Ph.D.
Department of Cardiovascular Medicine, Graduate School of Biomedical and Health Sciences, Hiroshima University, Hiroshima, Japan

H. Kubo, Ph.D.
Department of Cardiovascular Regeneration and Medicine, Research Institute for Radiation Biology and Medicine, Hiroshima University, 1-2-3 Kasumi, Minami-ku, Hiroshima 734-8551, Japan

Y. Higashi, M.D., Ph.D., F.A.H.A. (✉)
Department of Cardiovascular Regeneration and Medicine, Research Institute for Radiation Biology and Medicine, Hiroshima University, 1-2-3 Kasumi, Minami-ku, Hiroshima 734-8551, Japan

Division of Regeneration and Medicine, Medical Center for Translational and Clinical Research, Hiroshima University Hospital, Hiroshima, Japan
e-mail: yhigashi@hiroshima-u.ac.jp

Y. Higashi, T. Murohara (eds.), *Therapeutic Angiogenesis*,
DOI 10.1007/978-981-10-2744-4_11

Keywords Low-intensity pulsed ultrasound • Angiogenesis • Peripheral arterial disease

11.1 What Is PUS?

Ultrasound is an oscillating sound pressure wave. An ultrasound wave has three components: frequency, length, and velocity of the wave. Ultrasound frequency is divided into two categories: low-frequency ultrasound (20–200 kHz) and high-frequency ultrasound (1–20 MHz). Ultrasonic energy is also divided into two categories: low-intensity ultrasound (0.5–3000 mW/cm^2) and high-intensity ultrasound (5–15 kW/cm^2) [1]. A combination of ultrasound frequency and ultrasound intensity is used in various clinical settings. A combination of low-frequency ultrasound and low-intensity ultrasound is used for treatment of bone fracture healing and soft tissue healing, treatment of erectile dysfunction, and drug and gene delivery; a combination of low-frequency ultrasound and high-intensity ultrasound or high-frequency ultrasound and high-intensity ultrasound is used for treatment of cancer and surgery; and a combination of high-frequency ultrasound and low-intensity ultrasound is used in diagnostic imaging.

Although there is no strict definition of LIPUS, ultrasound frequency of 0.5–3 MHz, pulse wave of 100–400 μs in pulse burst and 1–3 kHz in repetition rate, and ultrasound energy of 0.5–3000 mW/cm^2 have been used in many previous studies as LIPUS [2–10].

11.2 Therapeutic Medicine with LIPUS

Ultrasound is an oscillating sound pressure wave that induces various cellular responses in different cells such as osteogenic transcription in osteoblasts [11], collagen synthesis in osteoblasts and fibroblasts [12], and calcium uptake in fibroblast [13]. Previous studies have shown that LIPUS stimulation induces the expression of the osteogenesis-related gene [3] and increases levels of different angiogenic factors such as interleukin-8 (IL-8) and basic fibroblast growth factor (bFGF) in osteoblasts and vascular endothelial growth factor (VEGF) in osteoblasts and monocytes [4].

In 1983, it was reported for the first time that LIPUS is effective for healing of nonunions after bone fractures in humans [2]. LIPUS has since become an established treatment for fresh bone fractures and refractory bone fractures and is now widely used in a clinical setting. LIPUS for treatment of fresh fractures, delayed unions, and nonunions has been based on the same protocol in most studies: ultrasound frequency of 1.5 MHz, ultrasound output power of 30 mW/cm^2, pulse duration of 200 μs, pulse repetition frequency of 1 kHz, and output duration of 20 min [4–8]. It is thought that LIPUS accelerates the regeneration of soft tissues including skin, skeletal muscles, ligaments, tendons, alveolar gingiva, and nerves [14, 15]. Interestingly, it has been shown that LIPUS might be useful for treatment of rheumatoid arthritis and osteoarthritis though inhibition of inflammatory responses [16]. Recently, it has been shown that LIPUS might be beneficial for treatment of erectile dysfunction [17].

11.3 LIPUS Therapy for Angiogenesis

Previous studies have shown positive effects of LIPUS and low-intensity ultrasound with a continuous wave on angiogenesis. In 1990, Young and Dyson [18] reported for the first time that therapeutic ultrasound of 5 min per day for 5 days at a spatial average temporal average intensity of 0.1 W/cm^2 with a frequency of either 0.75 or 3.0 MHz accelerated wound healing in the skin of adult rats through an increase in neovascularization. Barzelai et al. [5] showed that low-intensity ultrasound with a continuous wave of 2 MHz, 5 V peak-to-peak amplitude, and intensity of 0.05 W/cm^2 for 5 min per day over a 3-week period improved tissue blood flow and angiogenesis in a rat hind-limb ischemic model. Using low-intensity ultrasound with a continuous wave increased blood vessels in tropical fish [19]. Ramli et al. [20] showed that exposure to ultrasound with a 45 kHz wave at an intensity of 15 mW/cm^2 augmented vascularity of the chorioallantoic membrane of chicks. Abtahi et al. [6] showed that LIPUS for 5 min per day over a period of 2 weeks at an intensity of 0.3 W/cm^2 with a frequency of 3.0 MHz and pulse mode of 1:4 accelerated angiogenesis assessed by CD31 expression in transplanted ovaries of mice. LIPUS therapy increased capillary density and improved regional myocardial blood flow in a porcine model of chronic myocardial ischemia [7]. Although angiogenesis was not directly assessed in vivo, LIPUS for 20 min per day over a period of 2–4 weeks at an intensity of 30 mW/cm^2 with a frequency of 1.5 MHz and pulse duration of 200 μs accelerated bone fracture healing through enhancement of angiogenesis assessed by VEGF expression in rats with ovariectomy-induced osteoporosis [8], and LIPUS for 20 min per day over a period of 16 weeks at an intensity of 30 mW/cm^2 with a frequency of 1.5 MHz and pulse duration of 200 μs improved bone-tendon junction healing through enhancement of angiogenesis assessed by VEGF expression and enhancement of chondrogenesis in a rabbit patellectomy model [9]. These findings suggest that LIPUS therapy is promising for angiogenesis.

Although almost all studies have shown positive effects of LIPUS on angiogenesis, one study showed that acute pulsed ultrasound at an intensity of 2.5 or 5 W/cm^2 with a frequency of 1.0 MHz and pulse repeat rate of 5 kHz had no effect on angiogenesis [21].

We also confirmed the effects of LIPUS on angiogenesis in an ischemic limb mouse model. A hind-limb ischemia mouse model is widely used for assessing the angiogenic response and for testing the therapeutic potential of a new regimen for PAD [22, 23]. To verify the effects of LIPUS on ischemic wound

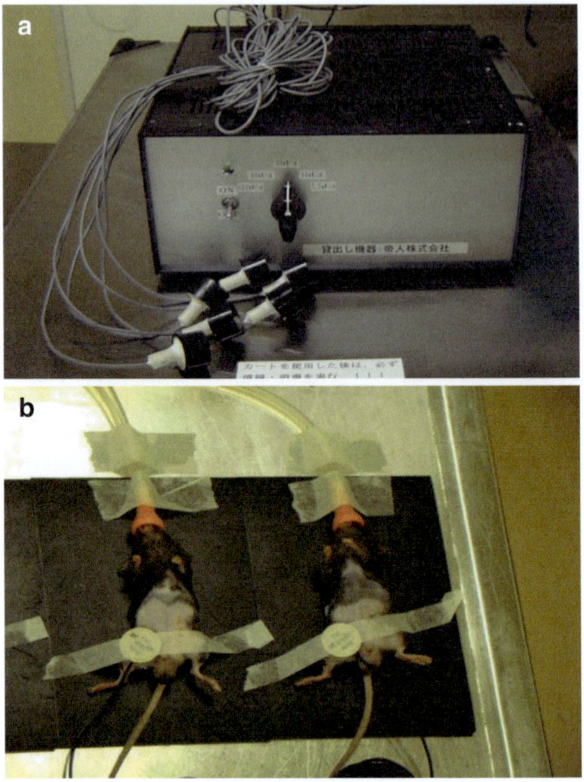

Fig. 11.1 (**a**) Low-intensity pulsed ultrasound (LIPUS) exposure system for experimental study. (**b**) Mice were anesthetized. Ultrasonic gel was put on an ultrasound transducer (Nippon Sigmax Co., Ltd., Tokyo, Japan). The transducer was placed over the skin of the ischemic area of the thigh region. LIPUS exposure was repeated for 20 min per day. LIPUS exposure conditions were as follows: effective area of the transducer was 1.68 cm^2, intensity was 30 mW/cm^2 with a 20% duty cycle, and pulse frequency was 2.0 MHz with a 1 kHz repeat rate

healing, hind-limb ischemia was surgically created in mice by femoral artery occlusion, and the mice were treated with LIPUS for 20 min per day over a period of 2 weeks at an intensity of 30 mW/cm^2 with a frequency of 2.0 MHz and pulse repeat rate of 1 kHz using a device developed for in vitro and in vivo study (Fig. 11.1). The blood flow in ischemic legs was measured by laser Doppler blood imaging, and blood flow rate index was calculated by Moor LDI software (Moor Instruments Ltd., Devon, United Kingdom). In the control group (sham group), the blood flow rate on the first day after the ischemic operation was reduced to 29% of the preoperative blood flow rate and slightly increased to 62% and 60% at day 7 and day 14, respectively. In the LIPUS exposure group (LIPUS group), the blood flow rate was reduced to 27% of the preoperative value at day 1 and then increased to 79% and 97% at day 7 and day 14, respectively. At day 14, the blood flow rate in the LIPUS group was significantly increased compared to that in the control group (Fig. 11.2). Hematoxylin-eosin staining of sections of control mice revealed that ischemia caused injury of muscle fibers and infiltration of fibrous tissue and immune cells, whereas infiltration of fibrous tissue and immune cells was inhibited in mice exposed to

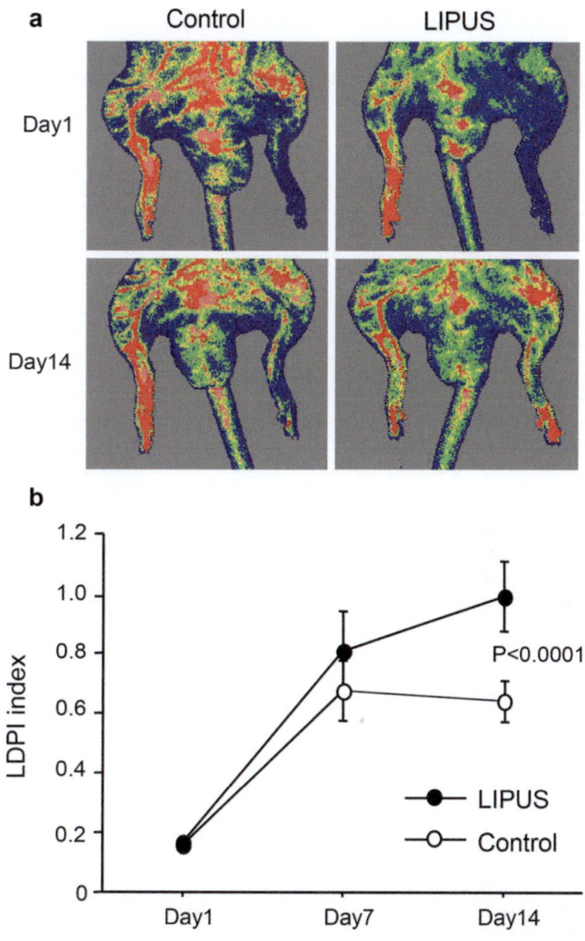

Fig. 11.2 Effect of low-intensity pulsed ultrasound (LIPUS) in the ischemic hind limb of mice: quantitative analysis of blood flow in ischemic hind limb after surgery. (**a**) Representative laser Doppler perfusion imaging in the ischemic right hind limb of sham mice and LIPUS-exposed mice. (**b**) Quantitative analysis of the ischemic/nonischemic limb laser Doppler perfusion imaging on postoperative days 1, 7, and 14 ($n = 8$ for each group). Results are shown as means ± SE. On day 14, there was a significant difference between the control and LIPUS groups

LIPUS. These findings suggest that LIPUS accelerated ischemic wound healing or prevented infiltration of fibrous tissue and immune cells.

It is known that angiogenesis is involved in recovery from ischemic damage. To determine whether the accelerated recovery of blood flow rate in the LIPUS group was caused by microvessel proliferation, we investigated the effects of LIPUS on capillary density of tissue sections stained with alkaline-phosphatase or anti-CD31 antibody. Capillary density ratios of alkaline-phosphatase staining were 1.14 for the LIPUS group and 0.94 for the control group (Fig. 11.3). In addition, the ratio of vascular endothelial cells was estimated by histological staining using anti-CD31 antibody. In the control group, the ratio was reduced to 0.61. The ratio of vascular endothelial cells was significantly increased in the LIPUS group. These results suggest that microvessel proliferation was stimulated by LIPUS, leading to recovery of peripheral blood flow.

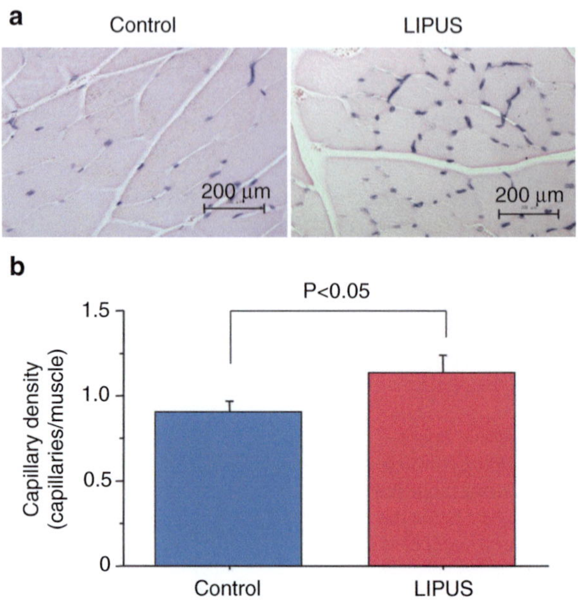

Fig. 11.3 Low-intensity pulsed ultrasound (LIPUS) exposure enhances neovascularization in ischemic limb tissues at 2 weeks after surgery: (**a**) representative anti-CD31-stained slides obtained quadriceps femoris muscle sections from sham mice and LIPUS-exposed mice. (**b**) Quantitative analysis of the ischemic/nonischemic limb capillary density ratio (*left*) in ischemic regions in the sham group and LIPUS group (*right*) ($n = 8$ for each group). Results are shown as means ± SE. Capillary density was significantly enhanced ($P < 0.05$) by LIPUS exposure

11.4 Mechanisms of LIPUS-Induced Angiogenesis

Several experimental studies have shown that LIPUS and low-intensity ultrasound with a continuous wave are beneficial for angiogenesis, osteogenesis, and wound healing [3, 10]. It is thought that LIPUS, in vitro and in vivo, induces angiogenesis through increases in angiogenic growth factors and angiogenic cytokines [24, 25]. Therapeutic ultrasound stimulates the production of angiogenic cytokines such as IL-8, bFGF, and VEGF [4].

Young and Dyson [18] reported that there were more blood vessels in equivalent regions of granulation tissue of ultrasound-treated wounds than in control wounds, suggesting that the ultrasound-treated wounds were at a more advanced stage in the repair process. Ultrasound is used as a noninvasive therapeutic method for treatment of nonunion fractures. The expression of VEGF was reported to be increased during bone fracture healing in LIPUS-treated rabbits and rats [3, 26]. VEGF is well known to be involved in angiogenesis. Ultrasound induced the expression of angiogenic factors in human fibroblasts, osteoblasts, and monocytes [27]. Ultrasound exposure increased VEGF expression in osteoporotic bone fracture healing [3] and patella-patella tendon junction healing [9]. Angiogenic transcription factors, angiogenic cytokines, and their receptors would contribute to therapeutic ultrasound-induced angiogenesis.

Ultrasound induces endothelial cell proliferation, migration, and sprouting [3–9]. The therapeutic range of ultrasound stimulates both nitric oxide (NO) synthesis and prostaglandin E_2 synthesis by human osteoblasts [28]. Exposure of endothelial cells to therapeutic ultrasound increases NO synthase activity and NO production, which could lead experimentally or therapeutically to vasodilatation [29]. Ultrasound treatment at an intensity of 1.6–2.0 W/cm^2 for 6 days enhanced NO and Ca^{2+} release from endothelial cells but did not promote endothelial cell growth [30]. In addition, ultrasound stimulation changed the cellular morphology and orientation and increased extracellular matrix secretion from endothelial cells [30]. These findings suggest that exposure to LIPUS induces angiogenesis through activation of NO synthesis.

Mesenchymal stem cells (MSCs) are multipotent stem cells that can differentiate into different cell types including osteoblasts, chondrocytes, myocytes, adipocytes, ß-pancreatic islets cells, and endothelial cells [31, 32]. The therapeutic potential of MSCs in cell-based tissue engineering has been widely studied. Therefore, we focused on the effects of LIPUS on MSCs derived from human bone marrow. After starvation for 16 h, MSCs were irradiated with LIPUS for 20 min, and the cells were harvested at 1, 3, 6, 9, and 12 h after LIPUS exposure. Total RNA was isolated from the MSCs, and the expression of mRNA was assessed by real-time PCR using primers for VEGF, FGF2, and angiopoietin 1 (ANGPT1), which are well-known factors for angiogenesis. Expression of VEGF was 0.97 at 1 h after LIPUS irradiation and was increased to 3.29 at 3 h after LIPUS exposure (Fig. 11.4a). FGF2 and ANGPT1 expression levels were significantly increased at 6 h after LIPUS exposure (Fig. 11.4b). Expression of the transcriptional factor PPAR gamma coactivator-1α (PGC-1α), but not that of hypoxia-induced factor-1α (HIF-1α), was significantly enhanced at 6 h after LIPUS exposure (Fig. 11.4c). These findings suggest that VEGF produced in MSCs may cause angiogenesis through paracrine signaling.

As mentioned above, transcriptional levels of the angiogenic factors VEGF, FGF2, and ANGPT1 were increased in MSCs after LIPUS exposure. The effects of LIPUS on hind-limb ischemia in mice might be caused by upregulation of these angiogenic factors. Immunostaining was performed with antibodies against VEGF, FGF2, and ANGPT1. In the sham group, VEGF, FGF2, and ANGPT1 were slightly expressed in stained sections. In the LIPUS group, the expression of VEGF, FGF2, and ANGPT1 was much stronger than that in contralateral nonischemic side sections, suggesting that the expression of VEGF, FGF2, and ANGPT1 was induced by LIPUS exposure in in vivo conditions. Arany et al. [33] reported that the VEGF gene is upregulated by the transcription factor transcriptional coactivator peroxisome PGC-1α. To determine whether PGC-1α is involved in upregulation of the VEGF gene after LIPUS exposure, immunostaining was performed with a PGC-1α antibody. PGC-1α was substantially expressed in the ischemic side in the LIPUS group but was not detected in either ischemic or counter sections of the sham group, suggesting that PGC-1α was upregulated in the LIPUS exposure area and might induce angiogenic factors in vivo.

The expression of VEGF, FGF2, and ANGPT1 mRNA was upregulated in MSCs by LIPUS exposure (Fig. 11.4). PGC-1α is known as a transcriptional coactivator and is known to play a role in the regulation of cellular energy metabolism. It was reported that PGC-1α induced VEGF independently of the

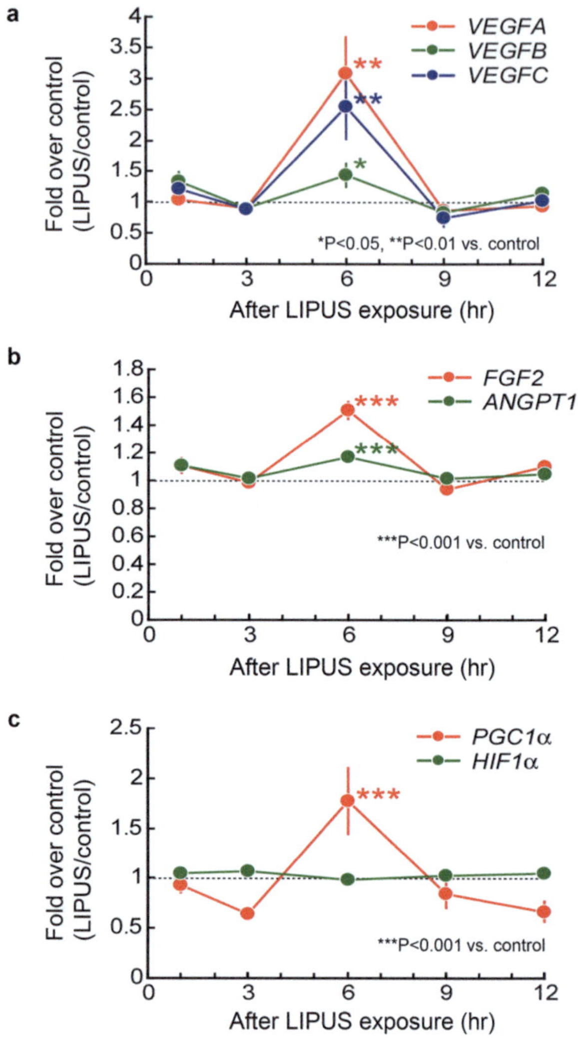

Fig. 11.4 Total RNA was isolated from collected mesenchymal stem cells (MSCs), and expression of mRNA was measured by real-time quantitative RT-PCR using primers for vascular endothelial growth factor (VEGF), fibroblast growth factor 2 (FGF2), angiopoietin 1 (ANGPT1), PPAR gamma coactivator-1α (PGC-1α), and hypoxia-induced factor-1α (HIF-1α), which are well known as angiogenesis-related genes. (**a**) Expression of VEGF was 0.97 at 3 h after low-intensity pulsed ultrasound (LIPUS) exposure and was increased to 3.29 at 6 h after LIPUS exposure. (**b**) FGF2 and ANGPT1 expression levels were significantly increased at 6 h after LIPUS exposure. (**c**) Expression of the transcriptional factor PGC-1α, but not that of HIF-1α, was significantly enhanced at 6 h after LIPUS exposure

transcription factor HIF-1 pathway [33]. Since LIPUS induced the expression of PGC-1α in the ischemic hind limb, LIPUS might induce an increase in VEGF via the PGC-1α pathway rather than the HIF-1α pathway. LIPUS upregulated the expression of VEGF, FGF2, and ANGPT1 in the ischemic hind limb, suggesting that LIPUS affects multiple pathways in several type cells and increased the expression of proteins involved in angiogenesis. It is well known that exercise and ischemia enhance VEGF expression by different mechanisms, leading to angiogenesis. Ischemia causes an increase in VEGF through the dual actions of

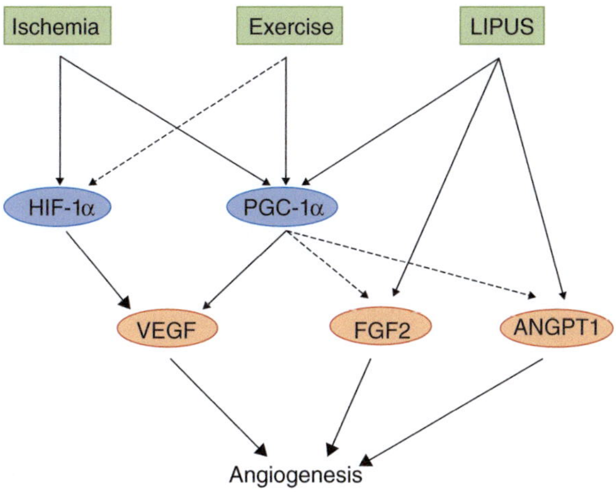

Fig. 11.5 Exercise-, ischemia-, and low-intensity pulsed ultrasound (LIPUS)-induced angiogenesis with vascular endothelial growth factor (VEGF) expression through different mechanisms of the enhancement of hypoxia-induced factor-1α (HIF-1α) and peroxisome proliferator-activated receptor γ coactivator-1α (PGC-1α)

enhancement of HIF-1α and PGC-1α expression [33]. Exercise causes an increase in VEGF through predominant enhancement of PGC-1α expression [33]. It is likely that the mechanism of LIPUS-induced angiogenesis is similar to that of exercise-induced angiogenesis (Fig. 11.5).

One of the possible mechanisms of the therapeutic benefits of MSC is paracrine release of cytokines. VEGF receptors were not expressed at detectable levels in MSCs exposed to LIPUS, and serum VEGF level was not affected by LIPUS exposure in ischemic mice. These findings suggest that the pro-angiogenic effects of LIPUS are paracrine effects of MSCs and are local effects; acceleration of angiogenesis occurs in the LIPUS exposure area, not in the whole body.

Recent studies have shown that LIPUS activated extracellular signal-regulated kinase 1/2 (ERK1/2) and p38, members of the mitogen-activated protein kinase (MAPK) family, in rat bone marrow-derived stromal cells and myoblasts [34, 35]. P38 MAPK responds to several stress stimuli [36]. ERK1/2 is activated rapidly by numerous extracellular signals and is ubiquitously distributed in various tissues [37].

Previous studies have shown that LIPUS induces phosphorylation of ERK1/2 in C2C12 cells [35], human embryonic palatal mesenchyme cells [38], MC3T3-E1 cells [39], human skin fibroblast cells [40], and mouse macrophage cells [41]. LIPUS exposure promoted proliferation via activation of focal adhesion components, such as integrin, integrin receptors and Src, and the Rho/Rho-associated kinase (ROCK)/Src/ERK signaling pathway in human skin fibroblasts [34, 35]. There is a possibility that LIPUS induces ERK1/2 phosphorylation via focal adhesion components in human umbilical vein endothelial cells. LIPUS enhances the

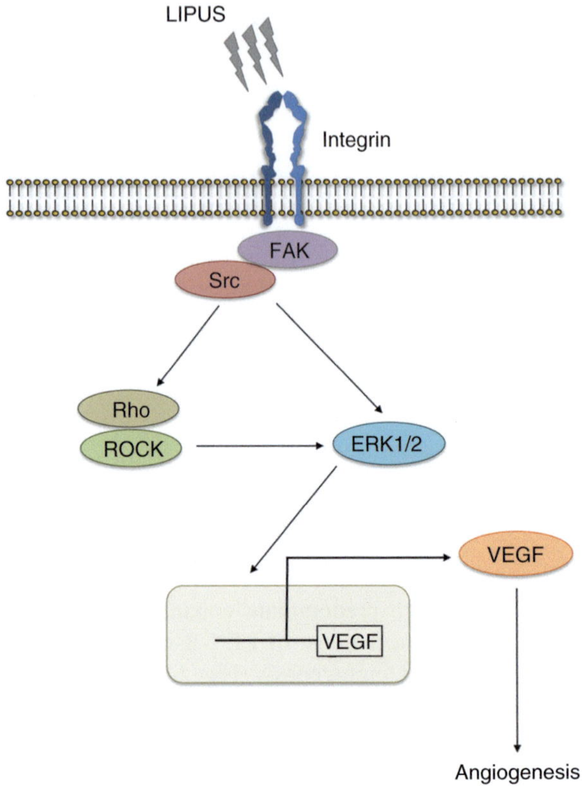

Fig. 11.6 Putative mechanism of low-intensity pulsed ultrasound (LIPUS)-induced angiogenesis through activation of the integrin/Src/Rho/Rho-associated kinase (ROCK)/extracellular signal-regulated kinase (ERK)/vascular endothelial growth factor (VEGF) signaling pathway

function of Src and FAK, which correlates with the activity of integrin. Previous studies have shown that the Rho/ROCK pathway is also an upstream regulator of ERK1/2 [40, 41].

Fluid shear stress stimulates ERK5 activity, and ERK5 inhibits apoptosis of endothelial cells [42]. It is likely that LIPUS exposure protects endothelial cells from apoptosis.

One possible mechanism of LIPUS-induced angiogenesis through activation of the integrin/Src/Rho/ROCK/ERK/VEGF signaling pathway is shown in Fig. 11.6.

Hogan et al. [43] reported that capillary density was significantly increased in cremaster muscle by 1 MHz pulsed ultrasound (1 MHz, 10 kHz 50% duty) at intensities in the range 1.25–10.0 W/cm^2. In contrast, Rubin et al. [21] reported that there was no significant effect on blood flow or capillary density of the mouse cremaster muscle of 1 MHz pulsed ultrasound (1 MHz, 5 kHz 50% duty, 2.5 and 5 W/cm^2). Since over 1 W/cm^2 levels of ultrasound intensity have thermal effects, we used a transducer of 30 mW/cm^2. We found that 1 MHz LIPUS

(1 MHz, 1 kHz 20% duty, and 30 mW/cm²) slightly increased blood flow in the ischemic leg, but there were no significant differences between the LIPUS group and sham group, suggesting that therapeutic angiogenesis for hind-limb ischemia is dependent on the ultrasound energy and that 2 MHz LIPUS is more suitable than I MHz LIPUS for acceleration of angiogenesis. In fact, it was reported that 2 MHz continuous ultrasound (50 mW/cm²) induced angiogenesis in rat hind-limb ischemia [5]. Further studies are needed to determine whether pulsed ultrasound is more efficacious than continuous ultrasound for the cure of hind-limb ischemia.

In addition, high-intensity ultrasound induces tissue warming, including warming of the vasculature [5]. Thermal effects may be a dual-edged sword for the vasculature. It has been shown that LIPUS has beneficial effects on the vasculature independently of thermal effects [31, 44]. Indeed, LIPUS of 27 kHz continuous wave at 0.25 W/cm² for 10 min increased NO synthase activity and NO production in human umbilical endothelial cells in a temperature-controlled water bath [44]. Although LIPUS of 1.4 W/cm² and a pulse wave with a duty cycle of 30% for 5 min induced coronary vasodilation in dogs, intracoronary temperature showed no change during exposure to LIPUS for 90 min [45].

11.5 Development of a LIPUS Device

From a clinical perspective, it is expected that LIPUS will be effective for angiogenesis in patients with PAD with limb ischemia ranging from a mild stage to a severe stage. We developed a new device of LIPUS therapy in humans (Fig. 11.7) [46]. The LIPUS device has a main body (size: width, 249 mm; depth, 298 mm; height, 142 mm; weight, 1.9 kg) with a power supply of 100 VAC–240 VAC ± 10% and electricity consumption of ≤150 VA. Ultrasound transducer element (Nippon Sigmax Co., Ltd., Tokyo, Japan) has a size (width, 34.68 mm; depth, 38.68 mm; height, 7.84 mm; weight, 17.0 g), eight elements (each transducer consists of four elements), ultrasound frequency of 2 MHz ± 10%, ultrasound output power of 30 mW/cm², beam nonuniformity ratio of 5 ± 2, pulse duration of 200 μs ± 5%, pulse repetition frequency of 1 kHz ± 5%, pulse duty of 20% ± 5%, and output duration of 20 min. Ultrasound control mode has a burst wave time-sharing system: each channel or group consists of four transducer elements. Within a channel transducer, elements are driven one by one with the power generator so that convolution of emitted ultrasound is unlikely. Ultrasound irradiation is applied over the skin of the ischemic area of the thigh region for 20 min. The gap between the transducer and skin is filled with ultrasonic gel. Eight transducers are attached to the skin over the gastrocnemius of each ischemic leg (Fig. 11.7).

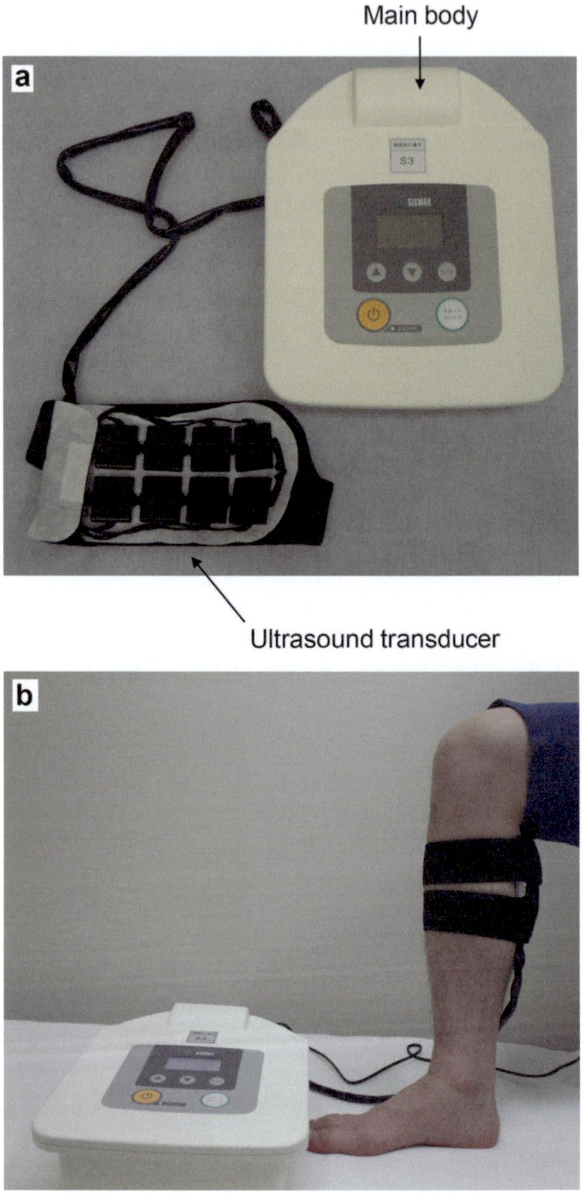

Fig. 11.7 (**a**) Appearance of the low-intensity pulsed ultrasound (LIPUS) device. (**b**) Application of transducers for LIPUS exposure in the gastrocnemius of the ischemic leg

11.6 Conclusions

In experimental hind-limb ischemic models, different pulse waves and intensities of ultrasound and different exposure times were used to find conditions effective for angiogenesis [13–16]. Unfortunately, an optimal ultrasound treatment method for therapeutic angiogenesis remains unclear. The establishment of optimal ultrasound treatment would enable more specific conclusion concerning the role of ultrasound in angiogenesis in humans to be drawn.

Further studies are needed to evaluate the effects of LIPUS on adverse effects and various events, including not only cardiovascular outcomes but also onset of malignancy, during a much longer follow-up period using a prospective randomized controlled study design. To evaluate the effect of LIPUS on clinical symptoms, we are currently performing a multicenter, double-blind, parallel randomized clinical trial using a new LIPUS device, SX-1001, in patients with Buerger disease with Fontaine III (UMIN Clinical Trial Registry; UMIN000014757) [46].

There are several therapies for PAD including exercise, pharmacologic medication, surgical bypass, endovascular intervention, and cell and gene therapies. In this chapter, we showed that LIPUS accelerated angiogenesis in an animal ischemic model. Since LIPUS has low energy and is noninvasive and easy to use, it will be beneficial for the treatment of PAD. We believe that the use of LIPUS will become a new strategy for angiogenesis in patients with PAD. The door is now open to expand ultrasound beyond the area of diagnostic imaging.

Acknowledgments We thank Megumi Wakisaka, Ki-ichiro Kawano, and Satoko Michiyama for their excellent secretarial assistance.

Sources of Founding: This study was supported in part by a Grant-in-Aid for Scientific Research from the Ministry of Education, Science and Culture of Japan (18590815 and 21590898).

References

1. Khanna A, Nelmes RT, Gougoulias N, Maffulli N, Gray J. The effects of LIPUS on soft-tissue healing: a review of literature. Br Med Bull. 2009;89:169–82. doi:10.1093/bmb/ldn040.
2. Duarte LR. The stimulation of bone growth by ultrasound. Arch Orthop Trauma Surg. 1983;101:153–9.
3. Cheung WH, Chow SK, Sun MH, Qin L, Leung KS. Low-intensity pulsed ultrasound accelerated callus formation, angiogenesis and callus remodeling in osteoporotic fracture healing. Ultrasound Med Biol. 2011;37:231–8. doi:10.1016/j.ultrasmedbio.2010.11.016.
4. Reher P, Doan N, Bradnock B, Meghji S, Harris M. Effect of ultrasound on the production of IL-8, basic FGF and VEGF. Cytokine. 1999;11:416–23.
5. Barzelai S, Sharabani-Yosef O, Holbova R, Castel D, Walden R, Engelberg S, et al. Low-intensity ultrasound induces angiogenesis in rat hind-limb ischemia. Ultrasound Med Biol. 2006;32:139–45.
6. Abtahi NS, Eimani H, Vosough A, Shahverdi A, Fathi R, Hayati N, et al. Effect of therapeutic ultrasound on folliculogenesis, angiogenesis and apoptosis after heterotopic mouse ovarian transplantation. Ultrasound Med Biol. 2014;40:1535–44. doi:10.1016/j.ultrasmedbio.2014.02.006.
7. Hanawa K, Ito K, Aizawa K, Shindo T, Nishimiya K, Hasebe Y, et al. Low-intensity pulsed ultrasound induces angiogenesis and ameliorates left ventricular dysfunction in a porcine model of chronic myocardial ischemia. PLoS One. 2014;9(8):e104863. doi:10.1371/journal.pone.0104863.
8. Cheung WH, Leung KS, Chow SK. Ultrasound and fragility fracture: is there a role? Injury. 2016;47(Suppl 1):S39–42. doi:10.1016/S0020-1383(16)30010-30019.
9. Lu H, Qin L, Cheung W, Lee K, Wong W, Leung K. Low-intensity pulsed ultrasound accelerated bone-tendon junction healing through regulation of vascular endothelial growth factor expression and cartilage formation. Ultrasound Med Biol. 2008;34:1248–60. doi:10.1016/j.ultrasmedbio.2008.01.009.

10. Hill G, Fenwick S, Matthews B, Chivers RA, Southgate J. The effect of low-intensity pulsed ultrasound on repair of epithelial cell monolayers in vitro. Ultrasound Med Biol. 2005;31:1701–6.
11. Chen YJ, Wang CJ, Yang KD, Chang PR, Huang HC, Huang YT, et al. Pertussis toxin-sensitive Galphai protein and ERK-dependent pathways mediate ultrasound promotion of osteogenic transcription in human osteoblasts. FEBS Lett. 2003;554:154–8.
12. Reher P, Doan N, Bradnock B, Meghji S, Harris M. Therapeutic ultrasound for osteoradionecrosis: an in vitro comparison between 1 MHz and 45 kHz machines. Eur J Cancer. 1998;34:1962–8.
13. Mortimer AJ, Dyson M. The effect of therapeutic ultrasound on calcium uptake in fibroblasts. Ultrasound Med Biol. 1988;14:499–506.
14. Fu SC, Shum WT, Hung LK, Wong MW, Qin L, Chan KM. Low-intensity pulsed ultrasound on tendon healing: a study of the effect of treatment duration and treatment initiation. Am J Sports Med. 2008;36:1742–9. doi:10.1177/0363546508318193.
15. Shiraishi R, Masaki C, Toshinaga A, Okinaga T, Nishihara T, Yamanaka N, et al. The effects of low-intensity pulsed ultrasound exposure on gingival cells. J Periodontol. 2011;82:1498–503. doi:10.1902/jop.2011.100627.
16. Zeng C, Li H, Yang T, Deng ZH, Yang Y, Zhang Y, et al. Effectiveness of continuous and pulsed ultrasound for the management of knee osteoarthritis: a systematic review and network meta-analysis. Osteoarthr Cartil. 2014;22:1090–9. doi:10.1016/j.joca.2014.06.028.
17. Xin Z, Lin G, Lei H, Lue TF, Guo Y. Clinical applications of low-intensity pulsed ultrasound and its potential role in urology. Transl Androl Urol. 2016;5:255–66. doi:10.21037/tau.2016.02.04.
18. Young SR, Dyson M. The effect of therapeutic ultrasound on angiogenesis. Ultrasound Med Biol. 1990;16:261–9.
19. Martin CJ, Pratt BM, Watmough DJ. A study of ultrasound-induced microstreaming in blood vessels of tropical fish. Br J Cancer Suppl. 1982;5:161–4.
20. Ramli R, Reher P, Harris M, Meghji S. The effect of ultrasound on angiogenesis: an in vivo study using the chick chorioallantoic membrane. Int J Oral Maxillofac Implants. 2009;24:591–6.
21. Rubin MJ, Etchison MR, Condra KA, Franklin Jr TD, Snoddy AM, et al. Acute effects of ultrasound on skeletal muscle oxygen tension, blood flow and capillary density. Ultrasound Med Biol. 1990;16:271–7.
22. Sata M, Nishimatsu H, Suzuki E, Sugiura S, Yoshizumi M, Ouchi Y, et al. Endothelial nitric oxide synthase is essential for the HMG-CoA reductase inhibitor cerivastatin to promote collateral growth in response to ischemia. FASEB J. 2001;15:2530–2.
23. Sumi M, Sata M, Hashimoto A, Imaizumi T, Yanaga K, Ohki T, et al. OPC-28326, a selective femoral arterial vasodilator, augments ischemia induced angiogenesis. Biomed Pharmacother. 2007;61:209–15.
24. Li J, Waugh LJ, Hui SL, Burr DB, Warden SJ. Low-intensity pulsed ultrasound and nonsteroidal anti-inflammatory drugs have opposing effects during stress fracture repair. J Orthop Res. 2007;25:1559–67.
25. Pittenger MF, Mackay AM, Beck SC, Jaiswal RK, Douglas R, Mosca JD, et al. Multilineage potential of adult human mesenchymal stem cells. Science. 1999;284:143–7.
26. Walsh W, Langdown A, Auld J, Stephens P, Yu Y, Vizesi F, et al. Effect of low intensity pulsed ultrasound on healing of an ulna defect filled with a bone graft substitute. J Biomed Mater Res B Appl Biomater. 2008;86:74–81.
27. Doan N, Reher P, Meghji S, Harris M. In vitro effects of therapeutic ultrasound on cell proliferation, protein synthesis, and cytokine production by human fibroblasts, osteoblasts, and monocytes. J Oral Maxillofac Surg. 1999;57:409–19.
28. Reher P, Harris M, Whiteman M, Hai HK, Meghji S. Ultrasound stimulates nitric oxide and prostaglandin E2 production by human osteoblasts. Bone. 2002;31:236–41.

29. Altland O, Dalecki D, Suchkova V, Francis CW. Low-intensity ultrasound increases endothelial cell nitric oxide synthase activity and nitric oxide synthesis. J Thromb Haemost. 2004;2:637–43.
30. Hsu S, Huang T. Bioeffect of ultrasound on endothelial cells in vitro. Biomol Eng. 2004;21:99–104.
31. Tang J, Wang J, Zheng F, Kong X, Guo L, Yang J, et al. Combination of chemokine and angiogenic factor genes and mesenchymal stem cells could enhance angiogenesis and improve cardiac function after acute myocardial infarction in rats. Mol Cell Biochem. 2010;399:107–18. doi:10.1007/s11010-009-0374-0.
32. Chen L, Jiang X, Yang L. Differentiation of rat marrow mesenchymal stem cells into pancreatic islet beta-cells. World J Gastroenterol. 2004;10:3016–20.
33. Arany Z, Foo SY, Ma Y, Ruas JL, Bommi-Reddy A, Girnun G, et al. HIF-indcpendent regulation of VEGF and angiogenesis by the transcriptional coactivator PGC-1alpha. Nature. 2008;451:1008–12. doi:10.1038/nature06613.
34. Sena K, Angle SR, Kanaji A, Aher C, Karwo DG, Sumner DR, et al. Low-intensity pulsed ultrasound (LIPUS) and cell-to-cell communication in bone marrow stromal cells. Ultrasonics. 2011;51:639–44. doi:10.1016/j.ultras.2011.01.007.
35. Ikeda K, Takayama T, Suzuki N, Shimada K, Otsuka K, Ito K. Effects of low-intensity pulsed ultrasound on the differentiation of C2C12 cells. Life Sci. 2006;79:1936–43.
36. Cuenda A, Rousseau S. p38 MAP-kinases pathway regulation, function and role in human diseases. Biochim Biophys Acta. 2007;1773:1358–75.
37. Boulton TG, Cobb MH. Identification of multiple extracellular signal-regulated kinases (ERKs) with antipeptide antibodies. Cell Regul. 1991;2:357–71.
38. Appleford MR, Oh S, Cole JA, Protivinský J, Ong JL. Ultrasound effect on osteoblast precursor cells in trabecular calcium phosphate scaffolds. Biomaterials. 2007;28:4788–94.
39. Bandow K, Nishikawa Y, Ohnishi T, Kakimoto K, Soejima K, Iwabuchi S, et al. Low-intensity pulsed ultrasound (LIPUS) induces RANKL, MCP-1, and MIP-1beta expression in osteoblasts through the angiotensin II type 1 receptor. J Cell Physiol. 2007;211:392–8.
40. Zhou S, Zhou S, Schmelz A, Seufferlein T, Li Y, Zhao J, et al. Molecular mechanisms of low intensity pulsed ultrasound in human skin fibroblasts. J Biol Chem. 2004;279:54463–9.
41. Zhou S, Bachem MG, Seufferlein T, Li Y, Gross HJ, Schmelz A. Low intensity pulsed ultrasound accelerates macrophage phagocytosis by a pathway that requires actin polymerization, Rho, and Src/MAPKs activity. Cell Signal. 2008;20:695–704. doi:10.1016/j.cellsig.2007.12.005.
42. Pi X, Yan C, Berk BC. Big mitogen-activated protein kinase (BMK1)/ERK5 protects endothelial cells from apoptosis. Circ Res. 2004;94:362–9.
43. Hogan RD, Burke KM, Franklin TD. The effect of ultrasound on microvascular hemodynamics in skeletal muscle: effects during ischemia. Microvasc Res. 1982;23:370–9.
44. Tokcaer-Keskin Z, Akar AR, Ayaloglu-Butun F, Terzioglu-Kara E, Durdu S, Ozyurda U, et al. Timing of induction of cardiomyocyte differentiation for in vitro cultured mesenchymal stem cells: a perspective for emergencies. Can J Physiol Pharmacol. 2009;87:143–50. doi:10.1139/Y08-111.
45. Miyamoto T, Neuman Y, Luo H, Jeon DS, Kobal S, Ikeno F, et al. Coronary vasodilation by noninvasive transcutaneous ultrasound: an in vivo canine study. J Am Coll Cardiol. 2003;41:1623–7.
46. Higashi Y, Azuma N, Takeishi Y, Minamino T, Kihara Y, Node K, et al. Effect of low-intensity pulsed ultrasound device, SX-1001, on clinical symptoms in Buerger disease with critical limb ischemia: a multicentre, double-blind, parallel randomized clinical trial. Int Heart J. 2015; 56:633–8. doi:10.1536/ihj.15-191.

Chapter 12
Low-Energy Extracorporeal Shock Wave Therapy

Kenta Ito, Tomohiko Shindo, and Hiroaki Shimokawa

Abstract Despite recent advances in medical knowledge and technology, ischemic heart disease is still one of the major causes of death, with the morbidity increasing worldwide. We have recently developed a new, noninvasive angiogenic therapy using low-energy shock waves. Low-energy extracorporeal cardiac shock wave therapy improves myocardial ischemia in a porcine model of chronic myocardial ischemia and in patients with refractory angina pectoris. Shock wave therapy also improves walking ability in patients with peripheral arterial disease and ameliorates digital skin ulcers in patients with systemic sclerosis. Furthermore, animal studies suggest that shock wave therapy may be effective to suppress left ventricular remodeling after acute myocardial infarction and to enhance locomotor recovery after spinal cord injury. Here, we summarize the studies in animals and humans and discuss the advantages and perspectives of low-energy shock wave therapy.

Keywords Shock wave therapy • Ischemic heart disease • Growth factors • New technology

12.1 Introduction

Despite recent advances in medical knowledge and technology, ischemic heart disease (IHD) is still one of the major causes of death in developed countries, with the morbidity increasing worldwide [1–4]. The standard management of IHD has three

K. Ito • T. Shindo • H. Shimokawa, M.D., Ph.D. (✉)
Department of Cardiovascular Medicine, Tohoku University Graduate School of Medicine, 1-1 Seiryomachi, Aoba-ku, Sendai, Miyagi 980-8574, Japan
e-mail: shimo@cardio.med.tohoku.ac.jp

Y. Higashi, T. Murohara (eds.), *Therapeutic Angiogenesis*,
DOI 10.1007/978-981-10-2744-4_12

major therapeutic options, including medication, percutaneous coronary intervention (PCI), and coronary artery bypass grafting (CABG). However, the number of severe IHD patients with multiple comorbidities has been increasing as the population is aging. Thus, new, noninvasive therapeutic strategies are urgently needed for aged and severely diseased patients.

Shock wave (SW) is a longitudinal acoustic wave that propagates through water, fat, and soft tissues as ultrasound does. SW is a single pressure pulse with a short needlelike positive spike <1 μs in duration and up to 100 MPa in amplitude, followed by a tensile portion lasting several microseconds with lower amplitude. Extracorporeal shock wave (SW) therapy was clinically introduced more than 30 years ago, which has markedly improved the treatment of urolithiasis. In extracorporeal SW lithotripsy, high-energy SW is used to break up stones in the urinary tract. We and others have demonstrated that low-energy SW (about 10% of the energy density that is used for urolithiasis) enhances the expression of vascular endothelial growth factor (VEGF) (Fig. 12.1) and nitric oxide (NO) release in cultured human umbilical vein endothelial cells (HUVEC) [5, 6]. Furthermore, we have demonstrated that low-energy cardiac SW therapy improves myocardial ischemia in a porcine model of chronic myocardial ischemia and in patients with refractory angina pectoris [5, 7, 8]. In this chapter, we summarize the studies in animals and humans and discuss the advantages and perspectives of low-energy SW therapy.

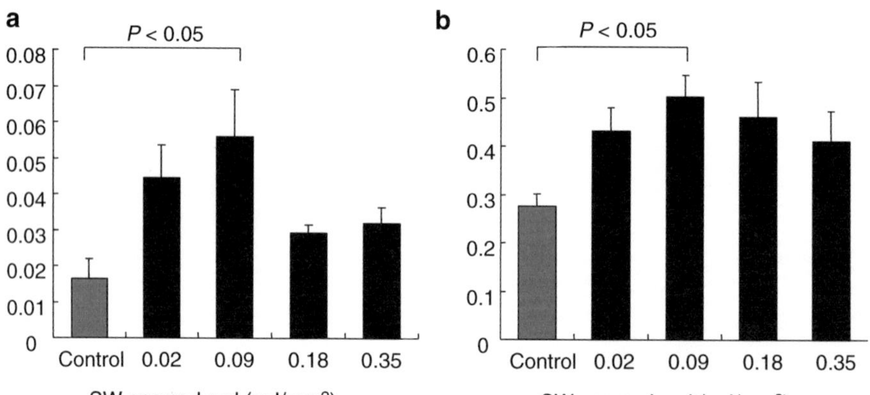

Fig. 12.1 Effects of SW on mRNA expression in HUVECs in vitro. SW therapy upregulated mRNA expression of VEGF (**a**) and Flt-1 (**b**) with a maximum effect noted at 0.09 mJ/mm^2, in which level is approximately 10% of that used for urolithiasis. Results are expressed as mean ± SEM (n = 10 each) (from [5] with permission)

12.2 Extracorporeal Cardiac SW Therapy for Angina Pectoris

12.2.1 Animal Studies

Based on our in vitro studies in HUVEC, we studied whether low-energy SW therapy ameliorates myocardial ischemia in a porcine model in vivo [5]. A porcine model of chronic myocardial ischemia was prepared by placing an ameroid constrictor at the proximal segment of the left circumflex coronary artery (Lcx). This constrictor gradually induced a total occlusion of the Lcx with sustained myocardial dysfunction but without myocardial infarction in 4 weeks. Four weeks after the implantation of the ameroid constrictor, we performed low-energy extracorporeal cardiac SW therapy in the SW group ($n = 8$) three times during the first week, whereas animals in the control group ($n = 8$) received the same anesthesia procedures three times a week but without the SW therapy. Low-energy SW was applied to nine spots (0.09 mJ/mm^2, 200 shots/spot) in the ischemic Lcx region in the left ventricle (LV) with a guidance of an echocardiogram equipped within a specially designed SW generator (Storz Medical AG, Kreuzlingen, Switzerland) with an R-wave-triggered manner. We evaluated cardiac function before (baseline) and at 4 and 8 weeks after the ameroid implantation. Four weeks after the implantation of the constrictor, wall motion of the Lcx region in the LV was reduced to the same extent in both the control and the SW group (Fig. 12.2a, c).

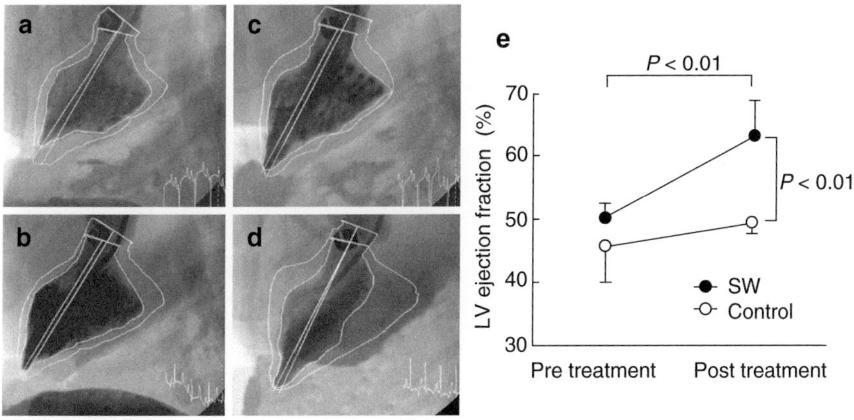

Fig. 12.2 Effects of the SW therapy on LV function in pigs in vivo. The extracorporeal cardiac SW therapy improved ischemia-induced myocardial dysfunction in vivo as evaluated by left ventriculography. Four weeks after the implantation of an ameroid constrictor, LV wall motion of the LCX (posterolateral) region was reduced in both the control (**a**) and the SW group (before the SW therapy) (**c**). Eight weeks after the implantation of an ameroid constrictor, no significant change in LV wall motion was noted in the control group (**b**), whereas marked recovery was noted in the SW group (**d**). (**e**) The SW therapy normalized LV ejection fraction in the SW group but not in the control group. Results are expressed as mean ± SEM ($n = 8$ each) (from [5] with permission)

However, 4 weeks after the SW therapy (8 weeks after the implantation of the constrictor), LV wall motion was marked improved only in the SW group (Fig. 12.2b, d, e). We also confirmed that the SW therapy upregulated the expression of VEGF, increased capillary density (Fig. 12.3), and normalized regional myocardial blood flow in the ischemic myocardium in vivo. No complications or adverse effects, such as tissue injury, hemorrhage, or arrhythmia, were noted during or after the SW therapy. These results suggest that the low-energy cardiac SW therapy enhances the endogenous angiogenic system in pigs in vivo. This was the first report that demonstrates the potential usefulness of low-energy extracorporeal cardiac SW therapy as a noninvasive angiogenic approach to chronic myocardial ischemia.

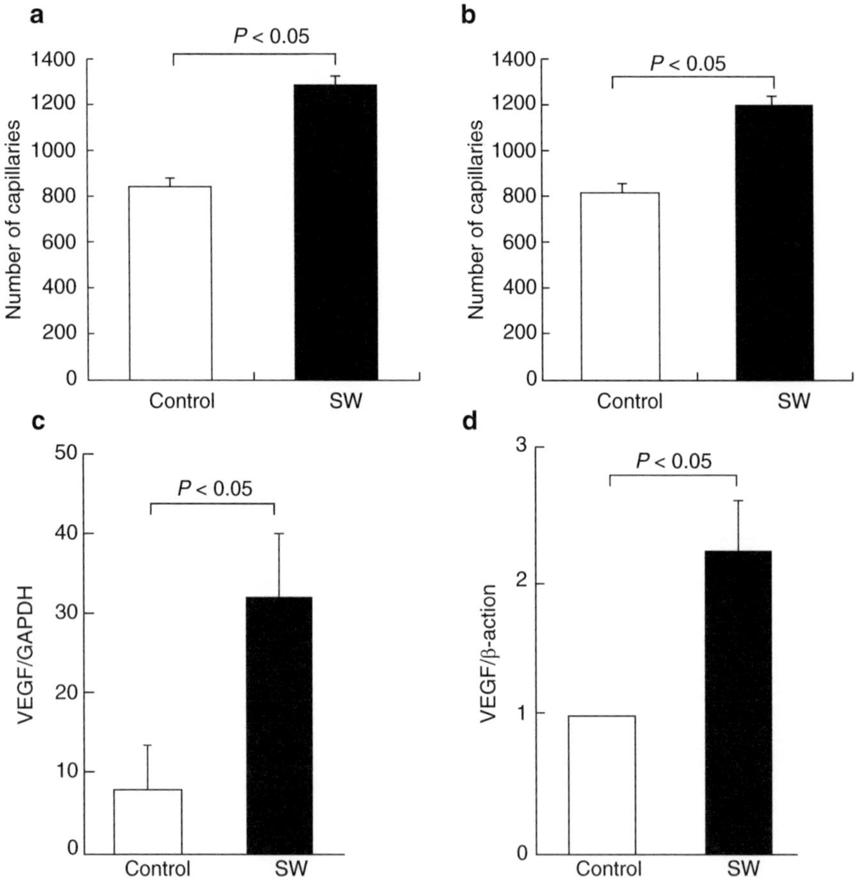

Fig. 12.3 Effects of the SW therapy on capillary density and VEGF expression in the ischemic myocardium in pigs in vivo. The extracorporeal cardiac SW therapy increased the density of factor VIII-positive capillaries and VEGF expression in the ischemic myocardium. Capillary density was significantly greater in the SW group (SW) than in the control group (control) in both the endocardium (**a**) and the epicardium (**b**). The mRNA expression (**c**) and the protein levels (**d**) of VEGF were significantly higher in the SW group than in the control group. Results are expressed as mean ± SEM ($n = 6$ each) (from [5] with permission)

12.2.2 Clinical Studies

Standard therapeutic approaches to ischemic heart disease (IHD) include medication, percutaneous coronary intervention (PCI), and coronary artery bypass grafting (CABG). However, the number of IHD patients who are resistant to those therapies is increasing. Based on the promising results in animal studies, we conducted the first clinical trial of low-energy extracorporeal cardiac SW therapy in nine patients with refractory angina pectoris without indication of PCI or CABG (55–82 years old, five men and four women) [7]. Low-energy SW was applied to 20–40 spots (0.09 mJ/mm^2, 200 shots/spot) in the ischemic area in the LV three times during the first week. During the therapy, a patient lays on the bed in a supine position without any anesthesia or analgesics (Fig. 12.4). The low-energy SW therapy significantly improved symptoms and reduced the use of nitroglycerin (Fig. 12.5) and also ameliorated myocardial perfusion as assessed by stress scintigraphy only in the ischemic area treated with the SW therapy (Fig. 12.6). No complications or adverse effects related to the SW therapy were noted. These results indicate that low-energy extracorporeal cardiac SW therapy is a safe, effective, and noninvasive therapeutic

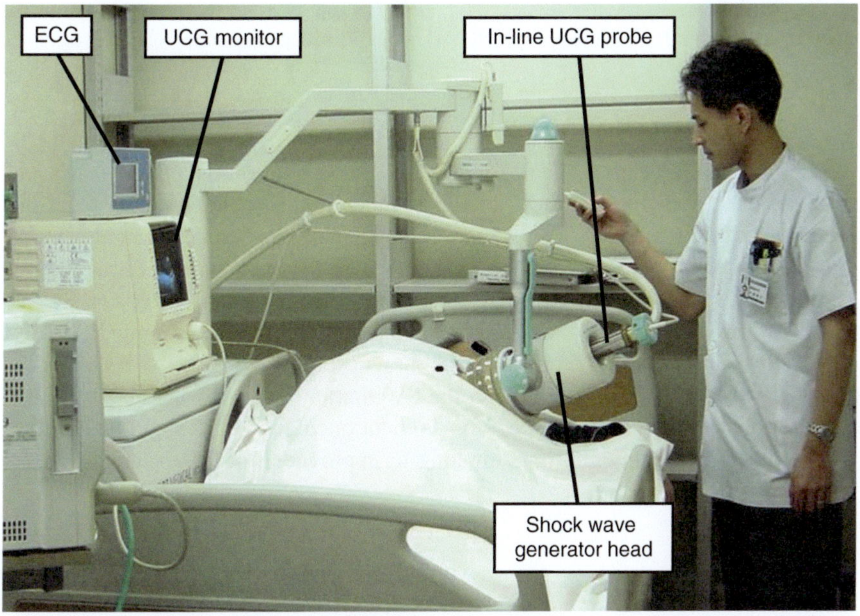

Fig. 12.4 Extracorporeal cardiac SW therapy in action in a patient with refractory angina pectoris. The machine is equipped with a SW generator head and in-line echocardiography probe. The SW generator is attached to the chest wall of the patient when used. The SW pulse is easily focused on the ischemic myocardium under the guidance of echocardiography. There is no need of anesthesia or analgesics

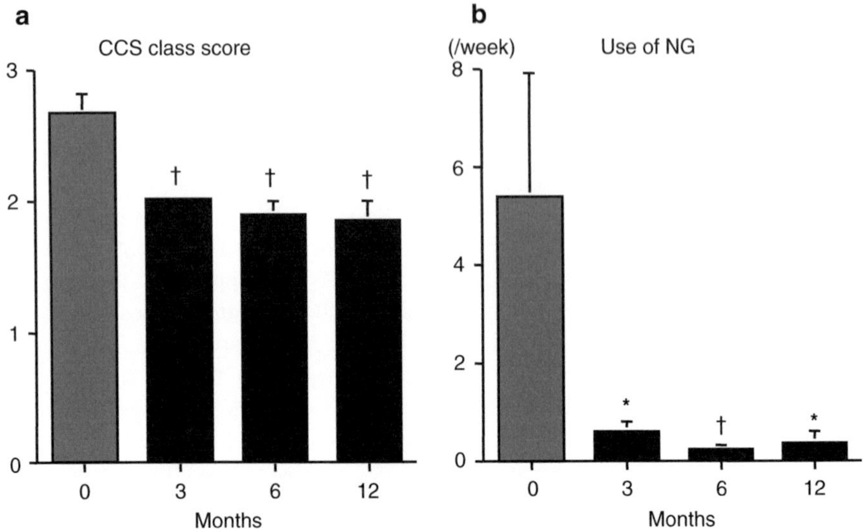

Fig. 12.5 Effects of the extracorporeal cardiac SW therapy on symptom and the use of nitroglycerin. Extracorporeal cardiac SW therapy significantly improved Canadian Cardiovascular Society (CCS) class scores (**a**) and the use of nitroglycerin (NG) (**b**) in patients with refractory angina pectoris. Results are expressed as mean ± SEM. $^{*}P < 0.05$ and $^{\dagger}P < 0.01$ vs. 0 month (statistically analyzed by post hoc test after one-way ANOVA) (from [7] with permission)

strategy for severe ischemic heart disease. To further confirm the effectiveness and safety of the SW therapy, we performed a second clinical trial in a randomized and placebo-controlled manner [8]. In this second clinical trial, we again demonstrated that the low-energy SW therapy not only improves symptoms and reduces the use of nitroglycerin but also improves LV function (Fig. 12.7), establishing cardiac SW therapy as an effective and safe angiogenic strategy for severe ischemic heart disease. Following our initial report, several clinical studies with positive results were reported worldwide [9–15]. Although the SW therapy improves the quality of life (QOL) in patients with angina pectoris as mentioned above, it should be clarified whether the SW therapy improves the long-term prognosis of those patients.

Fig. 12.7 Effects of the extracorporeal cardiac SW therapy in patients with refractory angina pectoris in the placebo-controlled and double-blind study. *CCS* Canadian Cardiovascular Society, *NTG* nitroglycerin, *Max. exercise* maximum exercise capacity in watts (W), *Peak* VO_2 peak oxygen uptake, *LVEF* left ventricular (LV) ejection fraction, *LVSV* LV stroke volume, *LVEDV* LV end-diastolic volume, *BNP* brain natriuretic peptide. Results are mean ± SE ($n = 8$ each) (from [8] with permission)

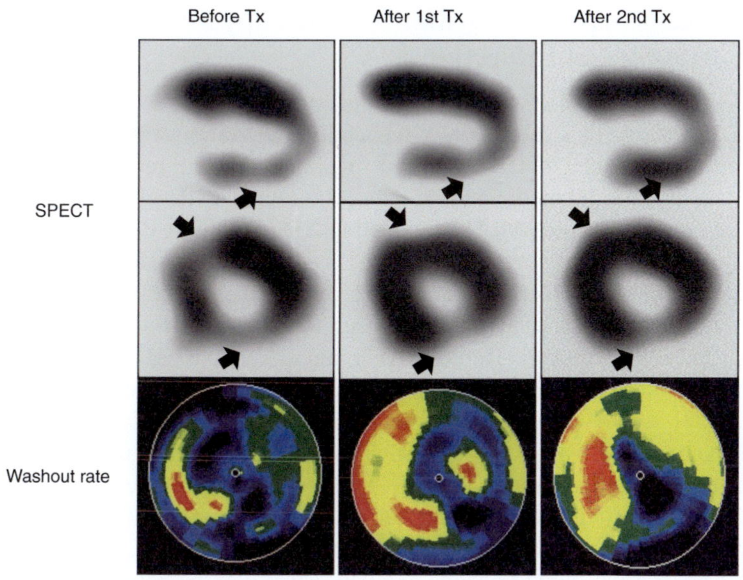

Fig. 12.6 Effects of the extracorporeal cardiac SW therapy on myocardial perfusion in patients with refractory angina pectoris. Dipyridamole stress thallium-201 single-photon emission computed tomography (SPECT) imaging and polar map in a patient with severe three-vessel coronary artery disease before and after the SW therapy. The results clearly demonstrated that the SW therapy ameliorated myocardial perfusion only where SW was applied, in the anteroseptal wall after the first treatment (1st Tx) and in the lateral wall after the second treatment (2nd Tx) (*arrows*) in a stepwise manner after the staged SW treatment. The areas treated with the SW therapy are indicated with dotted lines (from [7] with permission)

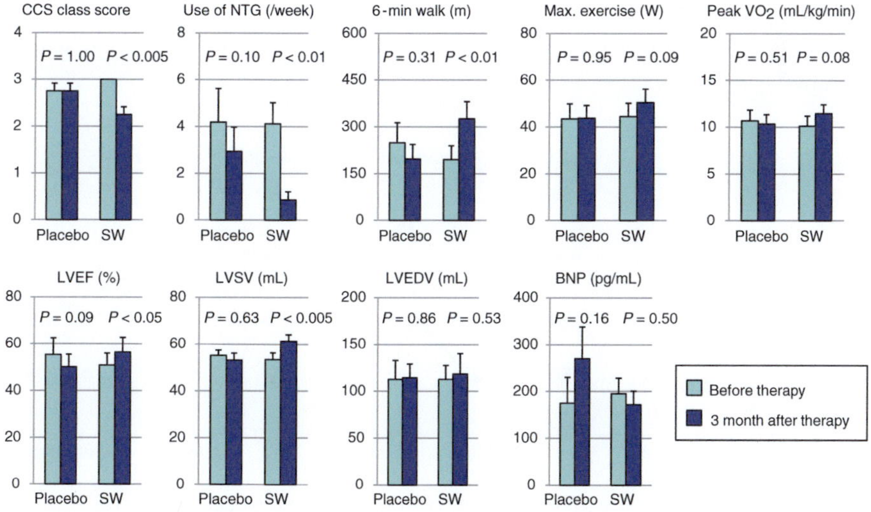

12.3 Extracorporeal Cardiac SW Therapy for Acute Myocardial Infarction

Although primary PCI substantially reduced the mortality of patients with acute myocardial infarction (AMI), LV remodeling after AMI still remains an important issue in cardiovascular medicine [16]. Thus, it is crucial to develop new therapeutic strategies to suppress post-MI LV remodeling. Since capillary density in the border zone adjacent to the infarcted myocardium is negatively correlated with infarct size 1 month after AMI [17], enhancing angiogenesis in the border zone is expected to ameliorate the progression of post-MI LV remodeling in patients. Thus, we studied whether the SW therapy is also effective to ameliorate post-MI LV remodeling in pigs in vivo. First, we created AMI by surgically excising the proximal segment of the Lcx [18]. Low-energy extracorporeal cardiac SW therapy was performed at 3, 5, and 7 days after AMI in the SW group. The animals in the control group were treated in the same manner but without the SW therapy. Four weeks after the therapy, LV ejection fraction and LV end-diastolic volume were significantly improved in the SW group compared with the control group (Fig. 12.8). Furthermore, regional myocardial blood flow and capillary density in the border zone were significantly improved in the SW group compared with the control group. Again, no procedural complications or adverse effects were noted. These results suggest that the low-energy extracorporeal cardiac SW therapy is an effective and noninvasive therapy to ameliorate post-MI LV remodeling as well. This was the first report that demonstrates the usefulness and safety of extracorporeal cardiac SW therapy as a noninvasive treatment of AMI. We also confirmed the beneficial effects and safety of the SW therapy in another porcine model of AMI due to myocardial ischemia (90 min)/reperfusion to mimic the clinical setting [19]. Based on the promising results in two types of AMI models in pigs, the first clinical trial in AMI patients is conducted to examine the feasibility, effectiveness, and safety of cardiac SW therapy. In this trial, low-energy SW is applied to the border zone around the infarcted area in AMI patients who are successfully treated with PCI as an adjunctive therapy.

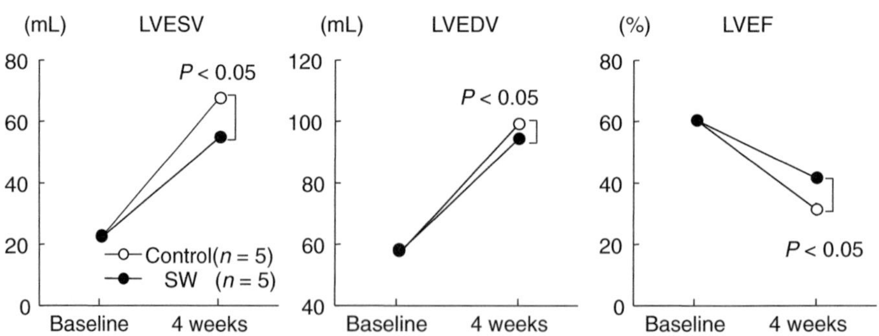

Fig. 12.8 Effects of the extracorporeal cardiac SW therapy on LV remodeling in pigs in vivo. The SW therapy significantly ameliorated LV remodeling characterized by the increase in LV end-systolic volume (LVESV) and end-diastolic volume (LVEDV) and reduced LV ejection fraction (LVEF) in a porcine model of AMI (from [18] with permission)

12.4 Additional Indications of Low-Energy Extracorporeal SW Therapy

12.4.1 SW Therapy for Peripheral Arterial Disease

Peripheral arterial disease (PAD) is often associated with IHD and the prognosis of patients with critical limb ischemia is quite poor [20–22]. Thus, we studied the effects of SW therapy on hindlimb ischemia in rabbits [23]. Hindlimb ischemia was induced by surgical excision of the entire unilateral femoral artery. One week after the operation, low-energy SW was applied to 30 spots (0.09 mJ/mm^2, 200 shots/spot) in the ischemic region three times a week for 3 consecutive weeks. Four weeks after the operation, blood flow, blood pressure, and capillary density were all significantly higher in the SW group than in the control group. Based on the results in animal studies, we conducted a clinical trial in 12 patients with PAD with intermittent claudication (Fontaine stage II; 60–86 years old, ten men and two women) [24]. Low-energy SW was applied to 40 spots (0.1 mJ/mm^2, 200 shots/spot) in the ischemic region three times a week for 3 consecutive weeks. Subjective walking ability was evaluated with a Walking Impairment Questionnaire (WIQ), and walking ability was evaluated with a treadmill test at 4, 8, 12, and 24 weeks after the SW therapy. The low-energy SW therapy significantly improved symptoms, maximum walking distance, and peripheral perfusion (Fig. 12.9). Tara et al. also reported the beneficial effects of low-energy SW therapy in PAD patients including Fontaine stage III and IV [25]. These results suggest that low-energy SW therapy is promising as a new, noninvasive angiogenic therapy for PAD.

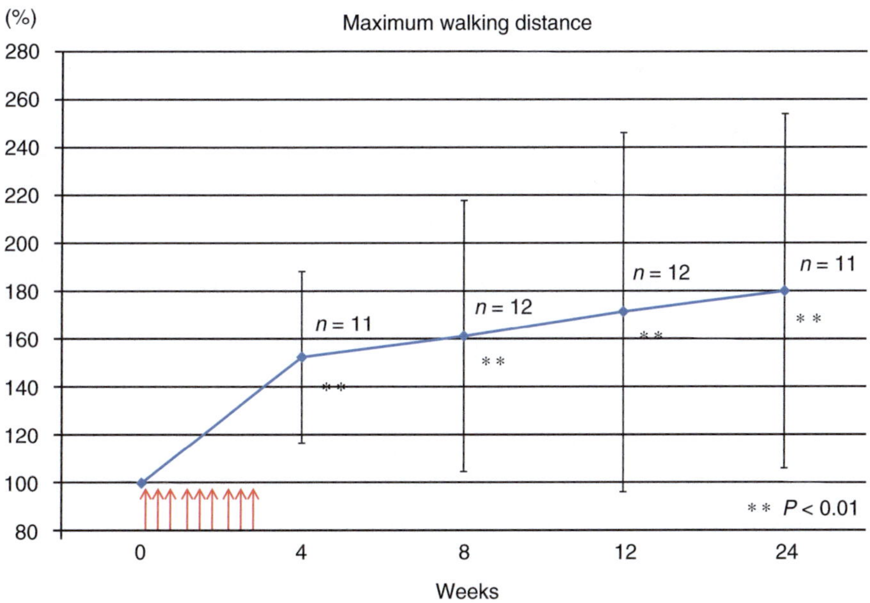

Fig. 12.9 Effects of the extracorporeal cardiac SW therapy on walking ability in patients with PAD and intermittent claudication. Maximum walking distance during the treadmill test was significantly increased and was maintained for 24 weeks after the SW therapy (from [24] with permission)

12.4.2 SW Therapy for Refractory Skin Ulcer

Raynaud's phenomenon and digital skin ulcers are severe complications of systemic sclerosis (SSc) which is related to immune activation, endothelial cell damage, and persistent vasospasms [26]. However, conventional immunosuppressive therapies, vasodilators, and anticoagulants are often ineffective. We and others have reported that low-energy SW therapy enhances wound healing in rodents [27–30] and in patients [31, 32]. We have demonstrated that the SW therapy facilitates wound healing in a mouse model of skin ulcers and that eNOS, VEGF, and angiogenesis play important roles in the repair process. Thus, we studied the effects of low-energy SW therapy on digital skin ulcers in nine patients with SSc [33]. Low-energy SW was applied to 20 areas on both hands and to 15 areas on both feet, totaling 7000 pulses once a week for 9 consecutive weeks with observations over 20 weeks. The low-energy SW therapy significantly improved digital ulcers in terms of size and number. No adverse effect was noted during the study period, demonstrating that this therapy can be safely repeated for a long period. These results suggest that the SW therapy may be added to standard treatments for refractory digital ulcers due to SSc.

12.4.3 SW Therapy for Other Disorders

We and others have reported the effects of low-energy SW therapy on secondary lymphedema in animals [34, 35]. We created a tail model of lymphedema in rats, and the tail was treated with low-energy SW on 2, 4, 6, and 8 days after the surgery [35]. Secondary lymphedema was sustained in the control group, which was significantly attenuated in the SW group. The lymphatic system function, the lymphatic vessel density, and the expression of VEGF-C and bFGF were all enhanced by the SW therapy. These results suggest that the low-energy SW therapy induces therapeutic lymphangiogenesis by upregulating VEGF-C and bFGF and that the SW therapy could be a noninvasive therapeutic strategy for lymphedema in humans. Low-energy SW has been widely used for the treatment of orthopedic diseases, such as bone nonunions, tendinosis calcarea, epicondylitis, and calcaneal spur through anti-inflammatory effects [36–39]. We have recently reported the effects of low-energy SW therapy on locomotor function after spinal cord injury in rats [40]. In this study, thoracic spinal cord contusion injury was inflicted using an impactor. Low-energy SW was applied to the injured spinal cord three times a week for 3 consecutive weeks. The SW therapy enhanced the expression of VEGF, attenuated neural tissue damage, and improved locomotor recovery. This study provides the first evidence that low-energy SW therapy can be a safe and promising therapeutic strategy for spinal cord injury.

12.5 Potential Mechanisms for the Beneficial Effects of SW Therapy

We and others have reported angiogenic effects of low-energy SW therapy in various animal models and in humans as mentioned above. Low-energy SW is reported to enhance VEGF release from bone marrow-derived mononuclear cells (BMDMNCs) and their differentiation into endothelial phenotype cells when applied to BMDMNCs [40]. Low-energy SW also activates proliferation and differentiation in cultured progenitors and precursors of cardiac cell lineages from the human heart [41, 42]. Furthermore, it has been reported that the beneficial effects of cell therapy were enhanced by pretreating BMDMNCs with SW before implantation into infarcted area in rabbits and that the pretreatment of ischemic leg with SW before cell therapy in a rat model of hindlimb ischemia enhanced the expression of stromal cell-derived factor 1 (SDF-1) in ischemic tissue and the resultant recruitment of endothelial progenitor cells [43, 44]. Thus, combination of cell therapy and SW therapy could be one of the potential approaches.

Recently, we have studied the effects of low-energy SW therapy on inflammatory responses in a rat model of AMI [45]. In this study, low-energy SW was applied to whole hearts at 1, 3, and 5 days after AMI. The SW therapy significantly suppressed the infiltration of inflammatory cells during acute phase in addition to enhanced angiogenesis in the border zone. These results suggest that the SW therapy suppressed post-MI LV remodeling through anti-inflammatory effects in addition to its angiogenic effects.

12.6 Conclusions

The beneficial effects of SW (angiogenesis, anti-inflammatory effects, neuroprotection, etc.) may be mediated by the enhancement of various intrinsic pathways. Although the precise intracellular mechanisms remain to be elucidated, low-energy extracorporeal SW therapy is promising as an effective, safe, and noninvasive approach to not only ischemic cardiovascular disorders but a wide range of disorders.

References

1. Jessup M, Brozena S. Heart failure. N Engl J Med. 2003;348:2007–18.
2. Japanese Coronary Artery Disease (JCAD) Study Investigators. Current status of the background of patients with coronary artery disease in Japan. Circ J. 2006;70:1256–62.
3. Ruff CT, Braunwald E. The evolving epidemiology of acute coronary syndromes. Nat Rev Cardiol. 2011;8:140–7. doi:10.1038/nrcardio.2010.199.

4. Hata J, Kiyohara Y. Epidemiology of stroke and coronary artery disease in Asia. Circ J. 2013; 77:1923–32.
5. Nishida T, Shimokawa H, Oi K, Tatewaki H, Uwatoku T, Abe K, et al. Extracorporeal cardiac shock wave therapy markedly ameliorates ischemia-induced myocardial dysfunction in pigs in vivo. Circulation. 2004;110:3055–61.
6. Mariotto S, Cavalieri E, Amelio E, Ciampa AR, de Prati AC, Marlinghaus E, et al. Extracorporeal shock waves: from lithotripsy to anti-inflammatory action by NO production. Nitric Oxide. 2005;12:89–96.
7. Fukumoto Y, Ito A, Uwatoku T, Matoba T, Kishi T, Tanaka H, et al. Extracorporeal cardiac shock wave therapy ameliorates myocardial ischemia in patients with severe coronary artery disease. Coron Artery Dis. 2006;17:63–70.
8. Kikuchi Y, Ito K, Ito Y, Shiroto T, Tsuburaya R, Aizawa K, et al. Double-blind and placebo-controlled study of the effectiveness and safety of extracorporeal cardiac shock wave therapy for severe angina pectoris. Circ J. 2010;74:589–91.
9. Khattab AA, Brodersen B, Schuermann-Kuchenbrandt D, Beurich H, Tölg R, Geist V, et al. Extracorporeal cardiac shock wave therapy: first experience in the everyday practice for treatment of chronic refractory angina pectoris. Int J Cardiol. 2007;121:84–5.
10. Prinz C, Lindner O, Bitter T, Hering D, Burchert W, Horstkotte D, et al. Extracorporeal cardiac shock wave therapy ameliorates clinical symptoms and improves regional myocardial blood flow in a patient with severe coronary artery disease and refractory angina. Case Rep Med. 2009;2009:639594. doi:10.1155/2009/639594.
11. Vasyuk YA, Hadzegova AB, Shkolnik EL, Kopeleva MV, Krikunova OV, Iouchtchouk EN, et al. Initial clinical experience with extracorporeal shock wave therapy in treatment of ischemic heart failure. Congest Heart Fail. 2010;16:226–30. doi:10.1111/j.1751-7133.2010.00182.x.
12. Wang Y, Guo T, Cai HY, Ma TK, Tao SM, Sun S, et al. Cardiac shock wave therapy reduces angina and improves myocardial function in patients with refractory coronary artery disease. Clin Cardiol. 2010;33:693–9. doi:10.1002/clc.20811.
13. Wang Y, Guo T, Ma TK, Cai HY, Tao SM, Peng YZ, et al. A modified regimen of extracorporeal cardiac shock wave therapy for treatment of coronary artery disease. Cardiovasc Ultrasound. 2012;10:35. doi:10.1186/1476-7120-10-35.
14. Yang P, Guo T, Wang W, Peng YZ, Wang Y, Zhou P, et al. Randomized and double-blind controlled clinical trial of extracorporeal cardiac shock wave therapy for coronary heart disease. Heart Vessel. 2013;28:284–91. doi:10.1007/s00380-012-0244-7.
15. Schmid JP, Capoferri M, Wahl A, Eshtehardi P, Hess OM. Cardiac shock wave therapy for chronic refractory angina pectoris. A prospective placebo-controlled randomized trial. Cardiovasc Ther. 2013;31:e1–6. doi:10.1111/j.1755-5922.2012.00313.x.
16. Takii T, Yasuda S, Takahashi J, Ito K, Shiba N, Shirato K, et al. MIYAGI-AMI study investigators. Trends in acute myocardial infarction incidence and mortality over 30 years in Japan: report from the MIYAGI-AMI registry study. Circ J. 2010;74:93–100.
17. Olivetti G, Ricci R, Beghi C, Guideri G, Anversa P. Response of the border zone to myocardial infarction in rats. Am J Pathol. 1986;125:476–83.
18. Uwatoku T, Ito K, Abe K, Oi K, Hizume T, Sunagawa K, et al. Extracorporeal cardiac shock wave therapy improves left ventricular remodeling after acute myocardial infarction in pigs. Coron Artery Dis. 2007;18:397–404.
19. Ito Y, Ito K, Shiroto T, Tsuburaya R, Yi GJ, Takeda M, et al. Cardiac shock wave therapy ameliorates left ventricular remodeling after myocardial ischemia-reperfusion injury in pigs in vivo. Coron Artery Dis. 2010;21:304–11.
20. Hiatt WR. Medical treatment of peripheral arterial disease and claudication. N Engl J Med. 2001;344:1608–21.
21. Wennberg PW. Approach to the patient with peripheral arterial disease. Circulation. 2013;128:2241–50. doi:10.1161/CIRCULATIONAHA.113.000502.
22. Jaff MR, White CJ, Hiatt WR, Fowkes GR, Dormandy J, Razavi M, et al. An update on methods for revascularization and expansion of the TASC lesion classification to include below-the-knee arteries: a supplement to the inter-society consensus for the management of peripheral arterial

disease (TASC II): the TASC steering committee. Catheter Cardiovasc Interv. 2015;86:611–25. doi:10.1002/ccd.26122.
23. Oi K, Fukumoto Y, Ito K, Uwatoku T, Abe K, Hizume T, et al. Extracorporeal shock wave therapy ameliorates hindlimb ischemia in rabbits. Tohoku J Exp Med. 2008;214:151–8.
24. Serizawa F, Ito K, Kawamura K, Tsuchida K, Hamada Y, Zukeran T, et al. Extracorporeal shock wave therapy improves the walking ability of patients with peripheral artery disease and intermittent claudication. Circ J. 2012;76:1486–93.
25. Tara S, Miyamoto M, Takagi G, Kirinoki-Ichikawa S, Tezuka A, Hada T, et al. Low-energy extracorporeal shock wave therapy improves microcirculation blood flow of ischemic limbs in patients with peripheral arterial disease: pilot study. J Nippon Med Sch. 2014;81:19–27.
26. Abraham DJ, Varga J. Scleroderma: from cell and molecular mechanisms to disease models. Trends Immunol. 2005;26:587–95.
27. Stojadinovic A, Elster EA, Anam K, et al. Angiogenic response to extracorporeal shock wave treatment in murine skin isografts. Angiogenesis. 2008;11:369–80. doi:10.1007/s10456-008-9120-6.
28. Yan X, Zeng B, Chai Y, et al. Improvement of blood flow, expression of nitric oxide, and vascular endothelial growth factor by low-energy shockwave therapy in random-pattern skin flap model. Ann Plast Surg. 2008;61:646–53. doi:10.1097/SAP.0b013e318172ba1f.
29. Hayashi D, Kawakami K, Ito K, Ishii K, Tanno H, Imai Y, et al. Low-energy extracorporeal shock wave therapy enhances skin wound healing in diabetic mice: a critical role of endothelial nitric oxide synthase. Wound Repair Regen. 2012;20:887–95. doi:10.1111/j.1524-475X.2012.00851.x.
30. Weihs AM, Fuchs C, Teuschl AH, Hartinger J, Slezak P, Mittermayr R, et al. Shock wave treatment enhances cell proliferation and improves wound healing by ATP release-coupled extracellular signal-regulated kinase (ERK) activation. J Biol Chem. 2014;289:27090–104. doi:10.1074/jbc.M114.580936.
31. Saggini R, Figus A, Troccola A, et al. Extracorporeal shock wave therapy for management of chronic ulcers in the lower extremities. Ultrasound Med Biol. 2008;34:1261–71. doi:10.1016/j.ultrasmedbio.2008.01.010.
32. Moretti B, Notarnicola A, Maggio G, Moretti L, Pascone M, Tafuri S, et al. The management of neuropathic ulcers of the foot in diabetes by shock wave therapy. BMC Musculoskelet Disord. 2009;10:54. doi:10.1186/1471-2474-10-54.
33. Saito S, Ishii T, Kamogawa Y, Watanabe R, Shirai T, Fujita Y, et al. Extracorporeal shock wave therapy for digital ulcers of systemic sclerosis: a phase 2 pilot study. Tohoku J Exp Med. 2016;238:39–47. doi:10.1620/tjem.238.39.
34. Kubo M, Li TS, Kamota T, Ohshima M, Shirasawa B, Hamano K. Extracorporeal shock wave therapy ameliorates secondary lymphedema by promoting lymphangiogenesis. J Vasc Surg. 2010;52:429–34. doi:10.1016/j.jvs.2010.03.017.
35. Serizawa F, Ito K, Matsubara M, Sato A, Shimokawa H, Satomi S. Extracorporeal shock wave therapy induces therapeutic lymphangiogenesis in a rat model of secondary lymphedema. Eur J Vasc Endovasc Surg. 2011;42:254–60. doi:10.1016/j.ejvs.2011.02.029.
36. Ogden JA, Alvarez RG, Levitt R, Marlow M. Shock wave therapy (Orthotripsy) in musculoskeletal disorders. Clin Orthop Relat Res. 2001;387:22–40.
37. Birnbaum K, Wirtz DC, Siebert CH, et al. Use of extracorporeal shock-wave therapy (ESWT) in the treatment of non-unions. A review of the literature. Arch Orthop Trauma Surg. 2002;122:324–30.
38. Wang CJ. Extracorporeal shockwave therapy in musculoskeletal disorders. J Orthop Surg Res. 2012;7:11. doi:10.1186/1749-799X-7-11.
39. Al-Abbad H, Simon JV. The effectiveness of extracorporeal shock wave therapy on chronic achilles tendinopathy: a systematic review. Foot Ankle Int. 2013;34:33–41. doi:10.1177/1071100712464354.
40. Yamaya S, Ozawa H, Kanno H, Kishimoto KN, Sekiguchi A, Tateda S, et al. Low-energy extracorporeal shock wave therapy promotes VEGF expression and neuroprotection and improves locomotor recovery after spinal cord injury. J Neurosurg. 2014;121:1514–25. doi:10.3171/2014.8.JNS132562.

41. Yip HK, Chang LT, Sun CK, Youssef AA, Sheu JJ, Wang CJ. Shock wave therapy applied to rat bone marrow-derived mononuclear cells enhances formation of cells stained positive for CD31 and vascular endothelial growth factor. Circ J. 2008;72:150–6.
42. Nurzynska D, Di Meglio F, Castaldo C, Arcucci A, Marlinghaus E, Russo S, et al. Shock waves activate in vitro cultured progenitors and precursors of cardiac cell lineages from the human heart. Ultrasound Med Biol. 2008;34:334–42.
43. Sheu JJ, Sun CK, Chang LT, Fang HY, Chung SY, Chua S, et al. Shock wave-pretreated bone marrow cells further improve left ventricular function after myocardial infarction in rabbits. Ann Vasc Surg. 2010;24:809–21. doi:10.1016/j.avsg.2010.03.027.
44. Aicher A, Heeschen C, Sasaki K, Urbich C, Zeiher AM, Dimmeler S. Low-energy shock wave for enhancing recruitment of endothelial progenitor cells: a new modality to increase efficacy of cell therapy in chronic hind limb ischemia. Circulation. 2006;114:2823–30.
45. Abe Y, Ito K, Hao K, Shindo T, Ogata T, Kagaya Y, et al. Extracorporeal low-energy shock-wave therapy exerts anti-inflammatory effects in a rat model of acute myocardial infarction. Circ J. 2014;78:2915–25.

Chapter 13
Granulocyte Colony-Stimulating Factor

Yasuyuki Fujita and Atsuhiko Kawamoto

Abstract Granulocyte colony-stimulating factor (G-CSF) is a potent hematopoietic protein that promotes the development and function of granulocytes and mobilizes stem/progenitor cells from the bone marrow. Recent studies have shown that G-CSF also directly influences the activity of some non-hematopoietic cells, such as cardiomyocytes, endothelial cells, and neurons via G-CSF receptor. This chapter provides an overview of the preclinical and clinical reports to demonstrate the usefulness and the current limitations of the therapeutic strategy using G-CSF for ischemic diseases.

Keywords CD34+ cells • Cerebrovascular disease (CVD) • Coronary artery disease (CAD) • Endothelial progenitor cells (EPCs) • Granulocyte colony-stimulating factor (G-CSF) • Peripheral artery disease (PAD)

Y. Fujita, M.D., Ph.D. • A. Kawamoto, M.D., Ph.D. (✉)
Division of Vascular Regeneration, Unit of Regenerative Medicine,
Institute of Biomedical Research and Innovation (IBRI) Hospital,
2-2 Minatojima-Minamimachi, Chuo-Ku,
Kobe, Hyogo 650-0047, Japan

Vascular Regeneration Research Group,
IBRI Research Center, Kobe, Hyogo, Japan

Translational Research Informatics Center,
Foundation for Biomedical Research and Innovation,
Kobe, Hyogo, Japan
e-mail: kawamoto@fbri.org

Y. Higashi, T. Murohara (eds.), *Therapeutic Angiogenesis*,
DOI 10.1007/978-981-10-2744-4_13

13.1 Granulocyte Colony-Stimulating Factor (G-CSF)

Granulocyte colony-stimulating factor (G-CSF) is a potent hematopoietic protein that promotes the maturation, proliferation, and differentiation of the precursor cells of neutrophilic granulocytes [1–3] and mobilizes stem/progenitor cells from the bone marrow (BM) into peripheral blood (PB) [4]. Clinically, G-CSF has been used for the treatment of patients suffering from chemotherapy-associated neutropenia and the collection of stem cells used in allogeneic or autologous PB stem cell (PBSC) transplantation.

G-CSF is a 19.6 kD glycoprotein encoded by a single gene located on human chromosome 17 at 17q11–12 [5]. Two variant forms of this protein including 174-amino acid protein and 177-amino acid protein can be derived from differential splicing of the pre-mRNA of G-CSF [2, 6]. The 174-amino acid G-CSF is more active at stimulating proliferation of progenitor cells than the 177-amino acid G-CSF [2]. Monocyte/macrophage lineage cells are the major source of G-CSF; however, vascular endothelial cells, mesothelial cells, and fibroblasts have been found to produce G-CSF in response to inflammatory cytokines [7–10]. Production of G-CSF can be induced in vitro by appropriate stimulation with inflammatory mediators such as lipopolysaccharide (LPS), tumor necrosis factor (TNF)-α, interferon (IFN)-β, vascular endothelial growth factor (VEGF), interleukin (IL)-17 and IL-1 in endothelial cells, macrophages, epithelial cells, and fibroblasts [11–13]. G-CSF primarily acts via activation of G-CSF receptor (G-CSFR), which is a type I receptor (amino terminus is extracellular) consisting of four different domains: an immunoglobulin-like domain, a cytokine receptor homologous domain, and three fibronectin type III domains in the extracellular region [14]. G-CSFRs are expressed on not only hematopoietic cells including precursors, mature neutrophils, platelets, lymphocytes, and monocytes but also endothelial cells, neurons, glial cell, and cardiomyocytes [15–21]. G-CSFR activates a variety of intracellular pathways, including the Janus kinase/signal transducer and activator of transcription (JAK-STAT) pathway [22, 23], the Ras/mitogen-activated protein kinase (MAPK) pathway, and phosphatidylinositol 3-kinase (PI3K)/Akt pathway [24, 25]. Activation of these pathways mediates inhibition of apoptosis, promotes cell survival, and enhances proliferative response to G-CSF [25]. Cumulating evidence suggests a much broader role of G-CSF signaling in the repair of range of tissues outside the hematopoietic system.

Based on its ability to mobilize stem/progenitor cells and tissue repair and protection properties, regenerative effects of G-CSF on ischemic diseases have been rapidly investigated.

13.2 G-CSF Therapy for Ischemic Diseases

13.2.1 G-CSF for Coronary Artery Disease (CAD)

13.2.1.1 Preclinical Studies

G-CSF

The safety and efficacy of G-CSF for myocardial repair have been reported in several preclinical studies. G-CSF-mobilized stem cells are recruited to ischemic myocardium and differentiated into specialized cells such as cardiomyocytes, endothelial cells, and smooth muscle cells [26–30]. Furthermore, these mobilized cells may accelerate the healing process by induction of matrix metalloproteinases and VEGF [31, 32].

G-CSF also directly affects the activity of some non-hematopoietic cells such as cardiomyocytes and endothelial cells [33]. When administered immediately after myocardial infarction (MI), G-CSF activates the JAK-STAT pathway, which promotes the production of several anti-apoptosis-related proteins, decreases cardiomyocyte death, and limits infarct size [18]. In a murine model of MI, G-CSF treatment was associated with improvement of left ventricular (LV) function and enhancement of arteriogenesis [34]. On the other hand, G-CSF can also stimulate the differentiation of lineage-committed progenitor cells into neutrophils and macrophages [35], which could worsen inflammation and cardiac remodeling [36].

13.2.1.2 Clinical Trials

The safety and efficacy of G-CSF therapy in patients with acute MI (AMI) have been evaluated in several clinical trials (Table 13.1). In the MAGIC Cell (Myocardial Regeneration and Angiogenesis in Myocardial Infarction with G-CSF and Intra-Coronary Stem Cell Infusion) trial, 27 patients with revascularized AMI were randomly assigned to G-CSF-mobilized cell infusion group, G-CSF alone group, or control group. Among the three groups, cell infusion group showed significant improvement of cardiac function at 6 months. However, the trial was halted prematurely because of high rate of in-stent restenosis at the culprit lesion in patients who received G-CSF [37]. On the other hand, in subsequent phase I trials, G-CSF administration after percutaneous coronary intervention (PCI) in patients with AMI revealed improvement of cardiac function [26, 39–41, 47]. In the FIRSTLINE-AMI (Front-Integrated Revascularization and Stem Cell Liberation in Evolving Acute Myocardial Infarction) randomized, open-label trial, 25 patients with ST-segment elevation MI

Table 13.1 Randomized controlled trials of G-CSF monotherapy for CAD

Trial name/author	Year	Study design	Disease	Number of patients			Product	Dose and route of administration	Follow-up duration	Outcomes
				Total	Treated	Control				
MAGIC CELL/Kang et al. [37, 38]	2004	RCT	Recent MI (>48 h after AMI)	27	G-CSF: 10	7	G-CSF	10 μg/kg/day for 4 days	6 months	(6 months)
										Exercise capacity (cell group) ↑
										Myocardial perfusion (cell group) ↑
										LVEF (cell group) ↑
					G-CSF-mobilized MNCs: 10			SC	2 years	High rate of in-stent restenosis at culprit lesion
										(2 years)
										LVEF (cell group) ↑
										ESV (cell group)↓
FIRSTLINE-AMI/Ince et al. [39, 40]	2005	RCT (open label)	STEMI	50	25	25	G-CSF	10 μg/kg/day for 6 days	4 and 12 months	Regional wall motion ↑
								SC		LV remodeling ↓

Valgimigli et al. [41]	2005	RCT	STEMI	20	10	10	G-CSF	5 μg/kg/day for 4 days	6 months	LVEF ↑
								SC		LVEDV ↓
STEMMI/Ripa et al. [42]	2006	RCT (double blind)	STEMI	78	39	39	G-CSF	10 μg/kg/day for 6 days	6 months	No effect
								SC		
REVIVAL-2/Zohlnhofer et al. [43]	2006	RCT (double blind)	STEMI	114	56	58	G-CSF	10 μg/kg/day for 5 days	4–6 months	No effect
								SC		
G-CSF-STEMI/ Engelmann et al. [44]	2006	RCT (open label)	STEMI	44	23	21	G-CSF	10 μg/kg/day for 5 days	12 months	No effect
								SC		
STEM-AMI/Achilli et al. [45, 46]	2010	RCT (single blind)	STEMI	60	30	30	G-CSF	5 μg/kg/day bid for 5 days	6 months	(6 months)
										LVEF (MRI) →,
										perfusion (SPECT) →
								SC	3 years	Infarct size ↓
										(3 years)
										Infarct size →
										LVEDV ↓

Abbreviations: *CAD* coronary artery disease, *G-CSF* granulocyte colony-stimulating factor, *IC* intracoronary, *LVEF* left ventricular ejection fraction, *LVEDV* left ventricular end-diastolic volume, *RCT* randomized controlled trial, *SC* subcutaneous

(STEMI) were randomly assigned to receive subcutaneous injection of G-CSF at 10 μg/kg/day for 6 days starting within 90 min after primary PCI for STEMI; the control group ($n = 25$) did not receive placebo injections but had standard post-interventional care. Four months [39] and 1 year [40] after the PCI procedure, improvement of LV ejection fraction (LVEF) were significantly greater in G-CSF-treated patients than the control group. However, these promising results were not reproduced in subsequent double-blind, placebo-controlled trials, such as the STEMMI (Stem Cells in Myocardial Infarction), REVIVAL-2 (Regenerate Vital Myocardium by Vigorous Activation of Bone Marrow Stem Cells), and G-CSF-STEMI (Granulocyte Colony-Stimulating Factor ST-Segment Elevation Myocardial Infarction) trials [42–44]. Again, meta-analyses reported mixed results and a high heterogeneity in published studies [48–50]. The lack of consensus about efficacy of G-CSF in improving LV performance after AMI may be due to small sample size studies and heterogeneity in a variety of factors including different doses and timings of G-CSF administration, infarct size, LV function, symptom-to-balloon time, and techniques used to assess LV remodeling. Abdel-Latif et al. suggested that early administration of G-CSF may potentially benefit patients with large MI and severely impaired LV function in meta-analyses [48]. In the STEM-AMI (STem cEll Mobilization in Acute Myocardial Infarction), randomized, single-blind, placebo-controlled, and phase II study, 60 patients with anterior STEMI undergoing primary PCI with symptom-to-reperfusion time of 2–12 h and EF $\leq$45% after PCI were randomized to G-CSF 5 μg/kg subcutaneously bis in die (bid) or placebo for 5 days, starting 12 h after PCI. There were no significant differences in LVEF as measured by magnetic resonance imaging (MRI) or ^{99m}Technetium Sestamibi single-photon emission computed tomography (SPECT) perfusion between the groups. However, a significant reduction in transmural late gadolinium enhancement segments was seen at 6 months in the G-CSF group versus placebo group [45]. The final 3-year follow-up of the STEM-AMI trial on 35 patients confirmed that G-CSF attenuated adverse LV remodeling also in the long term [46]. The phase III STEM-AMI OUTCOME trial is ongoing [51].

The safety and efficacy of G-CSF therapy for patients with chronic coronary artery disease (chronic CAD) have been also evaluated [52–55] (Table 13.1). In a phase I study of G-CSF therapy for 29 patients with severe occlusive CAD, 13 patients were treated with G-CSF (5 μg/kg/day for 6 days) and 16 patients served as controls. G-CSF therapy increased the number of circulating CD34+ cells and reduced nitroglycerin consumption and angina attacks without improvement of myocardial perfusion or function [53]. In another study, 16 patients with chronic CAD received G-CSF (10 μg/kg/day) for 5 days. G-CSF therapy increased the number of circulating CD34+/CD133+ cells but without improvement of LV function, perfusion, or exercise duration and with two severe adverse cardiac events including one case of non-ST-segment elevation MI (non-STEMI) and the other case of death after MI [52]. In the study by Ripa et al. the combination therapy of intramyocardial injection of VEGF-A165 plasmid followed by subcutaneous injection of G-CSF in patients with severe occlusive CAD did not improve myocardial perfusion or clinical symptoms [54].

13.2.2 G-CSF for Peripheral Artery Disease (PAD)

13.2.2.1 Preclinical Studies

As noted above, G-CSF is a growth factor that mobilizes CD34+ endothelial progenitor cells (EPCs) from the BM into the PB [4]. On the other hand, Ohki et al. found that G-CSF increased VEGF production from Gr-1 + CD11b- neutrophils in vitro, and intramuscular injection of G-CSF increased plasma levels of VEGF in nonischemic mice. They also demonstrated that local G-CSF administration into ischemic tissue increased the number of Gr-1 + VEGF+ double positive cells and enhanced neovascularization of ischemic tissue in a mouse hind limb ischemia model [33]. Furthermore, blockade of the VEGF/VEGFR1 pathway eliminated G-CSF-induced (200 μg/kg/day for 5 days) angiogenesis in a mouse hind limb ischemia model, suggesting that G-CSF-induced vasculogenesis is VEGF dependent and could be indirectly mediated by this mechanism [33]. Lee et al. demonstrated that G-CSF enhanced migration and tube formation of human umbilical vein endothelial cells (HUVECs) in vitro. In vivo, local injection of low doses (2, 10, and 20 μg/kg per day for 6 days) augmented neovascularization in a rat model of hind limb ischemia [56].

13.2.2.2 Clinical Trials

There is only one clinical study of G-CSF for patients with PAD [57]. In this study, 39 patients were randomly assigned to conventional drug therapy, conventional drug therapy plus BM transplantation (BMT), or conventional therapy plus G-CSF. Subjective symptoms, ankle-brachial pressure index (ABI), and transcutaneous oxygen pressure (TcPO2) were significantly improved in the G-CSF and BMT groups to the same degree 1 month after treatment, whereas such improvements were not observed in the conventional therapy group (Table 13.2).

Table 13.2 Randomized controlled trials of G-CSF monotherapy for PAD

Trial name/author	Year	Study design	Disease	Number of patients			Product	Dose and route of administration	Follow-up duration	Outcomes
				Total	Treated	Control				
Arai et al. [57]	2006	RCT	Atherosclerotic PAD (Fontaine III or IV)	39	G-CSF: 14	12	G-CSF	2–5 μg/kg/day for 10 days	1 month	ABPI ↑, TcPO2↑
					BM-MNCs: 13			SC		Ulcer healing↑

Abbreviations: *PAD* peripheral artery disease

13.2.3 G-CSF for Cerebrovascular Disease (CVD)

13.2.3.1 Preclinical Studies

Preclinical studies have demonstrated the neuroprotective and neuroregenerative properties of G-CSF in experimental cerebral ischemia. Intravenous administration of G-CSF to mice [58] or rats [59, 60] enhanced survival rate and decreased cerebral infarct volume in experimental cerebral ischemia models. Furthermore, G-CSF therapy improved motor function and cognition in a mouse cerebral ischemia model [61, 62] and improved neurological behavior in a rat cerebral ischemia model [59]. However, a systemic review revealed that these favorable effects occur in transient but not permanent models of cerebral ischemia [63]. In the acute phase of transient cerebral ischemia, potential mechanisms underlying neuroprotective effects of G-CSF include inhibition of glutamate release [64] and direct protection against glutamate-induced excitotoxicity by G-CSFR-mediated activation of STAT3 [60], reduction of inflammation [61, 62, 65–68], anti-apoptotic activity [69, 70], and suppression of edema formation [71]. On the other hand, a negative effect of G-CSF was reported in a mouse permanent cerebral ischemia model [72]. Mice treated with G-CSF displayed cortical atrophy and impaired behavioral function, associated with G-CSF-induced exaggeration of inflammatory response, which was based on infiltration of CD11b-positive and F4/80-positive cells in the peri-infarction area. However, administration of G-CSF is accompanied by significant increase of circulating neutrophils 2 days after cerebral ischemia, but leukocytosis is restricted to the vessel compartment without deleterious effect on lesion formation and functional recovery [73]. Several experimental studies demonstrated that G-CSF also induces functional recovery by stimulating neuronal progenitor cells [59, 62] and promotes angiogenesis [59, 66, 67] and neurogenesis [59, 67, 69].

13.2.3.2 Clinical Trials

In the first randomized controlled trial of G-CSF therapy for patients with acute stroke within 7 days after onset, ten patients were randomized to receive subcutaneous G-CSF injections (15 μg/kg per day) for 5 days or usual care. No severe adverse effects were found in patients receiving G-CSF. There was a greater improvement in neurologic function including National Institutes of Health Stroke Scale (NIHSS) and Barthel Index [74]. STEMS (The Stem Cell Trial of Recovery EnhanceMent After Stroke) is a dose-escalation, double-blind, placebo-controlled, phase IIa trial of G-CSF treatment in patients with subacute ischemic stroke [75]. Thirty-six patients with ischemic stroke (7–30 days after the onset) were randomized to receive either subcutaneous G-CSF (1–10 μg/kg subcutaneously, one or five daily doses) or placebo. G-CSF increased PB-CD34+ cell count in a dose-dependent manner and appeared to be safe and well tolerated. However, there was no significant difference in functional outcomes 90 days after treatment between the groups. Schäbitz et al.

performed a placebo-controlled, dose-escalation, and phase IIa trial (AXIS trial) of four intravenous dose regimens (30–180 μg/kg over the course of 3 days) of G-CSF for 44 patients with ischemic stroke within 12 h after onset [76]. No treatment-related serious adverse events were found. Although there was no significant difference in the clinical outcome between G-CSF group and placebo, there was a dose-dependent beneficial effect in patients with larger diffusion-weighted image (DWI) lesions >14–17 cm^3. In a phase IIb single-center randomized controlled trial (STEMS 2) for patients 3–30 days after ischemic or hemorrhagic stroke [77], 60 patients were randomly assigned to receive subcutaneous injection of G-CSF (10 μg/kg) or placebo for 5 days. The study results suggested that G-CSF may be safe when administered subcutaneously and may reduce ischemic lesion volume by MRI. In the phase III AXIS 2 trial [78], intravenous injection of G-CSF (135 μg/kg over 72 h) was tested against placebo in 328 patients with acute stroke localized in the middle cerebral artery territory within 9 h after onset. Primary and secondary endpoints were the modified Rankin Scale score and NIHSS score at day 90 after treatment, respectively. Disappointingly, G-CSF treatment failed to meet the primary and secondary endpoints of the trial (Table 13.3).

In patients with chronic ischemic stroke at least 4 months after onset, Floel et al. reported that subcutaneous G-CSF treatment (10 μg/kg for 10 days) was feasible, safe, and reasonably tolerable; however the primary efficacy endpoint, hand motor function, was not significantly improved [79] (Table 13.3).

13.3 G-CSF-Mobilized Crude Mononuclear Cells (MNCs) and Fractionated EPCs (CD34+/CD133+ Cells)

Although results from preclinical studies have been promising, outcomes of G-CSF monotherapy trials remain controversial. Accumulating knowledge of the potential of stem/progenitor cells as therapeutic agents in both animal studies and clinical trials has shifted the interest in regenerative medicine from molecular to cell-based approaches.

In 1997, EPCs were first identified in adult human PB as CD34+ mononuclear cells (MNCs) [80]. They are phenotypically characterized by expression of antigens associated with hematopoietic stem cells (HSCs) including CD133, CD34, c-kit, VEGFR-2, CD144 (vascular endothelial (VE)-cadherin), and Sca-1. The discovery of circulating EPCs changed the traditional paradigm that "vasculogenesis" occurs exclusively in the developing embryo. EPC levels in the PB are low under normal condition; however, EPCs residing in the BM are mobilized into PB in response to physiological and pathological stimuli, such as myocardial and peripheral ischemia [81, 82]. Mobilized EPCs recruit to the foci of neovascularization where they form structural components of the growing vasculature [83]. Accumulated recent insights into the mechanism of EPC-mediated neovascularization reveal that EPCs secrete paracrine factors including VEGF-A, VEGF-B, stromal cell-derived factor-1 (SDF-1),

Table 13.3 Randomized controlled trials of G-CSF monotherapy for CVD

Trial name/author	Year	Study design	Disease	Number of patients			Product	Dose and route of administration	Follow-up duration	Outcomes
				Total	Treated	Control				
Shyu et al. [74]	2006	RCT (double blind)	Acute cerebral infarction within 7 days	10	7	3	G-CSF	15 μg/kg/day for 5 days SC	12 months	Neurological function (NIHSS and Barthel Index) ↑
STEMS/Sprigg et al. [75]	2006	RCT (double blind)	Recent ischemic stroke 7–30 days after onset	36	24	12	G-CSF	Dose escalation with 6 blocks of 1–10 μg/kg 1 or 5 daily doses SC	90 days	G-CSF (5 days of 10 μg/kg) increased CD34+ cell count dose dependently No difference in SAEs between treatment groups
AXIS/Schabitz et al. [76]	2010	RCT (double blind)	Acute ischemic stroke within 12 h	44	30	14	G-CSF	Total cumulative dose of 30–180 μg/kg for 3 days IV	90 days	Clinical outcome → Dose-dependent beneficial effect of G-CSF in patients with diffusion-weighted image lesions >14–17 cm^3
STEMS 2/England et al. [77]	2012	RCT (double blind)	Ischemic or hemorrhagic stroke 3–30 days after onset	60	40	20	G-CSF	10 μg/kg/day for 5 days SC	90 days	MRI ischemic lesion volume →↓
AXIS 2/Ringelstein et al. [78]	2013	RCT (double blind)	Acute ischemic stroke within 9 h	324	161	163	G-CSF	Total cumulative dose of 135 μg/kg over 72 h IV	90 days	Modified Rankin Scale score (primary endpoint) → NIHSS score (secondary endpoint) →
Floel et al. [79]	2011	RCT (double blind)	Chronic ischemic stroke	41	21	20	G-CSF	10 μg/kg/day for 10 days SC	38 days	Hand motor function (primary efficacy endpoint) →

Abbreviations: *CVD* cerebrovascular disease, *IV* intravenous, *SAEs* serious adverse events and insulin-like growth factor-1 (IGF-1) [84]. The paracrine effect of EPCs inhibits cell death, enhances cell proliferation, activates resident stem/progenitor cells in the ischemic tissue, and recruits additional stem/progenitor cells to the ischemic site [85–89]. On the other hand, Gehling et al. reported that a cell population positive for AC133 (CD133), a more immature HSC marker, consists of progenitor and stem cells with not only hematopoietic potential but also the capacity of endothelial differentiation [90]. Accumulated studies revealed that PB-, BM-, and umbilical cord blood-derived CD34+ or CD133+ cells are enriched for endothelial lineage, can express endothelial markers, and form endothelial structure in vitro and in vivo [28, 91–94].

The discovery of EPCs guided to the development of stem/progenitor cell-based therapeutic strategies for ischemic cardiovascular diseases. Since then, BM-, PB-, or G-CSF-mobilized PB-MNCs including EPCs as well as the EPC-enriched fraction purified from the crude MNCs have been preclinically applied for ischemic diseases including CAD, PAD, and CVD. The promising results from these experimental studies in rodents promoted the initiation of clinical pilot trials.

13.3.1 G-CSF-Mobilized Crude MNCs for CAD

13.3.1.1 Clinical Trials

Crude MNCs

In the MAGIC Cell trial [37], 27 patients with revascularized AMI were randomly assigned to G-CSF-mobilized cell infusion group, G-CSF alone group, or control group. Among the three groups, cell infusion group showed significant improvement of cardiac function at 6 months. However, the trial was halted prematurely because of high rate of in-stent restenosis at the culprit lesion in patients who received G-CSF. At 2 years of follow-up evaluation, cell infusion improved LV systolic function and remodeling compared to baseline, but G-CSF alone did not [38]. In the MAGIC Cell-3-DES (Myocardial Regeneration and Angiogenesis in Myocardial Infarction With G-CSF and Intra-Coronary Stem Cell Infusion-3-Drug-Eluting Stents) trial [95], 96 patients with myocardial infarction who underwent coronary revascularization with drug-eluting stent (DES) for the culprit lesion were randomly assigned to four groups, AMI cell infusion group, AMI control group, old myocardial infarction (OMI) cell infusion group, and OMI control group. In cell infusion groups, PB-MNCs were mobilized by daily subcutaneous injection of G-CSF 10 μg/kg for 3 days and delivered to infarcted myocardium via intracoronary infusion. The AMI cell infusion group showed a significant improvement in LVEF and remodeling compared with controls. In OMI cell group, there was no significant change of LVEF and ventricular remodeling in spite of significant improvement of coronary flow reserve after cell infusion (Table 13.4).

Table 13.4 Randomized controlled trials of G-CSF-mobilized crude MNCs for CAD

Trial name/ author	Year	Study design	Disease	Number of patients: Total	Number of patients: Treated	Number of patients: Control	Product	Route of administration	Follow-up duration	Outcomes
MAGIC Cell/Kang et al. [37, 38]	2004	RCT	Recent MI (>48 h after AMI)	27	G-CSF: 10	7	G-CSF-mobilized MNCs	IC	6 months	(6 months)
										Exercise capacity (cell group) ↑
										Myocardial perfusion (cell group) ↑
					G-CSF-mobilized MNCs: 10					LVEF (cell group) ↑
										High rate of in-stent restenosis at culprit lesion
										(2 years)
										LVEF (cell group) ↑
MAGIC Cell-3-DES/ Kang et al. [95]	2006	RCT	AMI (≤14 days from onset)	82	AMI: 25	AMI: 25	G-CSF-mobilized MNCs	IC	6 months	(AMI)
										LVEF ↑
										LV remodeling ↓
			OMI (>14 days from onset)		OMI: 16	OMI: 16				(OMI)
										LVEF →
										LV remodeling →

13.3.2 G-CSF-Mobilized Crude MNCs for PAD

13.3.2.1 Clinical Trials

G-CSF-mobilized PB-MNCs have been also investigated in critical limb ischemia (CLI), which is the advanced stage of PAD representing ischemic rest pain and/or ulcer/necrosis. PB-MNCs are usually mobilized using several doses of subcutaneous G-CSF and harvested by plasmapheresis. Both intramuscular and intra-arterial injections of unfractionated mobilized PB-MNCs improved ABPI and maximum walking distance in small clinical trials [96–99]. A phase III clinical trial is ongoing (ClinicalTrials.gov NCT 01833585), and the final results may provide a definitive evidence of clinical usefulness of this cell-based therapy in CLI.

Huang et al. compared the therapeutic effect of intramuscular administration of BM-MNCs with G-CSF-mobilized PB-MNCs in patients with CLI [100]. One hundred fifty patients with CLI were randomized to receive intramuscular injection of BM-MNCs or G-CSF-mobilized MNCs. At 12 weeks after therapy, improvement of ABPI, skin temperature, and rest pain were significantly greater in PB-MNC group than BM-MNC group despite no difference in pain-free walking distance, ulcer healing, and amputation rates between the two groups (Table 13.5).

13.4 G-CSF-Mobilized Fractionated CD34+/CD133+ Cells

13.4.1 G-CSF-Mobilized Fractionated CD34+ Cells for CAD

13.4.1.1 Clinical Trials

The first pilot (phase I/IIa) study of intramyocardial injection of autologous and G-CSF-mobilized CD34+ stem cells in patients with refractory angina provided early evidence of feasibility, safety, and bioactivity of the stem cells [102]. The promising results encouraged the ACT34-CMI study, a prospective, double-blind, randomized, controlled phase II trial [103]. In this study, 167 patients with refractory angina were randomized to receive intramyocardial injection of 1×10^5 or 5×10^5 cells/kg of mobilized autologous CD34+ cells or an equal volume of diluent (placebo). Improvement of weekly angina frequency and exercise tolerance was significantly greater in low-dose patients, but not high-dose group, than placebo-treated patients at 6 months and 12 months. Three patients in the control group died in this study. No major adverse cardiovascular events were causatively related to the cell therapy (Table 13.6).

Table 13.5 Clinical trials of G-CSF-mobilized crude MNCs for PAD

Trial name/author	Year	Study design	Disease	Number of patients			Product	Route of administration	Follow-up duration	Outcomes
				Total	Treated	Control				
Huang et al. [101]	2004	Patient series	PAD	5	5	0	G-CSF-mobilized PB-MNCs	IM	3 months	ABPI ↑
										Laser Doppler blood perfusion ↑
Ishida et al. [97]	2005	Patient series	PAD	6	6	0	G-CSF-mobilized PB-MNCs	IM	24 weeks	ABPI →
										$TcPO_2$ →
										Maximal walking distance↑
Huang et al. [96]	2005	RCT	PAD	28	14	14	G-CSF-mobilized PB-MNCs	IM	3 months	ABPI ↑
										Laser Doppler blood perfusion↑
										Ulcer healing↑
										Angiographic score↑
Lenk et al. [99]	2005	Patient series	PAD	7	7	0	G-CSF-mobilized PB-MNCs	IA	12 weeks	ABPI ↑
										$TcPO_2$↑
										Pain-free walking distance ↑
										Flow-dependent vasodilation ↑
										Flow reserve in response to adenosine↑
										Endothelium-dependent vasodilation↑

Huang et al. [100]	2007	RCT	PAD	150	BM-MNCs: 74	0	BM-MNCs	IM	12 weeks	ABPI ↑
										Skin temperature ↑
										Rest pain scale ↓ in the G-CSF-mobilized PB-MNC-treated group
					G-CSF-mobilized PB-MNCs: 76		G-CSF-mobilized PB-MNCs			$TcPO_2$ →
										Pain-free walking distance →
										Ulcer healing and amputation rate →
Lara-Hernandez et al. [98]	2010	Patient series	PAD	28	28	0	G-CSF-mobilized PB-MNCs	IM	14 months	ABPI ↑, rest pain scaled ↓,
										limb salvage↑

Abbreviations: *IA* intra-arterial, *IM* intramuscular

Table 13.6 Clinical trials of G-CSF-mobilized fractionated CD34+ Cells for CAD

Trial name/ author	Year	Study design	Disease	Number of patients			Product	Route of administration	Follow-up duration	Outcomes
				Total	Treated	Control				
Losordo et al. [102]	2007	RCT (double blind)	Chronic CAD	24	18	6	G-CSF-mobilized CD34+ cells	IMC (TED)	6 months	ETT↑, symptom↑
ACT34-CMI/ Losordo et al. [103]	2011	RCT (double blind)	Chronic CAD	167	1×10^5 cells/kg: 55 5×10^5 cells/kg: 56	56	G-CSF-mobilized CD34+ cells	IMC (TED)	6 months	ETT↑, symptom↑

Abbreviations: *ETT* exercise tolerance test, *IMC* intramyocardial, *TED* transendocardial delivery

13.4.2 G-CSF-Mobilized Fractionated CD34+/CD133+ Cells for PAD

13.4.2.1 Clinical Trials

BM-derived EPCs comprise a small fraction (0.1–2%) of total MNCs. The advantage of the administration of fractionated EPCs is a higher concentration of EPCs compared with that of crude MNCs resulting in greater therapeutic potency. Onodera et al. reported that treatment with small number of harvested CD34+ cells was a negative independent predictor of amputation and death following either BM- or PB-MNC implantation in patients with CLI [104]. This finding suggests an important role of EPCs for therapeutic neovascularization and may provide a reasonable rationale for transplantation of CD34+ cells purified from crude MNCs in patients with CLI.

In a phase I/IIa clinical trial, our group evaluated the safety and feasibility of G-CSF-mobilized CD34+ cells in no-option patients with atherosclerotic PAD or Buerger's disease representing CLI [105]. CD34+ cells were isolated from the G-CSF-mobilized apheresis product using a magnetic cell sorting system and then intramuscularly transplanted in a dose-escalating manner into 17 patients (10^5 cells/kg, $n = 6$; 5×10^5 cells/kg, $n = 8$; or 10^6 cells/kg, $n = 3$). CD34+ cell therapy significantly improved Rutherford's category, pain scale, skin ulcer size, and blood perfusion at 12 weeks after treatment, although no significant dose-dependent relationship was observed. Furthermore, the safety and efficacy of CD34+ cell therapy were sustained for up to 4 years after cell therapy [106]. Furthermore, our phase II trial almost reproduced the clinical outcomes in the previous phase I/Iia trial, indicating the safety, feasibility, and potential effectiveness of CD34+ cell transplantation for CLI patients [107]. Recently, in the ACT34-CLI (Autologous CD34+ Cell Therapy for Critical Limb Ischemia Investigator) study, a double-blind, randomized, placebo-controlled, phase I/Iia pilot clinical trial, 28 patients with CLI were randomized to receive intramuscular injection of 1×10^5 (low-dose, $n = 7$) or 1×10^6 (high-dose, $n = 9$) cells/kg of mobilized CD34+ cells or an equal volume of diluent [108]. A favorable trend toward improvement of amputation-free survival rate was observed in the cell-treated groups, especially in the high-dose group, compared with control group at 6 and 12 months after treatment (Table 13.7).

In a phase I trial, Burt et al. evaluated the safety and feasibility of intramuscular implantation of autologous G-CSF-mobilized CD133+ cells, another EPC-enriched fraction, in nine patients with CLI [109]. In this uncontrolled study, leg amputation was observed in two out of nine patients at 12 months after treatment. The seven amputation-free patients showed significant improvement in QOL score at 3 and 6 months, but not 12 months. There was a favorable trend toward improvement in pain-free walking distance and exercise capacity at 12 months (Table 13.7).

Larger randomized clinical trials are warranted to clarify the efficacy of EPCs fractionated by various methods for the treatment of PAD patients. Following these favorable outcomes, a multicenter, phase II/III clinical trial for G-CSF-mobilized CD34+ cells in patients with CLI is in preparation in our institution.

Table 13.7 Clinical trials of G-CSF-mobilized fractionated CD34+/CD133+ cells for PAD

Trial name/ author	Year	Study design	Disease	Number of patients			Product	Route of administration	Follow-up duration	Outcomes
				Total	Treated	Control				
Kawamoto et al. [105]	2009	Patient series	PAD	17	17	0	G-CSF-mobilized CD34+ cells	IM	52 weeks	Efficacy score (TBPI, rest pain scale, total walking distance) ↑
Kinoshita et al. [106]										
ACT34-CLI/ Losordo et al. [108]	2012	RCT (double blind)	PAD	28	16	12	G-CSF-mobilized CD34+ cells	IM	12 months	Amputation rate ↓ ($p = 0.058$) compared with control
Fujita et al. [107]	2014	Patient series	PAD	11	11	0	G-CSF-mobilized CD34+ cells	IM	52 weeks	Rutherford's category ↑
										Rest pain scales ↓
										Skin perfusion pressure ↑
										TBPI ↑, $TcPO_2$ ↑
										Pain-free walking distance ↑
										Total walking distance ↑
Burt et al. [109]	2010	Patient series	PAD	9	9	0	G-CSF-mobilized CD133+ cells	IM	12 months	QOL ↑ (at 6 months)

Abbreviations: *QOL* quality of life, *TBPI* toe-brachial pressure index

13.5 Conclusions

Summarizing the results of the clinical trials regarding G-CSF monotherapy for CAD, G-CSF treatment is feasible and relatively safe. However, efficacy of this therapy remains controversial. The final result of an ongoing phase III clinical trial [51] may provide a definitive evidence of G-CSF treatment in STEMI. As for G-CSF monotherapy for PAD, there is only one early-phase clinical trial [57]. Randomized clinical trials are warranted to clarify the efficacy of G-CSF monotherapy for PAD. As for CVD, despite favorable results from animal studies, no significant effect of G-CSF monotherapy on major endpoints was observed in several randomized clinical trials.

Theoretically, G-CSF-mobilized stem/progenitor cell therapies may be superior over G-CSF monotherapy due to not only direct vasculogenic properties but also paracrine action by secreting multiple growth factors besides a single hematopoietic cytokine. Early-phase clinical trials reveled safety and feasibility of G-CFS-mobilized stem/progenitor cell therapy for CAD and PAD. As far as the authors searched, any clinical trials of G-CSF-mobilized stem/progenitor cell therapy for CVD have not been conducted. To prove the safety, feasibility, and efficacy of the cell-based therapies, well-designed, larger-scaled, and randomized clinical trials will be needed.

References

1. Metcalf D. The molecular control of cell division, differentiation commitment and maturation in haemopoietic cells. Nature. 1989;339:27–30. doi:10.1038/339027a0.
2. Nagata S, Tsuchiya M, Asano S, Kaziro Y, Yamazaki T, Yamamoto O, et al. Molecular cloning and expression of cDNA for human granulocyte colony-stimulating factor. Nature. 1986;319:415–8.
3. Welte K, Gabrilove J, Bronchud MH, Platzer E, Morstyn G. Filgrastim (r-metHuG-CSF): the first 10 years. Blood. 1996;88:1907–29.
4. Bussolino F, Ziche M, Wang JM, Alessi D, Morbidelli L, Cremona O, et al. In vitro and in vivo activation of endothelial cells by colony-stimulating factors. J Clin Invest. 1991;87:986–95. doi:10.1172/JCI115107.
5. Le Beau MM, Lemons RS, Carrino JJ, Pettenati MJ, Souza LM, Diaz MO, et al. Chromosomal localization of the human G-CSF gene to 17q11 proximal to the breakpoint of the t(15;17) in acute promyelocytic leukemia. Leukemia. 1987;1:795–9.
6. Nagata S, Tsuchiya M, Asano S, Yamamoto O, Hirata Y, Kubota N, et al. The chromosomal gene structure and two mRNAs for human granulocyte colony-stimulating factor. EMBO J. 1986;5:575–81.
7. Koeffler HP, Gasson J, Ranyard J, Souza L, Shepard M, Munker R. Recombinant human TNF alpha stimulates production of granulocyte colony-stimulating factor. Blood. 1987;70:55–9.
8. Kaushansky K, Lin N, Adamson JW. Interleukin 1 stimulates fibroblasts to synthesize granulocyte-macrophage and granulocyte colony-stimulating factors. Mechanism for the hematopoietic response to inflammation. J Clin Invest. 1988;81:92–7. doi:10.1172/JCI113316.
9. Vellenga E, Rambaldi A, Ernst TJ, Ostapovicz D, Griffin JD. Independent regulation of M-CSF and G-CSF gene expression in human monocytes. Blood. 1988;71:1529–32.

10. Zsebo KM, Yuschenkoff VN, Schiffer S, Chang D, McCall E, Dinarello CA, et al. Vascular endothelial cells and granulopoiesis: interleukin-1 stimulates release of G-CSF and GM-CSF. Blood. 1988;71:99–103.
11. Demetri GD, Griffin JD. Granulocyte colony-stimulating factor and its receptor. Blood. 1991;78:2791–808.
12. Fossiez F, Djossou O, Chomarat P, Flores-Romo L, Ait-Yahia S, Maat C, et al. T cell interleukin-17 induces stromal cells to produce proinflammatory and hematopoietic cytokines. J Exp Med. 1996;183:2593–603.
13. Sano E, Ohashi K, Sato Y, Kashiwagi M, Joguchi A, Naruse N. A possible role of autogenous IFN-Beta for cytokine productions in human fibroblasts. J Cell Biochem. 2007;100:1459–76. doi:10.1002/jcb.21128.
14. Fukunaga R, Ishizaka-Ikeda E, Pan CX, Seto Y, Nagata S. Functional domains of the granulocyte colony-stimulating factor receptor. EMBO J. 1991;10:2855–65.
15. Boneberg EM, Hareng L, Gantner F, Wendel A, Hartung T. Human monocytes express functional receptors for granulocyte colony-stimulating factor that mediate suppression of monokines and interferon-gamma. Blood. 2000;95:270–6.
16. Bussolino F, Wang JM, Defilippi P, Turrini F, Sanavio F, Edgell CJ, et al. Granulocyte- and granulocyte-macrophage-colony stimulating factors induce human endothelial cells to migrate and proliferate. Nature. 1989;337:471–3. doi:10.1038/337471a0.
17. Hanazono Y, Hosoi T, Kuwaki T, Matsuki S, Miyazono K, Miyagawa K, et al. Structural analysis of the receptors for granulocyte colony-stimulating factor on neutrophils. Exp Hematol. 1990;18:1097–103.
18. Harada M, Qin Y, Takano H, Minamino T, Zou Y, Toko H, et al. G-CSF prevents cardiac remodeling after myocardial infarction by activating the Jak-Stat pathway in cardiomyocytes. Nat Med. 2005;11:305–11. doi:10.1038/nm1199.
19. Kuhlmann MT, Kirchhof P, Klocke R, Hasib L, Stypmann J, Fabritz L, et al. G-CSF/SCF reduces inducible arrhythmias in the infarcted heart potentially via increased connexin43 expression and arteriogenesis. J Exp Med. 2006;203:87–97. doi:10.1084/jem.20051151.
20. Morikawa K, Morikawa S, Nakamura M, Miyawaki T. Characterization of granulocyte colony-stimulating factor receptor expressed on human lymphocytes. Br J Haematol. 2002;118:296–304.
21. Shimoda K, Okamura S, Harada N, Kondo S, Okamura T, Niho Y. Identification of a functional receptor for granulocyte colony-stimulating factor on platelets. J Clin Invest. 1993;91:1310–3. doi:10.1172/JCI116330.
22. Tian SS, Lamb P, Seidel HM, Stein RB, Rosen J. Rapid activation of the STAT3 transcription factor by granulocyte colony-stimulating factor. Blood. 1994;84:1760–4.
23. Shimoda K, Feng J, Murakami H, Nagata S, Watling D, Rogers NC, et al. Jak1 plays an essential role for receptor phosphorylation and Stat activation in response to granulocyte colony-stimulating factor. Blood. 1997;90:597–604.
24. Dong F, Larner AC. Activation of Akt kinase by granulocyte colony-stimulating factor (G-CSF): evidence for the role of a tyrosine kinase activity distinct from the Janus kinases. Blood. 2000;95:1656–62.
25. Hunter MG, Avalos BR. Phosphatidylinositol 3′-kinase and SH2-containing inositol phosphatase (SHIP) are recruited by distinct positive and negative growth-regulatory domains in the granulocyte colony-stimulating factor receptor. J Immunol. 1998;160:4979–87.
26. Orlic D, Kajstura J, Chimenti S, Limana F, Jakoniuk I, Quaini F, et al. Mobilized bone marrow cells repair the infarcted heart, improving function and survival. Proc Natl Acad Sci U S A. 2001;98:10344–9. doi:10.1073/pnas.181177898.
27. Kocher AA, Schuster MD, Szabolcs MJ, Takuma S, Burkhoff D, Wang J, et al. Neovascularization of ischemic myocardium by human bone-marrow-derived angioblasts prevents cardiomyocyte apoptosis, reduces remodeling and improves cardiac function. Nat Med. 2001;7:430–6. doi:10.1038/86498.

28. Kawamoto A, Murayama T, Kusano K, Ii M, Tkebuchava T, Shintani S, et al. Synergistic effect of bone marrow mobilization and vascular endothelial growth factor-2 gene therapy in myocardial ischemia. Circulation. 2004;110:1398–405. doi:10.1161/01.CIR.0000141563.71410.64.
29. Kajstura J, Rota M, Whang B, Cascapera S, Hosoda T, Bearzi C, et al. Bone marrow cells differentiate in cardiac cell lineages after infarction independently of cell fusion. Circ Res. 2005;96:127–37. doi:10.1161/01.RES.0000151843.79801.60.
30. Leri A, Kajstura J, Anversa P. Cardiac stem cells and mechanisms of myocardial regeneration. Physiol Rev. 2005;85:1373–416. doi:10.1152/physrev.00013.2005.
31. Minatoguchi S, Takemura G, Chen XH, Wang N, Uno Y, Koda M, et al. Acceleration of the healing process and myocardial regeneration may be important as a mechanism of improvement of cardiac function and remodeling by postinfarction granulocyte colony-stimulating factor treatment. Circulation. 2004;109:2572–80. doi:10.1161/01.CIR.0000129770.93985.3E.
32. Ohki Y, Heissig B, Sato Y, Akiyama H, Zhu Z, Hicklin DJ, et al. Granulocyte colony-stimulating factor promotes neovascularization by releasing vascular endothelial growth factor from neutrophils. FASEB J. 2005;19:2005–7. doi:10.1096/fj.04-3496fje.
33. Kuethe F, Figulla HR, Herzau M, Voth M, Fritzenwanger M, Opfermann T, et al. Treatment with granulocyte colony-stimulating factor for mobilization of bone marrow cells in patients with acute myocardial infarction. Am Heart J. 2005;150:115. doi:10.1016/j.ahj.2005.04.030.
34. Deindl E, Zaruba MM, Brunner S, Huber B, Mehl U, Assmann G, et al. G-CSF administration after myocardial infarction in mice attenuates late ischemic cardiomyopathy by enhanced arteriogenesis. FASEB J. 2006;20:956–8. doi:10.1096/fj.05-4763fje.
35. Suda T, Suda J, Kajigaya S, Nagata S, Asano S, Saito M, et al. Effects of recombinant murine granulocyte colony-stimulating factor on granulocyte-macrophage and blast colony formation. Exp Hematol. 1987;15:958–65.
36. Vandervelde S, van Luyn MJ, Tio RA, Harmsen MC. Signaling factors in stem cell-mediated repair of infarcted myocardium. J Mol Cell Cardiol. 2005;39:363–76. doi:10.1016/j.yjmcc.2005.05.012.
37. Kang H-J, Kim H-S, Zhang S-Y, Park K-W, Cho H-J, Koo B-K, et al. Effects of intracoronary infusion of peripheral blood stem-cells mobilised with granulocyte-colony stimulating factor on left ventricular systolic function and restenosis after coronary stenting in myocardial infarction: the MAGIC cell randomised clinical trial. Lancet. 2004;363:751–6. doi:10.1016/S0140-6736(04)15689-4.
38. Kang HJ, Kim HS, Koo BK, Kim YJ, Lee D, Sohn DW, et al. Intracoronary infusion of the mobilized peripheral blood stem cell by G-CSF is better than mobilization alone by G-CSF for improvement of cardiac function and remodeling: 2-years follow-up results of the Myocardial Regeneration and Angiogenesis in Myocardial Infarction with G-CSF and Intra-Coronary Stem Cell Infusion (MAGIC Cell) 1 trial. Am Heart J. 2007;153:237. doi:10.1016/j.ahj.2006.11.004.e1-8
39. Ince H, Petzsch M, Kleine HD, Schmidt H, Rehders T, Korber T, et al. Preservation from left ventricular remodeling by front-integrated revascularization and stem cell liberation in evolving acute myocardial infarction by use of granulocyte-colony-stimulating factor (FIRSTLINE-AMI). Circulation. 2005;112:3097–106. doi:10.1161/CIRCULATIONAHA.105.541433.
40. Ince H, Petzsch M, Kleine HD, Eckard H, Rehders T, Burska D, et al. Prevention of left ventricular remodeling with granulocyte colony-stimulating factor after acute myocardial infarction: final 1-year results of the Front-integrated Revascularization and Stem Cell Liberation in Evolving Acute Myocardial Infarction by Granulocyte Colony-Stimulating Factor (FIRSTLINE-AMI) trial. Circulation. 2005;112:I73–80. doi:10.1161/CIRCULATIONAHA.104.524827.
41. Valgimigli M, Rigolin GM, Cittanti C, Malagutti P, Curello S, Percoco G, et al. Use of granulocyte-colony stimulating factor during acute myocardial infarction to enhance bone marrow stem cell mobilization in humans: clinical and angiographic safety profile. Eur Heart J. 2005;26:1838–45. doi:10.1093/eurheartj/ehi289.
42. Ripa RS, Jorgensen E, Wang Y, Thune JJ, Nilsson JC, Sondergaard L, et al. Stem cell mobilization induced by subcutaneous granulocyte-colony stimulating factor to improve cardiac

regeneration after acute ST-elevation myocardial infarction: result of the double-blind, randomized, placebo-controlled stem cells in myocardial infarction (STEMMI) trial. Circulation. 2006;113:1983–92. doi:10.1161/CIRCULATIONAHA.105.610469.
43. Zohlnhofer D, Ott I, Mehilli J, Schomig K, Michalk F, Ibrahim T, et al. Stem cell mobilization by granulocyte colony-stimulating factor in patients with acute myocardial infarction: a randomized controlled trial. JAMA. 2006;295:1003–10. doi:10.1001/jama.295.9.1003.
44. Engelmann MG, Theiss HD, Hennig-Theiss C, Huber A, Wintersperger BJ, Werle-Ruedinger AE, et al. Autologous bone marrow stem cell mobilization induced by granulocyte colony-stimulating factor after subacute ST-segment elevation myocardial infarction undergoing late revascularization: final results from the G-CSF-STEMI (Granulocyte Colony-Stimulating Factor ST-Segment Elevation Myocardial Infarction) trial. J Am Coll Cardiol. 2006;48:1712–21. doi:10.1016/j.jacc.2006.07.044.
45. Achilli F, Malafronte C, Lenatti L, Gentile F, Dadone V, Gibelli G, et al. Granulocyte colony-stimulating factor attenuates left ventricular remodelling after acute anterior STEMI: results of the single-blind, randomized, placebo-controlled multicentre Stem Cell Mobilization in Acute Myocardial Infarction (STEM-AMI) Trial. Eur J Heart Fail. 2010;12:1111–21. doi:10.1093/eurjhf/hfq150.
46. Achilli F, Malafronte C, Maggiolini S, Lenatti L, Squadroni L, Gibelli G, et al. G-CSF treatment for STEMI: final 3-years follow-up of the randomised placebo-controlled STEM-AMI trial. Heart. 2014;100:574–81. doi:10.1136/heartjnl-2013-304955.
47. Suarez de Lezo J, Torres A, Herrera I, Pan M, Romero M, Pavlovic D, et al. Effects of stem-cell mobilization with recombinant human granulocyte colony stimulating factor in patients with percutaneously revascularized acute anterior myocardial infarction. Rev Esp Cardiol. 2005;58:253–61.
48. Abdel-Latif A, Bolli R, Zuba-Surma EK, Tleyjeh IM, Hornung CA, Dawn B. Granulocyte colony-stimulating factor therapy for cardiac repair after acute myocardial infarction: a systematic review and meta-analysis of randomized controlled trials. Am Heart J. 2008;156:216–26. e9 doi:10.1016/j.ahj.2008.03.024.
49. Ince H, Valgimigli M, Petzsch M, de Lezo JS, Kuethe F, Dunkelmann S, et al. Cardiovascular events and re-stenosis following administration of G-CSF in acute myocardial infarction: systematic review and meta-analysis. Heart. 2008;94:610–6. doi:10.1136/hrt.2006.111385.
50. Zohlnhofer D, Dibra A, Koppara T, de Waha A, Ripa RS, Kastrup J, et al. Stem cell mobilization by granulocyte colony-stimulating factor for myocardial recovery after acute myocardial infarction: a meta-analysis. J Am Coll Cardiol. 2008;51:1429–37. doi:10.1016/j.jacc.2007.11.073.
51. Achilli F, Malafronte C, Cesana F, Maggiolini S, Mauro C, De Ferrari GM, et al. Granulocyte-colony stimulating factor for large anterior ST-elevation myocardial infarction: rationale and design of the prospective randomized phase III STEM-AMI OUTCOME trial. Am Heart J. 2015;170:652–8. e7. doi:10.1016/j.ahj.2015.07.005.
52. Hill JM, Syed MA, Arai AE, Powell TM, Paul JD, Zalos G, et al. Outcomes and risks of granulocyte colony-stimulating factor in patients with coronary artery disease. J Am Coll Cardiol. 2005;46:1643–8. doi:10.1016/j.jacc.2005.01.067.
53. Wang Y, Tagil K, Ripa RS, Nilsson JC, Carstensen S, Jorgensen E, et al. Effect of mobilization of bone marrow stem cells by granulocyte colony stimulating factor on clinical symptoms, left ventricular perfusion and function in patients with severe chronic ischemic heart disease. Int J Cardiol. 2005;100:477–83. doi:10.1016/j.ijcard.2004.12.006.
54. Ripa RS, Wang Y, Jorgensen E, Johnsen HE, Hesse B, Kastrup J. Intramyocardial injection of vascular endothelial growth factor-A165 plasmid followed by granulocyte-colony stimulating factor to induce angiogenesis in patients with severe chronic ischaemic heart disease. Eur Heart J. 2006;27:1785–92. doi:10.1093/eurheartj/ehl117.
55. Huttmann A, Duhrsen U, Stypmann J, Noppeney R, Nuckel H, Neumann T, et al. Granulocyte colony-stimulating factor-induced blood stem cell mobilisation in patients with chronic heart

failure—feasibility, safety and effects on exercise tolerance and cardiac function. Basic Res Cardiol. 2006;101:78–86. doi:10.1007/s00395-005-0556-1.

56. Lee M, Aoki M, Kondo T, Kobayashi K, Okumura K, Komori K, et al. Therapeutic angiogenesis with intramuscular injection of low-dose recombinant granulocyte-colony stimulating factor. Arterioscler Thromb Vasc Biol. 2005;25:2535–41. doi:10.1161/01.atv.0000190609.28293.17.
57. Arai M, Misao Y, Nagai H, Kawasaki M, Nagashima K, Suzuki K, et al. Granulocyte colony-stimulating factor: a noninvasive regeneration therapy for treating atherosclerotic peripheral artery disease. Circ J. 2006;70:1093–8.
58. Six I, Gasan G, Mura E, Bordet R. Beneficial effect of pharmacological mobilization of bone marrow in experimental cerebral ischemia. Eur J Pharmacol. 2003;458:327–8.
59. Shyu WC, Lin SZ, Yang HI, Tzeng YS, Pang CY, Yen PS, et al. Functional recovery of stroke rats induced by granulocyte colony-stimulating factor-stimulated stem cells. Circulation. 2004;110:1847–54. doi:10.1161/01.CIR.0000142616.07367.66.
60. Schabitz WR, Kollmar R, Schwaninger M, Juettler E, Bardutzky J, Scholzke MN, et al. Neuroprotective effect of granulocyte colony-stimulating factor after focal cerebral ischemia. Stroke. 2003;34:745–51. doi:10.1161/01.STR.0000057814.70180.17.
61. Gibson CL, Bath PM, Murphy SP. G-CSF administration is neuroprotective following transient cerebral ischemia even in the absence of a functional NOS-2 gene. J Cereb Blood Flow Metab. 2010;30:739–43. doi:10.1038/jcbfm.2010.12.
62. Gibson CL, Bath PM, Murphy SP. G-CSF reduces infarct volume and improves functional outcome after transient focal cerebral ischemia in mice. J Cereb Blood Flow Metab. 2005;25:431–9. doi:10.1038/sj.jcbfm.9600033.
63. England TJ, Gibson CL, Bath PM. Granulocyte-colony stimulating factor in experimental stroke and its effects on infarct size and functional outcome: a systematic review. Brain Res Rev. 2009;62:71–82. doi:10.1016/j.brainresrev.2009.09.002.
64. Han JL, Blank T, Schwab S, Kollmar R. Inhibited glutamate release by granulocyte-colony stimulating factor after experimental stroke. Neurosci Lett. 2008;432:167–9. doi:10.1016/j.neulet.2007.07.056.
65. Komine-Kobayashi M, Zhang N, Liu M, Tanaka R, Hara H, Osaka A, et al. Neuroprotective effect of recombinant human granulocyte colony-stimulating factor in transient focal ischemia of mice. J Cereb Blood Flow Metab. 2006;26:402–13. doi:10.1038/sj.jcbfm.9600195.
66. Lee ST, Chu K, Jung KH, Ko SY, Kim EH, Sinn DI, et al. Granulocyte colony-stimulating factor enhances angiogenesis after focal cerebral ischemia. Brain Res. 2005;1058:120–8. doi:10.1016/j.brainres.2005.07.076.
67. Sehara Y, Hayashi T, Deguchi K, Zhang H, Tsuchiya A, Yamashita T, et al. Potentiation of neurogenesis and angiogenesis by G-CSF after focal cerebral ischemia in rats. Brain Res. 2007;1151:142–9. doi:10.1016/j.brainres.2007.01.149.
68. Solaroglu I, Cahill J, Tsubokawa T, Beskonakli E, Zhang JH. Granulocyte colony-stimulating factor protects the brain against experimental stroke via inhibition of apoptosis and inflammation. Neurol Res. 2009;31:167–72. doi:10.1179/174313209X393582.
69. Schneider A, Kruger C, Steigleder T, Weber D, Pitzer C, Laage R, et al. The hematopoietic factor G-CSF is a neuronal ligand that counteracts programmed cell death and drives neurogenesis. J Clin Invest. 2005;115:2083–98. doi:10.1172/JCI23559.
70. Solaroglu I, Tsubokawa T, Cahill J, Zhang JH. Anti-apoptotic effect of granulocyte-colony stimulating factor after focal cerebral ischemia in the rat. Neuroscience. 2006;143:965–74. doi:10.1016/j.neuroscience.2006.09.014.
71. Sevimli S, Diederich K, Strecker JK, Schilling M, Klocke R, Nikol S, et al. Endogenous brain protection by granulocyte-colony stimulating factor after ischemic stroke. Exp Neurol. 2009;217:328–35. doi:10.1016/j.expneurol.2009.03.018.
72. Taguchi A, Wen Z, Myojin K, Yoshihara T, Nakagomi T, Nakayama D, et al. Granulocyte colony-stimulating factor has a negative effect on stroke outcome in a murine model. Eur J Neurosci. 2007;26:126–33. doi:10.1111/j.1460-9568.2007.05640.x.

73. Strecker JK, Sevimli S, Schilling M, Klocke R, Nikol S, Schneider A, et al. Effects of G-CSF treatment on neutrophil mobilization and neurological outcome after transient focal ischemia. Exp Neurol. 2010;222:108–13. doi:10.1016/j.expneurol.2009.12.012.
74. Shyu WC, Lin SZ, Lee CC, Liu DD, Li H. Granulocyte colony-stimulating factor for acute ischemic stroke: a randomized controlled trial. CMAJ. 2006;174:927–33. doi:10.1503/cmaj.051322.
75. Sprigg N, Bath PM, Zhao L, Willmot MR, Gray LJ, Walker MF, et al. Granulocyte-colony–stimulating factor mobilizes bone marrow stem cells in patients with subacute ischemic stroke: the stem cell trial of recovery enhancement after stroke (STEMS) pilot randomized, controlled trial (ISRCTN 16784092). Stroke. 2006;37:2979–83. doi:10.1161/01.STR.0000248763.49831.c3.
76. Schabitz WR, Laage R, Vogt G, Koch W, Kollmar R, Schwab S, et al. AXIS: a trial of intravenous granulocyte colony-stimulating factor in acute ischemic stroke. Stroke. 2010;41:2545–51. doi:10.1161/STROKEAHA.110.579508.
77. England TJ, Abaei M, Auer DP, Lowe J, Jones DRE, Sare G, et al. Granulocyte-colony stimulating factor for mobilizing bone marrow stem cells in subacute stroke: the stem cell trial of recovery enhancement after stroke 2 randomized controlled trial. Stroke. 2012;43:405–11. doi:10.1161/strokeaha.111.636449.
78. Ringelstein EB, Thijs V, Norrving B, Chamorro A, Aichner F, Grond M, et al. Granulocyte colony–stimulating factor in patients with acute ischemic stroke: results of the AX200 for ischemic stroke trial. Stroke. 2013;44:2681–7. doi:10.1161/strokeaha.113.001531.
79. Floel A, Warnecke T, Duning T, Lating Y, Uhlenbrock J, Schneider A, et al. Granulocyte-colony stimulating factor (G-CSF) in stroke patients with concomitant vascular disease—a randomized controlled trial. PLoS One. 2011;6:e19767. doi:10.1371/journal.pone.0019767.
80. Asahara T, Murohara T, Sullivan A, Silver M, van der Zee R, Li T, et al. Isolation of putative progenitor endothelial cells for angiogenesis. Science. 1997;275:964–7.
81. Masuda H, Kalka C, Takahashi T, Yoshida M, Wada M, Kobori M, et al. Estrogen-mediated endothelial progenitor cell biology and kinetics for physiological postnatal vasculogenesis. Circ Res. 2007;101:598–606. doi:10.1161/CIRCRESAHA.106.144006.
82. Takahashi T, Kalka C, Masuda H, Chen D, Silver M, Kearney M, et al. Ischemia- and cytokine-induced mobilization of bone marrow-derived endothelial progenitor cells for neovascularization. Nat Med. 1999;5:434–8. doi:10.1038/7434.
83. Orlic D, Kajstura J, Chimenti S, Jakoniuk I, Anderson SM, Li B, et al. Bone marrow cells regenerate infarcted myocardium. Nature. 2001;410:701–5. doi:10.1038/35070587.
84. Narmoneva DA, Vukmirovic R, Davis ME, Kamm RD, Lee RT. Endothelial cells promote cardiac myocyte survival and spatial reorganization: implications for cardiac regeneration. Circulation. 2004;110:962–8. doi:10.1161/01.CIR.0000140667.37070.07.
85. Kamihata H, Matsubara H, Nishiue T, Fujiyama S, Amano K, Iba O, et al. Improvement of collateral perfusion and regional function by implantation of peripheral blood mononuclear cells into ischemic hibernating myocardium. Arterioscler Thromb Vasc Biol. 2002;22:1804–10.
86. Kinnaird T, Stabile E, Burnett MS, Lee CW, Barr S, Fuchs S, et al. Marrow-derived stromal cells express genes encoding a broad spectrum of arteriogenic cytokines and promote in vitro and in vivo arteriogenesis through paracrine mechanisms. Circ Res. 2004;94:678–85. doi:10.1161/01.RES.0000118601.37875.AC.
87. Kinnaird T, Stabile E, Burnett MS, Shou M, Lee CW, Barr S, et al. Local delivery of marrow-derived stromal cells augments collateral perfusion through paracrine mechanisms. Circulation. 2004;109:1543–9. doi:10.1161/01.CIR.0000124062.31102.57.
88. Rehman J, Li J, Orschell CM, March KL. Peripheral blood "endothelial progenitor cells" are derived from monocyte/macrophages and secrete angiogenic growth factors. Circulation. 2003;107:1164–9.
89. Wollert KC, Drexler H. Clinical applications of stem cells for the heart. Circ Res. 2005;96:151–63. doi:10.1161/01.RES.0000155333.69009.63.
90. Gehling UM, Ergun S, Schumacher U, Wagener C, Pantel K, Otte M, et al. In vitro differentiation of endothelial cells from AC133-positive progenitor cells. Blood. 2000;95:3106–12.

91. Kawamoto A, Iwasaki H, Kusano K, Murayama T, Oyamada A, Silver M, et al. CD34-positive cells exhibit increased potency and safety for therapeutic neovascularization after myocardial infarction compared with total mononuclear cells. Circulation. 2006;114:2163–9. doi:10.1161/CIRCULATIONAHA.106.644518.
92. Masuda H, Alev C, Akimaru H, Ito R, Shizuno T, Kobori M, et al. Methodological development of a clonogenic assay to determine endothelial progenitor cell potential. Circ Res. 2011;109:20–37. doi:10.1161/CIRCRESAHA.110.231837.
93. Ramos AL, Darabi R, Akbarloo N, Borges L, Catanese J, Dineen SP, et al. Clonal analysis reveals a common progenitor for endothelial, myeloid, and lymphoid precursors in umbilical cord blood. Circ Res. 2010;107:1460–9. doi:10.1161/CIRCRESAHA.110.223669.
94. Taguchi A, Soma T, Tanaka H, Kanda T, Nishimura H, Yoshikawa H, et al. Administration of CD34+ cells after stroke enhances neurogenesis via angiogenesis in a mouse model. J Clin Invest. 2004;114:330–8. doi:10.1172/JCI20622.
95. Kang H-J, Lee H-Y, Na S-H, Chang S-A, Park K-W, Kim H-K, et al. Differential effect of intracoronary infusion of mobilized peripheral blood stem cells by granulocyte colony–stimulating factor on left ventricular function and remodeling in patients with acute myocardial infarction versus old myocardial infarction: the MAGIC Cell-3-DES randomized, controlled trial. Circulation. 2006;114:I-145–I-51. doi:10.1161/circulationaha.105.001107.
96. Huang P, Li S, Han M, Xiao Z, Yang R, Han ZC. Autologous transplantation of granulocyte colony-stimulating factor-mobilized peripheral blood mononuclear cells improves critical limb ischemia in diabetes. Diabetes Care. 2005;28:2155–60.
97. Ishida A, Ohya Y, Sakuda H, Ohshiro K, Higashiuesato Y, Nakaema M, et al. Autologous peripheral blood mononuclear cell implantation for patients with peripheral arterial disease improves limb ischemia. Circ J. 2005;69:1260–5.
98. Lara-Hernandez R, Lozano-Vilardell P, Blanes P, Torreguitart-Mirada N, Galmes A, Besalduch J. Safety and efficacy of therapeutic angiogenesis as a novel treatment in patients with critical limb ischemia. Ann Vasc Surg. 2010;24:287–94. doi:10.1016/j.avsg.2009.10.012.
99. Lenk K, Adams V, Lurz P, Erbs S, Linke A, Gielen S, et al. Therapeutical potential of blood-derived progenitor cells in patients with peripheral arterial occlusive disease and critical limb ischaemia. Eur Heart J. 2005;26:1903–9. doi:10.1093/eurheartj/ehi285.
100. Huang PP, Yang XF, Li SZ, Wen JC, Zhang Y, Han ZC. Randomised comparison of G-CSF-mobilized peripheral blood mononuclear cells versus bone marrow-mononuclear cells for the treatment of patients with lower limb arteriosclerosis obliterans. Thromb Haemost. 2007;98:1335–42.
101. Huang PP, Li SZ, Han MZ, Xiao ZJ, Yang RC, Qiu LG, et al. Autologous transplantation of peripheral blood stem cells as an effective therapeutic approach for severe arteriosclerosis obliterans of lower extremities. Thromb Haemost. 2004;91:606–9. doi:10.1160/TH03-06-0343.
102. Losordo DW, Schatz RA, White CJ, Udelson JE, Veereshwarayya V, Durgin M, et al. Intramyocardial transplantation of autologous CD34+ stem cells for intractable angina: a phase I/IIa double-blind, randomized controlled trial. Circulation. 2007;115:3165–72. doi:10.1161/CIRCULATIONAHA.106.687376.
103. Losordo DW, Henry TD, Davidson C, Sup Lee J, Costa MA, Bass T, et al. Intramyocardial, autologous CD34+ cell therapy for refractory angina. Circ Res. 2011;109:428–36. doi:10.1161/CIRCRESAHA.111.245993.
104. Onodera R, Teramukai S, Tanaka S, Kojima S, Horie T, Matoba S, et al. Bone marrow mononuclear cells versus G-CSF-mobilized peripheral blood mononuclear cells for treatment of lower limb ASO: pooled analysis for long-term prognosis. Bone Marrow Transplant. 2011;46:278–84. doi:10.1038/bmt.2010.110.
105. Kawamoto A, Katayama M, Handa N, Kinoshita M, Takano H, Horii M, et al. Intramuscular transplantation of G-CSF-mobilized CD34(+) cells in patients with critical limb ischemia: a phase I/IIa, multicenter, single-blinded, dose-escalation clinical trial. Stem Cells. 2009;27:2857–64. doi:10.1002/stem.207.
106. Kinoshita M, Fujita Y, Katayama M, Baba R, Shibakawa M, Yoshikawa K, et al. Long-term clinical outcome after intramuscular transplantation of granulocyte colony stimulating factor-

mobilized CD34 positive cells in patients with critical limb ischemia. Atherosclerosis. 2012;224:440–5. doi:10.1016/j.atherosclerosis.2012.07.031.
107. Fujita Y, Kinoshita M, Furukawa Y, Nagano T, Hashimoto H, Hirami Y, et al. Phase II clinical trial of CD34+ cell therapy to explore endpoint selection and timing in patients with critical limb ischemia. Circ J. 2014;78:490–501.
108. Losordo DW, Kibbe MR, Mendelsohn F, Marston W, Driver VR, Sharafuddin M, et al. A randomized, controlled pilot study of autologous CD34+ cell therapy for critical limb ischemia. Circ Cardiovasc Interv. 2012;5:821–30. doi:10.1161/CIRCINTERVENTIONS.112.968321.
109. Burt RK, Testori A, Oyama Y, Rodriguez HE, Yaung K, Villa M, et al. Autologous peripheral blood CD133+ cell implantation for limb salvage in patients with critical limb ischemia. Bone Marrow Transplant. 2010;45:111–6. doi:10.1038/bmt.2009.102.

Chapter 14
Waon Therapy: Effect of Thermal Stimuli on Angiogenesis

Masaaki Miyata, Mitsuru Ohishi, and Chuwa Tei

Abstract Tei et al. developed a form of thermal therapy, namely, Waon therapy, which uses a dry sauna with temperature maintained at 60 °C and differs from the traditional sauna. We have previously reported that Waon therapy improves cardiac and vascular function in patients with chronic heart failure (CHF) and demonstrated that the molecular mechanism by which Waon therapy improves vascular flow and endothelial function is through increased endothelial nitric oxide synthase (eNOS) expression.

In a mouse model of hindlimb ischemia, we demonstrated that eNOS is a critical regulator of angiogenesis by Waon therapy. Furthermore, we reported that Waon therapy upregulates the heat shock protein 90 which contributes to the activation of the Akt/eNOS/NO pathway and induces angiogenesis in mice with hindlimb ischemia.

In addition, we have demonstrated that repeated Waon therapy is safe for patients with severe peripheral artery disease (PAD) and potentially effective as evidenced by substantial decrease in the pain score, by increases in ankle-brachial pressure index and blood flow assessed by laser Doppler perfusion imaging, and by formation of new collateral vessels on angiography. Ischemic ulcers heal or improve markedly. Moreover, we reported that Waon therapy mobilized circulating endothelial progenitor cells and improved limb ischemia in patients with PAD.

In conclusion, the thermal stimulus by Waon therapy is one of noninvasive therapies of therapeutic angiogenesis for PAD.

Keywords Waon therapy • Peripheral arterial disease • Nitric oxide • eNOS • Endothelial progenitor cell

M. Miyata (✉) • M. Ohishi
Department of Cardiovascular Medicine and Hypertension, Graduate School of Medical and Dental Sciences, Kagoshima University, 8-35-1 Sakuragaoka, Kagoshima 890-8520, Japan
e-mail: miyatam@m3.kufm.kagoshima-u.ac.jp

C. Tei
Waon Therapy Research Institute, Tokyo, Japan

Y. Higashi, T. Murohara (eds.), *Therapeutic Angiogenesis*,
DOI 10.1007/978-981-10-2744-4_14

14.1 Introduction

Peripheral artery disease (PAD) is associated with decreased functional capacity and quality of life. Furthermore, patients with PAD have a higher rate of limb amputation and an increased risk of death compared with the general population. Patients with chronic critical limb ischemia have a 20% mortality rate per year with many patients being poor candidates for revascularization procedures [1]. Therapeutic angiogenesis and vasculogenesis may provide a treatment option for those patients with critical limb ischemia who are not suited to conventional revascularization therapy [2].

Dr. Tei et al. developed a form of thermal therapy, namely, Waon therapy, which uses a dry sauna with temperature maintained at 60 °C and differs from the traditional sauna with high temperature and humidity [3]. This thermal therapy increases the deep body temperature by 1 °C. We reported that the repeated Waon therapy improves hemodynamics and ameliorates symptoms in patients with chronic heart failure (CHF) [4, 5]. Furthermore, we have reported that Waon therapy improves vascular endothelial dysfunction in patients with CHF and in patients with coronary risk factors such as hypertension, hyperlipidemia, diabetes mellitus, and smoking [6, 7]. We demonstrated that the molecular mechanism by which Waon therapy improves vascular flow and endothelial function is through increased endothelial nitric oxide synthase (eNOS) expression [8, 9].

Nitric oxide (NO), constitutively produced by eNOS, is a mediator of angiogenesis [10]. Therefore, we drew attention to the effect of NO induced by Waon therapy on angiogenesis and demonstrated that Waon therapy increases eNOS protein expression, blood flow, and capillary density in a mouse model of hindlimb ischemia [11]. Furthermore, we have reported that Waon therapy is safe and improves symptoms, status, and blood flow in patients with PAD [12]. In this section, we summarize the beneficial effect of Waon therapy as the therapeutic angiogenesis for patients with PAD.

14.2 Waon Therapy

Waon therapy uses a far-infrared-ray dry sauna room, which is evenly maintained at 60 °C and differs from the traditional sauna with high temperature and mist. Figure 14.1 shows how to perform Waon therapy. The patients were placed in a supine or sitting position in the dry room at 60 °C for 15 min and, once removed, allowed to rest on a bed in a supine position with a blanket to keep them warm for an additional 30 min. They were weighed before and after Waon therapy, and oral hydration with water was used to compensate for weight loss by perspiration [3, 13].

We have treated many CHF patients with Waon therapy, and none of the patients so far have shown any deterioration in their condition. However, Waon therapy does not appear to be indicated for CHF patients with severe aortic stenosis or obstructive

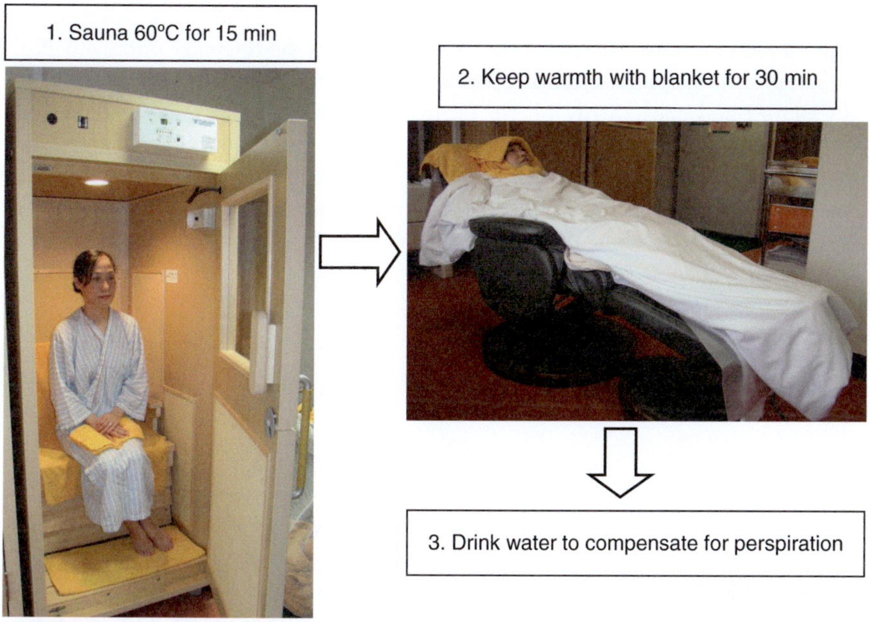

Fig. 14.1 Protocol of Waon therapy. The patients were placed in a supine or sitting position in the dry room at 60 °C for 15 min (1) and allowed to rest on a bed in a supine position with a blanket to keep them warm for an additional 30 min (2). They were weighed before and after Waon therapy, and oral hydration with water was used to compensate for weight loss by perspiration (3)

hypertrophic cardiomyopathy, because the pressure gradient might be increased during Waon therapy. Patients with infectious disease are also excluded from Waon therapy.

14.3 Waon Therapy for CHF

For the better understanding of therapeutic angiogenesis by Waon therapy for PAD, first, we would like to summarize the effects of Waon therapy on CHF and its mechanism.

14.3.1 Effects of Waon Therapy on CHF

Regarding the acute effect of Waon therapy, we have reported that 60 °C dry sauna therapy for 15 min improved acute hemodynamics in patients with CHF, including cardiac index, mean pulmonary wedge pressure, systemic and pulmonary resistance, and cardiac function [4]. Subsequently, we examined the chronic effects of repeated Waon therapy on clinical symptoms and cardiac function in patients with CHF and have reported that 4 weeks of Waon therapy significantly improved clinical symptoms, increased ejection fraction, and decreased cardiac size on the echocardiogram and chest X-ray [4, 5]. Furthermore, we demonstrated that repeated Waon therapy decreased the ventricular arrhythmias and the oxidative stress and improved the autonomic nervous activity and the prognosis in patients with CHF [14–17]. In two prospective multicenter studies, we have confirmed that Waon therapy is safe, improves clinical symptoms and cardiac function, and decreases cardiac size in CHF patients [18, 19].

14.3.2 Mechanism of the Effect of Waon Therapy on CHF

We then investigated the vascular endothelial function to clarify the mechanisms of the effect of Waon therapy on CHF, since vascular endothelial function had been reported to be impaired in CHF. We have reported that 2 weeks of a daily Waon therapy significantly reduced brain natriuretic peptide (BNP) concentrations and improved endothelial function in patients with CHF. There was a significant correlation between the change in flow-mediated dilatation (%FMD) and the percent improvement in BNP concentrations suggesting that Waon therapy improves cardiac function with the betterment of vascular function [6].

In order to confirm the effect of Waon therapy on CHF and clarify its mechanism, we performed experimental studies using TO-2 cardiomyopathic hamsters with heart failure. We reported that the repeated Waon therapy improved survival in TO-2 cardiomyopathic hamsters with heart failure [20]. Furthermore, we clarified that one of the molecular mechanisms by which repeated Waon therapy improved endothelial function was an increase in mRNA and protein of endothelial nitric oxide synthase (eNOS) in Syrian golden hamsters and TO-2 cardiomyopathic hamsters [8, 9]. We believe that eNOS upregulation induced by repeated Waon therapy is caused by an increase in cardiac output and blood flow, which in turn results in increased shear stress, although thermal stimulation might upregulate arterial eNOS directly (Fig. 14.2).

14.4 Waon Therapy Upregulates eNOS and Augments Angiogenesis

NO, constitutively produced by eNOS, plays roles in angiogenesis. Therefore, we drew attention to the effect of NO induced by Waon therapy on angiogenesis. Having reported that Waon therapy upregulated the expression of arterial eNOS in

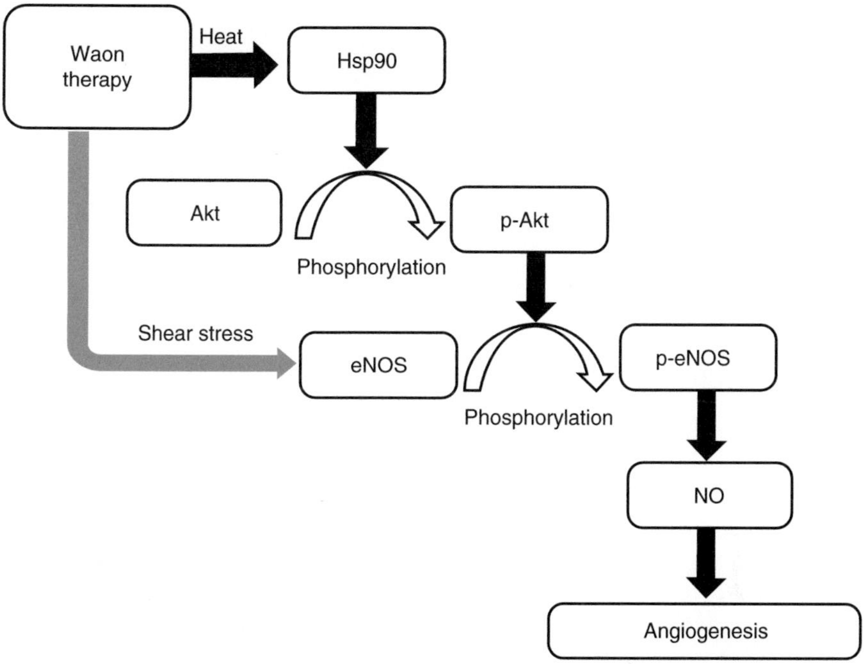

Fig. 14.2 Underlying pathway of angiogenesis upregulated by Waon therapy. Repeated Waon therapy upregulates eNOS expression through shear stress of the vessel wall and Hsp90. *Hsp* heat shock protein, *NO* nitric oxide, *eNOS* endothelial nitric oxide synthase

hamsters, we investigated whether this therapy increased angiogenesis in mice with hindlimb ischemia. Unilateral hindlimb ischemia was induced in apolipoprotein E-deficient mice divided into control and Waon therapy groups. The latter mice were placed in a far-infrared dry sauna at 41 °C for 15 min and then at 34 °C for 20 min, which increases the core temperature by 1 °C and is the same setting as patients, once a day for 5 weeks. Laser Doppler perfusion imaging demonstrated that the ischemic limb/normal side blood flow ratio in Waon therapy group was significantly increased beyond that in controls (Fig. 14.3). Significantly greater capillary density was seen in Waon therapy group. Western blotting showed Waon therapy markedly increased hindlimb eNOS expression (Fig. 14.4). To study possible involvement of eNOS in thermally induced angiogenesis, we gave Waon therapy to mice with hindlimb ischemia with or without N^G-nitro-L-arginine methyl ester (L-NAME) administration for 5 weeks. L-NAME treatment abolished angiogenesis induced by Waon therapy. In addition, Waon therapy did not increase angiogenesis in eNOS-deficient mice. From these results, we concluded that eNOS is a critical regulator of angiogenesis by Waon therapy [11].

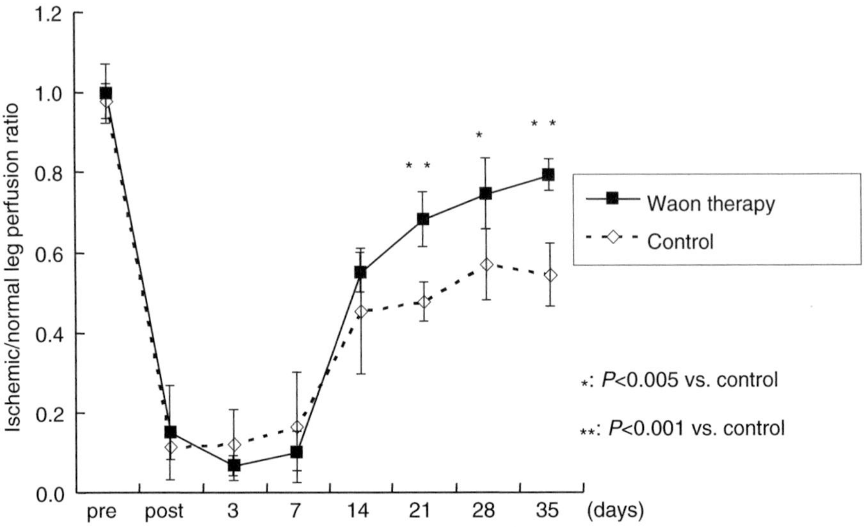

Fig. 14.3 Time course of ischemic/normal leg blood perfusion ratio in apolipoprotein E-deficient mice monitored by laser Doppler perfusion imaging. Computer-assisted quantitative analysis demonstrated a significantly greater increase of ischemic/normal hindlimb perfusion ratio in the Waon therapy group ($n = 7$) than in the control group ($n = 7$). $^*P < 0.005$, $^{**}P < 0.001$ vs. control group. Adapted from [11]

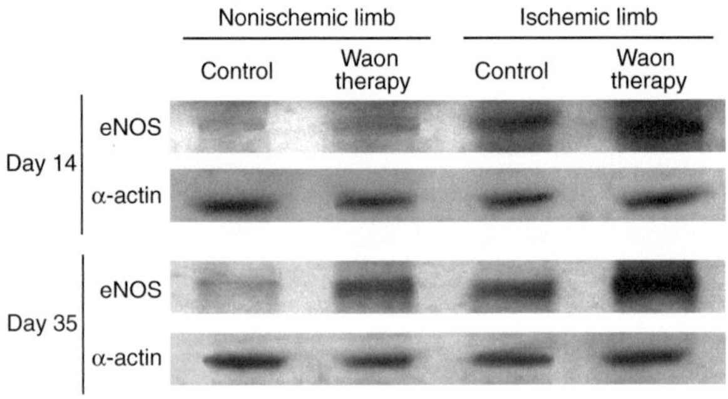

Fig. 14.4 Western blotting of eNOS protein in the muscle of mouse hindlimb. Waon therapy markedly increased eNOS protein expression not only in the ischemic hindlimb but also in the nonischemic hindlimb at days 14 and 35. *eNOS* endothelial nitric oxide synthase. Adapted from [11]

14.4.1 Heat Shock Protein and Akt/eNOS/NO Pathway

Using the same unilateral hindlimb ischemia model, we demonstrated that heat shock protein 90 (Hsp90), phosphorylated-Akt, and phosphorylated-eNOS were detected in arterial endothelial cells of ischemic hindlimbs and all were upregulated by Waon therapy. The effect of Waon therapy on angiogenesis through the activation of the Hsp90/Akt/eNOS pathway was attenuated by the administration of an Hsp90 inhibitor [21]. Therefore, we concluded that Waon therapy upregulates Hsp90 which contributes to the activation of the Akt/eNOS/NO pathway and induces angiogenesis in mice with hindlimb ischemia (Fig. 14.2).

14.5 Effects of Waon Therapy on PAD Patients

We evaluate the effect of repeated Waon therapy on 20 patients with PAD, including 15 patients with bilateral limb ischemia and 5 patients with unilateral limb ischemia [12]. Waon therapy was performed once a day for 5 days per week for a period of 10 weeks. All patients completed the study without any adverse events.

Leg pain was scored by a visual analogue scale, using a marked 10-cm line extending from "no pain, 0," to "severe pain, 10." The pain score significantly decreased after 10 weeks of Waon therapy (Fig. 14.5a). Exercise performance, evaluated by the 6-min walking distance, improved in all 18 patients after 10 weeks of Waon therapy (Fig. 14.5b). We measured ankle-brachial pressure index (ABI) in 31 limbs, and the mean ABI significantly increased after Waon therapy (Fig. 14.5c). Serial assessment of leg blood flow was performed with a laser Doppler imaging. It showed that blood flow increased after 10 weeks (Fig. 14.5d). Digital subtraction angiography was performed to evaluate new collateral vessel formation in 20 legs (13 patients), which showed a dramatic increase in the visible collateral vessels in

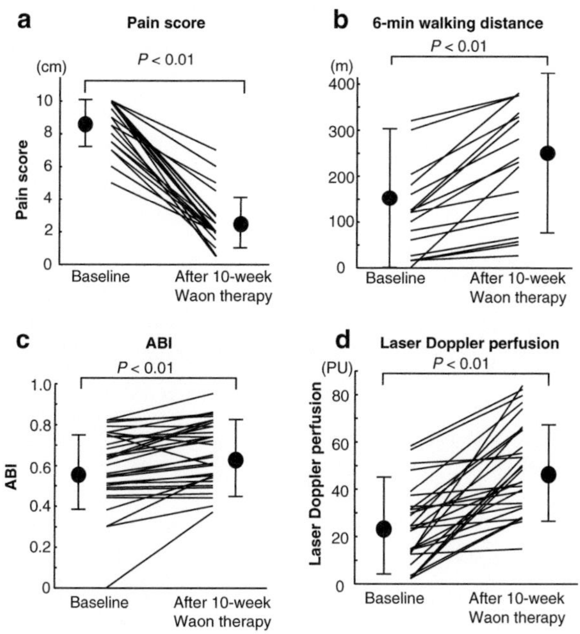

Fig. 14.5 Effect of Waon therapy on PAD patients. (**a**) Improvement in pain score evaluated by a visual analogue scale (10, severe pain; 0, no pain) after 10 weeks of Waon therapy. $n = 20$, $*P < 0.01$ vs. baseline. (**b**) Improvement in 6-min walking distance after 10 weeks of Waon therapy. $n = 18$, $*P < 0.01$ vs. baseline. (**c**) Improvement in ankle-brachial pressure index (ABI), after 10 weeks of Waon therapy. $n = 31$, $*P < 0.01$ vs. baseline. (**d**) Improvement in laser Doppler perfusion imaging after 10 weeks of Waon therapy. $n = 28$, $*P < 0.01$ vs. baseline. Adapted from [12]

12 ischemic legs after 10 weeks of Waon therapy. In addition, ischemic ulcers healed in all of seven limbs, resulting in successful limb salvage.

We described one of the impressive patients [22]. Although this patient had undergone femoropopliteal bypass surgery, all toes on his right foot, except for the fifth toe, had to be amputated. Furthermore, the patient developed a severe foot ulcer with intolerable pain, and therefore a below-knee amputation was thus being considered. He received Waon therapy without any changes in medication. The skin ulcer healed completely in 15 weeks, limb amputation was avoided, and he was discharged. After being discharged, he continued to receive Waon therapy twice per week at our outpatient clinic and no recurrence of the skin ulcer has been seen during 5 years of follow-up (Fig. 14.6). Patients followed at our out-patient clinic after

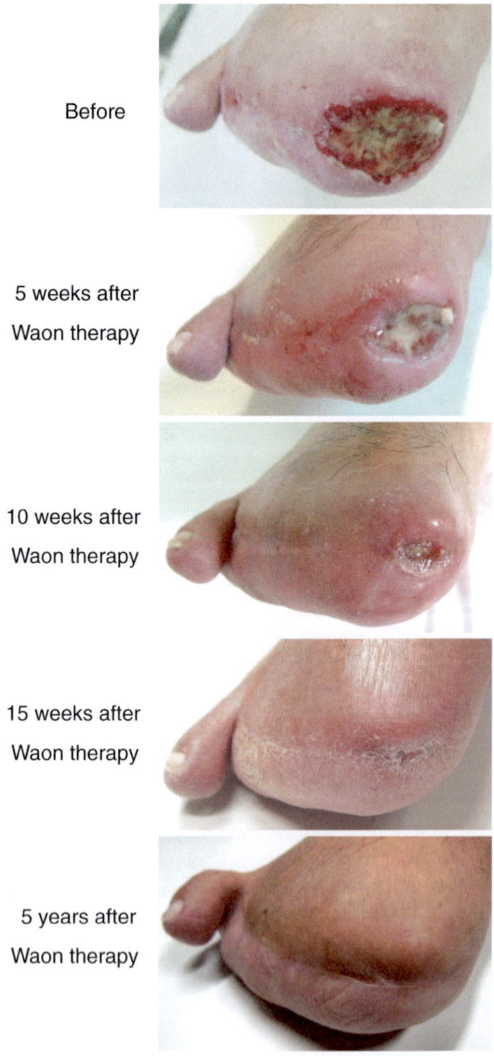

Fig. 14.6 Limb salvage by Waon therapy. The skin ulcer healed completely in 15 weeks after Waon therapy, and he was discharged. After being discharged, he continued to receive thermal therapy twice per week and no recurrence of the skin ulcer has been seen during 5 years of follow-up

discharge continued Waon therapy at least twice per week, and none of them showed the worsening symptoms of PAD. Therefore, to maintain the effect of Waon therapy, we believe that it should be continued at least twice per week after discharge.

14.6 Mechanism of the Effect of Waon Therapy on PAD

To investigate the underlying mechanism of Waon therapy for the treatment of patients with PAD, we determine whether Waon therapy can mobilize blood-derived progenitor cells [23]. Twenty-one consecutive PAD patients received standard medications and were randomly divided into control ($n = 10$) and Waon therapy groups ($n = 11$). The Waon therapy group received Waon therapy daily for 6 weeks. Frequency of circulating CD34+ cells increased after 6 weeks of Waon therapy, while it remained unchanged in the control group (Fig. 14.7). We confirmed serum nitrate and nitrite levels increased significantly after Waon therapy, but not in the control group. These results suggested that Waon therapy mobilized circulating endothelial progenitor cells and improved limb ischemia in patients with PAD.

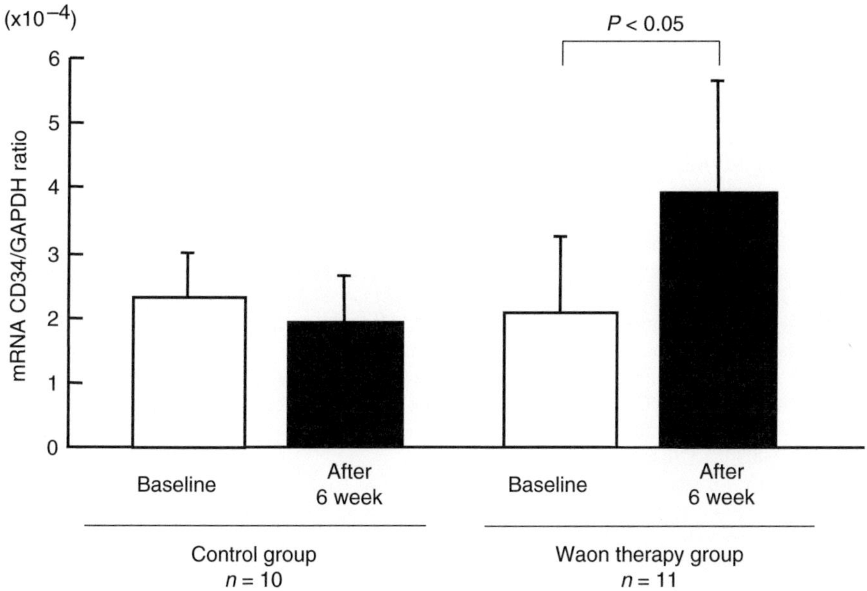

Fig. 14.7 Changes in frequency of mobilized circulatory CD34+ cells. Frequency of circulating CD34+ cells increased after 6 weeks of Waon therapy, while it remained unchanged in the control group. *Open bars*, at baseline; *solid bars*, after 6 weeks; *GAPDH* glyceraldehyde-3-phosphate dehydrogenase. Adapted from [23]

14.7 Conclusion

We reported that angiogenesis was induced via eNOS using Waon therapy in mice with hindlimb ischemia. Repeated Waon therapy upregulates eNOS expression through shear stress of the vessel wall and Hsp90. Furthermore, we demonstrated that repeated Waon therapy improved symptoms, status, and blood flow in patients with PAD. Waon therapy therefore is the noninvasive and highly promising strategy for treating PAD with therapeutic angiogenesis.

References

1. Norgren L, Hiatt WR, Dormandy JA, Nehler MR, Harris KA, Fowkes FG, on behalf of the TASC II Working Group. Inter-society consensus for the management of peripheral arterial disease (TASC II). J Vasc Surg. 2007;45:S5A–67A.
2. Isner JM, Asahara T. Angiogenesis and vasculogenesis as therapeutic strategies for postnatal neovascularization. J Clin Invest. 1999;10:1231–6.
3. Tei C. Waon therapy: soothing warmth therapy. J Cardiol. 2007;49:301–4.
4. Tei C, Horikiri Y, Park JC, Jeong JW, Chang KS, Toyama Y, et al. Acute hemodynamic improvement by thermal vasodilation in congestive heart failure. Circulation. 1995; 91:2582–90.
5. Tei C, Tanaka N. Thermal vasodilation as a treatment of congestive heart failure: a novel approach. J Cardiol. 1996;27:29–30.
6. Kihara T, Biro S, Imamura M, Yoshifuku S, Takasaki K, Ikeda Y, et al. Repeated thermal therapy treatment improves vascular endothelial and cardiac function in patients with chronic heart failure. J Am Coll Cardiol. 2002;39:754–9.
7. Imamura M, Biro S, Kihara T, Yoshifuku S, Takasaki K, Otsuji Y, et al. Repeated thermal therapy improves impaired vascular endothelial function in patients with coronary risk factors. J Am Coll Cardiol. 2001;38:1083–8.
8. Ikeda Y, Biro S, Kamogawa Y, Yoshifuku S, Eto H, Orihara K, et al. Repeated thermal therapy upregulates arterial endothelial nitric oxide synthase expression in Syrian golden hamsters. Jpn Circ J. 2001;65:434–8.
9. Ikeda Y, Biro S, Kamogawa Y, Yoshifuku S, Eto H, Orihara K, et al. Repeated sauna therapy increases arterial endothelial nitric oxide synthase expression and nitric oxide production in cardiomyopathic hamsters. Circ J. 2005;69:722–9.
10. Aicher A, Heeschen C, Mildner-Rihm C, Urbich C, Ihling C, Technau-Ihling K, et al. Essential role of endothelial nitric oxide synthase for mobilization of stem and progenitor cells. Nat Med. 2003;9:1370–6.
11. Akasaki Y, Miyata M, Eto H, Shirasawa T, Hamada N, Ikeda Y, et al. Repeated thermal therapy up-regulates endothelial nitric oxide synthase and augments angiogenesis in a mouse model of hindlimb ischemia. Circ J. 2006;70:463–70.
12. Tei C, Shinsato T, Miyata M, Kihara T, Hamasaki S. Waon therapy improves peripheral arterial disease. J Am Coll Cardiol. 2007;50:2169–71.
13. Miyata M, Tei C. Waon therapy for cardiovascular disease: innovative therapy for the 21st century. Circ J. 2010;74:617–21.
14. Kihara T, Biro S, Ikeda Y, Fukudome T, Shinsato T, Masuda A, et al. Effects of repeated sauna treatment on ventricular arrhythmias in patients with chronic heart failure. Circ J. 2004;68: 1146–51.

15. Fujita S, Ikeda Y, Miyata M, Shinsato T, Kubozono T, Kuwahata S, et al. Effect of Waon therapy on oxidative stress in chronic heart failure. Circ J. 2011;75:348–56.
16. Kuwahata S, Miyata M, Fujita S, Kubozono T, Shinsato T, Ikeda Y, et al. Improvement of autonomic nervous activity by Waon therapy in patients with chronic heart failure. J Cardiol. 2011;57:100–6.
17. Kihara T, Miyata M, Fukudome T, Ikeda Y, Shinsato T, Kubozono T, et al. Waon therapy improves the prognosis of patients with chronic heart failure. J Cardiol. 2009;53:214–8.
18. Miyata M, Kihara T, Kubozono T, Ikeda Y, Shinsato T, Izumi T, et al. Beneficial effects of Waon therapy on patients with chronic heart failure: results of a prospective multicenter study. J Cardiol. 2008;52:79–85.
19. Tei C, Imamura T, Kinugawa K, Inoue T, Masuyama T, Inoue H, et al. Waon therapy for managing chronic heart failure: results from a multicenter prospective randomized WAON-CHF Study. Circ J. 2016;80:827–34.
20. Ikeda Y, Biro S, Kamogawa Y, Yoshifuku S, Kihara T, Minagoe S, et al. Effect of repeated sauna therapy on survival in TO-2 cardiomyopathic hamsters with heart failure. Am J Cardiol. 2002;90:343–5.
21. Miyauchi T, Miyata M, Ikeda Y, Akasaki Y, Hamada N, Shirasawa T, et al. Waon therapy upregulates Hsp90 and leads to angiogenesis through the Akt-eNOS pathway in mouse hindlimb ischemia. Circ J. 2012;76:1712–21.
22. Tei C, Shinsato T, Kihara T, Miyata M. Successful thermal therapy for end-stage peripheral artery disease. J Cardiol. 2006;47:163–4.
23. Shinsato T, Miyata M, Kubozono T, Ikeda Y, Fujita S, Kuwahata S, et al. Waon therapy mobilizes CD34+ cells and improves peripheral arterial disease. J Cardiol. 2010;56:361–6.

Chapter 15
Exercise

Tatsuya Maruhashi, Yasuki Kihara, and Yukihito Higashi

Abstract Supervised exercise training has been recommended as first-line therapy for treatment of peripheral artery disease (PAD) patients with intermittent claudication. Capillary growth in skeletal muscle by angiogenesis is thought to be one of the possible mechanisms underlying the improvement of exercise performance and walking ability by exercise training in patients with PAD. Under normal conditions, exercise training has been demonstrated to increase the capillarity of active muscle. Exercise-induced increase in blood flow and tissue stretch has been proposed as important stimuli for angiogenesis in skeletal muscle. It is postulated that an increase in shear stress with increased blood flow and passive stretch primarily induces angiogenesis by longitudinal division of the vessel and sprouting angiogenesis, respectively. Vascular endothelial growth factor (VEGF) is recognized as the central pro-angiogenic factor strongly associated with angiogenesis. Exercise training induces angiogenesis through activation of various factors, including nitric oxide (NO), hypoxia-inducible factor-1α, and peroxisome proliferator-activated receptor-γ coactivator 1α (PGC-1α), through VEGF expression. However, PAD has been reported to be associated with reduced NO bioavailability represented by endothelial dysfunction or reduced PGC-1α expression in skeletal muscle. Therefore, it remains unclear whether similar mechanisms or angiogenic factors are involved in exercise-induced angiogenesis under normal and pathological conditions.

Keywords Exercise training • Angiogenesis • Peripheral artery disease • Skeletal muscle

---

T. Maruhashi (✉) • Y. Kihara
Department of Cardiovascular Medicine, Graduate School of Biomedical and Health Sciences, Hiroshima University, 1-2-3 Kasumi, Minami-ku, Hiroshima 734-8551, Japan
e-mail: 55maruchin@gmail.com

Y. Higashi
Department of Cardiovascular Regeneration and Medicine, Research Institute for Radiation Biology and Medicine, Hiroshima University,
1-2-3 Kasumi, Minami-ku, Hiroshima 734-8553, Japan

Y. Higashi, T. Murohara (eds.), *Therapeutic Angiogenesis*,
DOI 10.1007/978-981-10-2744-4_15

15.1 Exercise Training for Peripheral Artery Disease (PAD) Patients with Intermittent Claudication

In patients with PAD, intermittent claudication is the classical and main symptom characterized by muscular discomfort, including fatigue, aching, numbness, and cramping, in the lower limb that is reproducibly manifested by exercise and relieved by a short rest. Intermittent claudication is associated with functional impairment, including limited exercise performance and walking ability, leading to reduced physical activity and quality of life. Therefore, one of the main goals of treatment for PAD patients with intermittent claudication is to relieve symptoms and improve exercise performance and daily functional activity. In addition to risk factor modification and pharmacotherapy, supervised exercise training has been recommended as first-line therapy for PAD patients with intermittent claudication [1]. Compared with nonsupervised exercise training, it was shown that supervised exercise training significantly increased maximal treadmill walking distance by 50–200% in patients with PAD [2–4]. In addition, a randomized controlled clinical trial demonstrated that supervised exercise training and endovascular revascularization equally improved maximum walking distance after 6 and 12 months in PAD patients with intermittent claudication [5].

Although the mechanisms underlying the improvement in exercise performance and walking ability by supervised exercise training are still not completely understood, increase in collateral flow, improvement in endothelial function, enhancement of skeletal muscle metabolism and mitochondrial function, and increase in capillarity of skeletal muscle by the process of angiogenesis are thought to be the possible mechanisms of the beneficial effect of exercise training in patients with PAD. In this review, the current understanding of exercise-induced angiogenesis in skeletal muscle under normal and pathological conditions is discussed.

15.2 Importance of Angiogenesis in Skeletal Muscle in Response to Muscle Activity

15.2.1 Peripheral Adaptation in Response to Exercise

In addition to central cardiovascular hemodynamic factors, such as cardiac output, peripheral mechanisms for oxygen uptake and utilization, such as peripheral oxygen delivery, distribution, extraction, and mitochondrial respiration, are major determinants of exercise tolerance. Although peak oxygen uptake is increased after exercise training during constant oxygen supply to skeletal muscle, it is decreased after immobilization, indicating a role for peripheral factors in oxygen uptake in skeletal muscle and performance during exercise [6, 7]. Saltin et al. [8] demonstrated that after one-legged endurance training, peak oxygen uptake during exercise with the trained leg is higher than that with the contralateral untrained leg.

The capacity of central hemodynamics to perfuse the exercising leg is not enhanced by one-legged exercise. Therefore, these results demonstrate that peripheral factors are important determinants of skeletal muscle oxygen uptake and performance during exercise and that peripheral adaptations mediated by increases in oxygen extraction and utilization may contribute to the increased exercise capacity [8]. The capacity of maximum performance depends not only on central cardiovascular hemodynamic factors but also on peripheral adaptations for oxygen extraction and utilization through increasing capillarity and oxidative capacity.

Both oxidative activity and capillarity in peripheral tissues are enhanced by endurance training. Peripheral adaptations in response to exercise include increase in capillarization and increases in mitochondria content and enzymatic activity involved in oxidative and glycolytic metabolism in skeletal muscle, leading to improvement in oxygen diffusion capacity and utilization [9]. A capillary serves as the interface between blood and tissue for delivery of oxygen and removal of metabolites. Therefore, it is postulated that increased capillarity improves oxygen diffusion and metabolite clearance capacity by increasing the surface area, reducing the diffusion distance and increasing mean transient time, leading to improvement in endurance and aerobic capacity. Krogh demonstrated the importance of the capillary bed for oxygen delivery to muscle in his pioneering work showing that the capillary and muscle interface represents a limiting factor for oxygen supply [10, 11]. It has also been reported that there is a significant correlation between exercise performance or maximal aerobic capacity and skeletal muscle capillary density in animals [12].

15.2.2 Capillarization in Response to Exercise

Exercise training has been demonstrated to increase capillarity of active skeletal muscle both in animal studies [13–15] and human studies. Although differences in the degree of exercise-induced increase in capillarization assessed by capillary-to-fiber ratio exist among the studies, endurance training for 4–24 weeks increases capillarization by 10–40% [16–21]. It has also been reported that exercise-induced increase in capillarization rapidly disappears after cessation of training [22, 23]. Although there is little information on the relationships of degree of increase in capillarization with training intensity and duration, low-intensity training is minimally effective for increase in capillarity. In addition, high-intensity resistance training, such as weight lifting, often results in decreased capillarity as a result of fiber hypertrophy without angiogenesis [24, 25]. Moderate-intensity and high-intensity training have been reported to result in similar increases in skeletal muscle capillarizaion after 4–8 weeks [20, 22, 26, 27]. Recent animal studies using other techniques, such as electron microscopy and measurements of activated endothelial cells labeled with bromodeoxyuridine, for the assessment of angiogenic response to muscle activity support a role of exercise training for increase in skeletal muscle capillarity [28, 29]. These techniques have also provided mechanistic insights into capillarization in response to various stimuli to skeletal muscle.

15.2.3 Angiogenesis in Response to Exercise

Angiogenesis is defined as the formation of new capillaries from preexisting capillaries in response to alterations in a mechanical, metabolic, or hypoxic environment. The vasculature is normally quiescent in adults. However, the occurrence of angiogenesis is observed in both physiological and pathological conditions, such as during the proliferative phase of the menstrual cycle in the uterus, in the processes of wound healing and tumor progression, and in the adaptation of skeletal muscle to exercise training. Angiogenesis is a very complex process strictly regulated by a balance between pro- and anti-angiogenic factors through increases in synthesis or release from storage sites (Fig. 15.1). The growth process of angiogenesis is broadly divided into three categories: sprouting angiogenesis, longitudinal splitting, and intussusception [30].

Sprouting angiogenesis is thought to be the most common form of capillary growth. It is thought to be initiated by proteolytic degradation of the basement membrane surrounding a capillary, which allows endothelial cells to branch out from the parent vessel, migrate into the extracellular matrix, develop a lumen, and connect to existing vessels or other sprouts to form anastomoses. The newly formed vessel is initially leaky and subsequently becomes matured and functional through reconstruction of a basement membrane and investment by pericytes. Harmonization of several energetic processes, including degradation of the basement membrane and extracellular matrix, proliferation and migration of endothelial cells, and maturation of the new capillary, is required for sprouting angiogenesis.

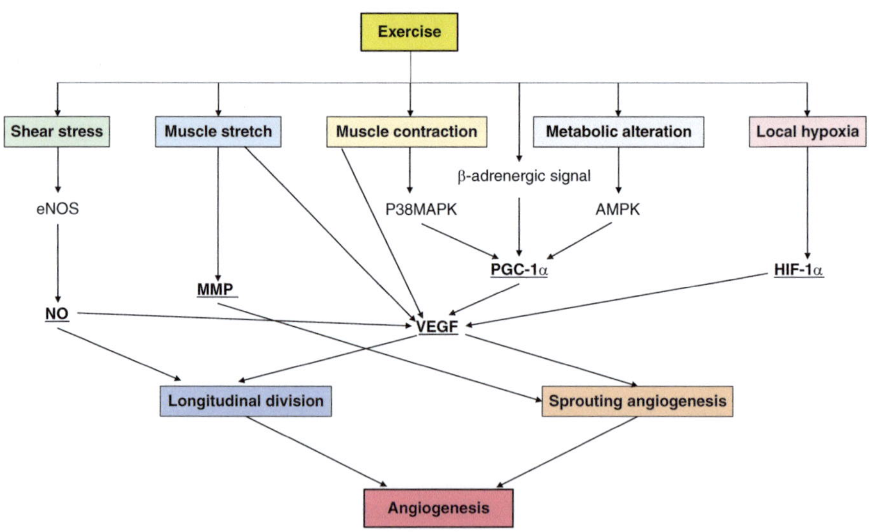

Fig. 15.1 Putative mechanisms of exercise-induced angiogenesis in skeletal muscle. *NO* indicates nitric oxide, *eNOS* endothelial nitric oxide synthase, *VEGF* vascular endothelial growth factor, *MMP* matrix metalloproteinase, *AMPK* AMP-activated kinase, *PGC-1α* peroxisome proliferator-activated receptor-γ coactivator 1α, *HIF-1α* hypoxic-inducible factor-1α

Longitudinal splitting refers to a process in which an activated endothelial cell extends a filopodial ridge into the lumen to split an existing vessel into two functionally separate vessels. The extending filopodium makes contact with the opposite surface of the vessel, and a cellular bridge across the lumen is consequently formed. This bridge continues to extend along the lumen, and two parallel flow channels are created [31]. This internal separation wall remodels and eventually separates the existing vessel into two distinct functional capillaries.

Intussusception also refers to an angiogenic process in which an existing vessel splits into two separate vessels. Intussusception is initiated by inward protrusion of opposing vessel walls from the abluminal surface into the vessel lumen. This process pinches the vessel wall and results in contact with each opposite wall, leading to the formation of a contact zone between endothelial cells. Then endothelial bilayer is perforated centrally and invaded by interstitial tissue, such as fibroblasts, pericytes, and collagen fibrils, resulting in the creation of a transluminal pillar and separation of the existing vessel into two lumens [32].

Although there is a possibility that longitudinal splitting and intussusception represent different aspects of the same angiogenic process, both types of capillary remodeling eventually separate a preexisting vessel into two functional parallel capillaries, resulting in redistribution of blood flow and a consequent decrease in shear stress. Compared with sprouting angiogenesis, capillary growth by longitudinal splitting and intussusception seems to be a faster, less complex, and less energy-demanding process, involving less degradation of the basement membrane and extracellular matrix, less proliferation of endothelial cells, and less compromised function of the capillary during the growth process [32, 33].

15.3 Physiological Mechanical Factors Related to Angiogenesis in Skeletal Muscle

Exercise-induced increases in blood flow and tissue stretch have been proposed as important stimuli for angiogenesis in skeletal muscle [12, 34]. Endothelial cells are exposed to increased shear stress, transmural pressure, and cyclic stretch by exercise. These mechanical stimuli activate mechanosensors on the endothelial cell surface, including integrins, receptor tyrosine kinase, G proteins and G protein-coupled receptors, Ca^{2+} channel, membrane lipids, and glycocalyx, consequently activating signal transduction involving phosphorylation of kinases and Rho family GTPases [35]. In endothelial cells, activation of these mechanotransduction pathways results in alterations in the expression of genes related to vascular growth, regulation of proliferation and migration, and rearrangement of the actin cytoskeleton. Increases in internal shear stress and external stretch, two main stimuli to the endothelium, are thought to promote physiological angiogenesis by different mechanisms [30]. It is thought that increased shear stress induces capillary growth by longitudinal division, whereas passive stretch induces capillary growth by sprouting angiogenesis [30, 36].

15.3.1 Shear Stress

Shear stress is a frictional force produced by blood flow that acts on the luminal surface of the endothelium. A normal range of shear stress plays a pivotal role in maintenance of vascular health by promoting the activation of vasodilatory and anti-thrombotic factors and inhibiting the activation of growth and pro-inflammatory factors [37–39]. In order to maintain shear stress in an appropriate range, arteries and arterioles alter their vessel diameter in a feedback manner, referred to as flow-mediated vasodilation, through various signaling pathways activated by shear stress in response to blood flow. Therefore, arteries and arterioles tend to become larger in response to elevated blood flow to maintain shear stress in a normal range. However, capillaries of skeletal muscle are less capable of modulating their diameter in response to elevated blood flow due to limited compliance. Therefore, the endothelium in capillaries is exposed to a continuously elevated shear stress induced by chronic vasodilation of arterioles. Chronic administration of the vasodilator dipyridamole has been shown to induce endothelial cell proliferation and increase the number of capillaries in skeletal muscle [40]. Similarly, chronic administration of the α_1-adrenergic receptor blocker prazosin, which induces vasodilation and high shear stress, has been shown to increase the number of capillaries in rat skeletal muscle [41]. Electron microscopic findings have further revealed that capillary growth in response to high shear stress induced by elevated blood flow occurs by longitudinal division of the vessel by filopodial extensions [30]. Capillaries adapt to chronically elevated shear stress by formation of new capillaries through the process of longitudinal division. Capillary remodeling by longitudinal division results in lower shear stress by redistributing blood flow.

However, it was shown that pump-controlled high blood flow without muscle contraction (passive hyperperfusion) resulted in no increase in VEGF or fibroblast growth factor-2 gene expression in skeletal muscle of the canine leg, whereas electrical nerve stimulation, which induces both blood flow increase and muscle contraction, caused an increase in VEGF mRNA [42]. These findings suggest that an increase in blood flow must be coupled with other factors, such as low tissue oxygen tension and metabolite concentration changes, or an increase in muscle contraction for the increase in mRNA expression of angiogenic factors. Nitric oxide (NO) produced by endothelial NO synthase (eNOS) activation through shear stress appears to play a pivotal role in the process of capillary growth by longitudinal division induced by increased blood flow. Angiogenesis via longitudinal division in response to increased blood flow induced by administration of prazosin was inhibited by co-treatment with an NOS inhibitor or by genetic knockout of eNOS, indicating that shear stress-induced eNOS activation and consequent NO production are necessary for angiogenesis via longitudinal division [43]. Matrix metalloproteases (MMPs) are thought to mediate proteolytic degradation of the basement membrane surrounding a capillary, which may be necessary for the process of sprouting angiogenesis but not required for luminal division during capillary growth. Previous studies demonstrated that production of MMPs is not increased in microvascular endothelial cells exposed to high shear stress in culture [44, 45] and in an in vivo model of prazosin-induced angiogenesis in rat skeletal muscle [44, 46], suggesting that shear stress induces angiogenesis not via sprouting angiogenesis but via longitudinal division.

15.3.2 Muscle Stretch

Muscle stretch is another potent angiogenic stimulus as a physiological mechanical factor. Capillaries are attached to the extracellular matrix and connective tissue components. Therefore, capillaries are tethered to myocytes during muscle contraction, relaxation, or lengthening through surrounding tissues, resulting in application of tensile forces to the capillaries. Application of stretch to cultured endothelial cells promotes upregulation of angiogenic factors, including VEGE and angiopoetins, activation of MMPs, and increase in migration and proliferation of endothelial cells [47–49]. Muscle stretch evoked by surgical extirpation of a functional synergist induced capillary growth [29, 50]. In that animal model, the expression of VEGF and MMPs was upregulated, and electron microscopic observations showed signs of basement membrane degradation, migration and proliferation of endothelial cells, and capillary sprouting. These findings suggest that capillary growth provoked by muscle stretch is primarily induced by sprouting angiogenesis [30, 43]. However, in this experimental animal model, the possibility of effects of additional metabolic signals induced by prolonged muscle overload cannot be excluded, making it difficult to establish a distinctive role of physiological mechanical stretch alone in angiogenesis.

In humans, a passive movement model is used to examine the role of physiological mechanical factors alone in capillary growth in skeletal muscle without the effect of metabolic factors. Passive limb movement by extension and flexion of the knee leads to an increase in blood flow-induced shear stress and stretch of muscle fibers with a negligible effect on muscle metabolism [51]. Ninety minutes of passive leg movement at a rate of 60 cycles/min has been demonstrated to increase interstitial VEGF protein concentration and eNOS mRNA levels in human skeletal muscle, suggesting that physiological mechanical stimuli alone can induce an angiogenic response without significant alterations of metabolic factors [51]. In addition, 4-week passive leg movement has been demonstrated to increase proliferation of endothelial cells and upregulation of angiogenic factors [52], supporting a role of physiological mechanical factors in the initiation of angiogenesis. However, compared with active exercise training, angiogenic signals evoked by passive movement appear not to be strong enough to induce an increase in capillarization. A 4-week period of passive movement training did not lead to a significant increase in capillary growth in healthy young subjects [52]. Alterations in metabolism and/or local oxygen tension appear to be necessary for exercise-induced angiogenesis in humans.

15.4 Local Hypoxia and Metabolic Alterations Related to Angiogenesis in Skeletal Muscle

Reduced oxygen tension and/or elevated metabolic activity have been shown to be related to exercise-induced angiogenesis [53]. Angiogenesis may lead to enhancement of oxygen delivery to skeletal muscle. Therefore, it is postulated that a relative lack of oxygen supply or low tissue oxygen tension serves as a key regulator of the process of angiogenesis. Several animal and human studies have demonstrated that increase in exercise-induced angiogenesis is augmented in trained skeletal muscle under the condition of restricted oxygen supply by limited blood flow [54, 55]. Compared with exercise in normoxia, exercise under a condition of hypoxia further increased VEGF mRNA expression in rats [56]. However, human studies have yielded conflicting results regarding the effect of exercise under a condition of hypoxia on the expression of VEGF mRNA. Although one study showed a further increase in VEGF mRNA expression by exercise under a condition of limited oxygen supply to the muscle by blood flow restriction compared with exercise under a normal condition, other studies have shown that there was no further increase in VEGF mRNA expression during exercise under a hypoxic condition compared with exercise under a normal condition [57, 58].

15.4.1 Hypoxia-Inducible Factor-1 (HIF-1)

HIF-1 is known to activate transcription of VEGF under hypoxic conditions. Under a condition of reduced oxygen availability, enzymatic activities of prolyl hydroxylase domain proteins (PHDs) and factor-inhibiting HIF-1 (FIH-1) are reduced, and then hydroxylation of HIF-1α by PHD and FIH-1 and degradation of HIF-1α by the proteasome are consequently inhibited, leading to stabilization and accumulation of HIF-α. HIF-α dimerizes with constitutively expressed HIF-1β, recruits p300, and binds to hypoxia response elements, resulting in upregulation of VEGF expression [59]. Exercise induces a significant reduction in intramuscular oxygen pressure [60], which is lowered enough to activate HIF-1α. Ameln et al. [61] demonstrated that HIF-1α protein level was increased after a 45-min bout of one-legged knee-extension exercise, indicating a role of exercise-induced hypoxia in the process of angiogenesis through an increase in HIF-1α protein level. However, they also reported that there was no further increase in exercise-induced expression of HIF-1α protein in a hypoxic condition with restriction of blood flow compared with exercise in a normal condition without restriction, whereas VEGF mRNA level was further increased by exercise training in a hypoxic condition compared with that in a normal condition [61]. These findings suggest that not HIF-1α but other transcriptional factors are associated with the further increase in VEGF mRNA levels induced by exercise under a hypoxic condition. As for the insignificant difference between HIF-1α protein levels with exercise training under hypoxic and normal condition, further studies are needed to determine whether a further reduction in oxygen tension induced by blood flow restriction is too small to promote further increases in HIF-1α protein levels and whether activation of HIF-mediated angiogenesis occurs in a threshold-dependent manner and this threshold is already exceeded during exercise under a normal condition.

The role of hypoxia in the process of exercise-induced angiogenesis is still controversial. In an animal study, myocyte-specific HIF-1α knockout mice had a higher baseline capillary-to-fiber ratio than that in control mice, and exercise training did not induce a further increase in the capillary-to-fiber ratio [62]. In human studies, in response to chronic high altitude hypoxia, muscle capillary density is remained unchanged or increased slightly as a result of a decrease in muscle fiber size rather than being due to an increase in capillary growth [63–65]. In addition, HIF-1α protein levels in skeletal muscle were not changed after prolonged exposure to high altitude hypoxia [66]. Involvement of HIF-1α in exercise-induced angiogenesis appears not to be as straightforward as originally expected.

15.4.2 Peroxisome Proliferator-Activated Receptor-γ Coactivator 1α (PGC-1α)

PGC-1α has been identified as one of the key transcriptional coactivators that modify the expression of genes involved in energy metabolism and mitochondrial biogenesis [67]. Recent studies have demonstrated that PGC-1α is also an important regulator of exercise-induced angiogenesis [68]. PGC-1α knockout mice showed lower basal VEGF protein expression and lower capillary-to-fiber ratio in skeletal muscle than those in control mice and showed no increase in VEGF protein after exercise [69]. Myocyte-specific PGC-1α knockout mice also did not show an exercise-induced increase in capillary density [70, 71]. These findings suggest that PGC-1α in skeletal muscle is essential for exercise-induced upregulation of VEGF expression and angiogenesis.

Expression and activity of PGC-1α are regulated by several pathways. p38γ mitogen-activated protein kinase (MAPK) is one of the important mediators of PGC-1α. Myocyte-specific p38γ MAPK knockout mice showed decreases in PGC-1α mRNA and VEGF mRNA with blunted mitochondrial biogenesis and angiogenesis in skeletal muscle following motor nerve stimulation [72]. Increases in cytosolic calcium and activated calcium/calmodulin-dependent kinase during muscle contraction are thought to activate p38 MAPK [73]. AMP-activated kinase (AMPK) is also thought to regulate PGC-1α activity. AMPK acts as an important cellular energy sensor regulating cell metabolism, including exercise-induced changes in glucose and fatty acid metabolism in skeletal muscle [74]. AMPK is activated by contractile activity [75–77]. AMPK induces the expression of PGC-1α mRNA and directly phosphorylates PGC-1α protein [78]. It has been shown that AMPK activation led to an increase in VEGF protein expression in control mice but not in PGC-1α knockout mice, indicating that AMPK stimulates VEGF expression through a PGC-1α-dependent mechanism [69]. AMPK has also been demonstrated to stimulate phosphorylation of p38 MAPK with increases in VEGF mRNA and protein levels in mouse skeletal muscle [79]. Recently, the β-adrenergic signal has been shown to be involved in exercise-induced angiogenesis through PGC-1α activation [80]. In an animal study, β-adrenergic stimulation by exercise or a long-acting β2 agonist increased the expression of PGC-1α mRNA from an alternative promoter (PGC-1α2) with an increase in VEGF mRNA, whereas exercise-induced expression of both PGC-1α and VEGF mRNA was blunted in mice treated with a β-adrenergic receptor blocker [70]. Expression of PGC-1α splice variants induced by exercise was also observed in human skeletal muscle [81, 82]. Translation of alternative PGC-1α2 mRNA produces two PGC-1α variants (PGC-1α2 and PGC-1α3) with a slight difference from the canonical PGC-1α at their NH_2-terminus [70]. PGC-1α2 appears to be the most exercise-responsive PGC-1α isoform, which is rapidly elevated by a single exercise bout in human skeletal muscle [82]. PGC-1α coactivates estrogen-related receptor-α (ERR-α) on conserved binding sites found in an enhancer within the first intron of the VEGF gene, inducing VEGF expression in an HIF-independent fashion [68, 70].

15.5 Exercise-Induced Angiogenesis in Patients with PAD

As discussed above, it has become apparent that angiogenesis is a very complex process that is tightly regulated through the balance between pro- and anti-angiogenic factors. However, it remains unclear whether similar mechanisms or angiogenic factors are involved in exercise-induced angiogenesis under normal and pathological conditions.

Endothelial function is impaired in patients with PAD. Flow-mediated vasodilation in the brachial and popliteal arteries is attenuated in patients with PAD, indicating reduced NO bioavailability in systemic circulation [83–85]. Shear stress-induced NO production in endothelial cells has been shown to be necessary for angiogenesis through the process of longitudinal division. Therefore, this raises a possibility that angiogenesis via longitudinal division in response to increased blood flow induced by exercise does not occur in PAD patients with endothelial dysfunction. However, endothelial function has been demonstrated to be improved by exercise training in patients with PAD [3, 86]. Exercise training is expected to improve endothelial function, leading to an increase in NO bioavailability and exercise-induced capillary growth through longitudinal division in patients with PAD.

VEGF is recognized as the central pro-angiogenic factor strongly associated with angiogenesis. Although there are a number of closely related VEGF isoforms (A-D) that have been characterized in human tissue, VEGF-A has been shown to play a predominant role in skeletal muscle angiogenesis [87]. Several splice variants of human VEGF-A with different molecular weights have been identified, among which $VEGF_{165}$ is the most physiologically active isoform in skeletal muscle

[88, 89]. In patients with PAD, VEGF-A levels have been reported to be paradoxically elevated compared with levels in subjects without PAD [90, 91]. It is unclear why adequate angiogenesis and collateralization do not occur despite increased VEGF-A levels in patients with PAD. Recent studies have revealed the existence of anti-angiogenic splice variants of human VEGF-A, generated from differential splicing of exon 8 [92, 93]. VEGF-A_{165}a protein has a pro-angiogenic effect, whereas VEGF-A_{165}b exerts an anti-angiogenic effect through inhibition of canonical VEGF-A signaling by competing with pro-angiogenic isoforms for binding to the VEGF receptor-2 (VEGFR-2)-neuropilin-1 receptor complex, leading to the attenuation of downstream signaling of VEGFR2. Kikuchi et al. demonstrated that total circulating VEGF-A levels are significantly higher in PAD patients than in control subjects, with increased levels of anti-angiogenic VEGF-A_{165}b splice isoform and corresponding decreased levels of pro-angiogenic VEGF-A_{165}a splice isoform in patients with PAD [94]. Expression of anti-angiogenic VEGF-A_{165}b is upregulated in ischemic tissue as a result of secreted frizzled-related protein 5 deficiency and Wnt5a activation in the setting of metabolic disease. It is unclear whether exercise-induced angiogenesis occurs by similar mechanisms in healthy individuals and PAD patients with dominant anti-angiogenic VEGF-A isoform over pro-angiogenic VEGF-A isoform.

Diabetes mellitus has been demonstrated to be associated with the development of PAD. The incidence of intermittent claudication is about two-times higher in diabetes patients than in non-diabetic patients [95]. As described above, PGC-1α is thought to be an important regulator in the process of exercise-induced angiogenesis. However, in diabetic animal models and human patients with diabetes mellitus, basal expression of PGC-1α mRNA has been reported to be reduced in skeletal muscle [96–98]. Moreover, PGC-1α mRNA expression in skeletal muscle is not induced by acute exercise in patients with type 2 diabetes mellitus [99]. Further studies are needed to investigate the mechanisms underlying exercise-induced angiogenesis in patients with PAD.

15.6 Conclusions

Tissue stretch and increased blood flow induced by exercise training are thought to be important stimuli for angiogenesis in skeletal muscle. Synthesis or release of angiogenic factors is activated by mechanical stimuli in skeletal muscle. VEGF appears to be the central pro-angiogenic factor. VEGF expression is stimulated by exercise training through various factors including NO, HIF-1α, and PGC-1α in a normal condition. Supervised exercise training has been recommended as first-line therapy for PAD patients with intermittent claudication, but the mechanisms underlying the improvement in exercise performance and walking ability by exercise training are not completely understood. Exercise-induced increase in capillarity of skeletal muscle through angiogenesis is thought to be one of the possible mechanisms of the beneficial effect of exercise training in patients with PAD. However, it remains to be elucidated whether similar mechanisms are involved in the process of exercise-induced angiogenesis under normal and pathological conditions. A better-defined understanding of the process of exercise-induced angiogenesis in normal and pathological conditions may lead to novel treatment and prevention of PAD.

References

1. Hirsch AT, Haskal ZJ, Hertzer NR, Bakal CW, Creager MA, Halperin JL, et al. ACC/AHA 2005 Practice Guidelines for the management of patients with peripheral arterial disease (lower extremity, renal, mesenteric, and abdominal aortic): a collaborative report from the American Association for Vascular Surgery/Society for Vascular Surgery, Society for Cardiovascular Angiography and Interventions, Society for Vascular Medicine and Biology, Society of Interventional Radiology, and the ACC/AHA Task Force on Practice Guidelines (Writing Committee to Develop Guidelines for the Management of Patients With Peripheral Arterial Disease): endorsed by the American Association of Cardiovascular and Pulmonary Rehabilitation; National Heart, Lung, and Blood Institute; Society for Vascular Nursing; TransAtlantic Inter-Society Consensus; and Vascular Disease Foundation. Circulation. 2006;113(11):e463–654. doi:10.1161/CIRCULATIONAHA.106.174526.
2. Fakhry F, van de Luijtgaarden KM, Bax L, den Hoed PT, Hunink MG, Rouwet EV, et al. Supervised walking therapy in patients with intermittent claudication. J Vasc Surg. 2012;56(4):1132–42. doi:10.1016/j.jvs.2012.04.046.
3. McDermott MM, Ades P, Guralnik JM, Dyer A, Ferrucci L, Liu K, et al. Treadmill exercise and resistance training in patients with peripheral arterial disease with and without intermittent claudication: a randomized controlled trial. JAMA. 2009;301(2):165–74. doi:10.1001/jama.2008.962.
4. Watson L, Ellis B, Leng GC. Exercise for intermittent claudication. Cochrane Database Syst Rev. 2008;(4):CD000990. doi:10.1002/14651858.CD000990.pub2.
5. Spronk S, Bosch JL, den Hoed PT, Veen HF, Pattynama PM, Hunink MG. Intermittent claudication: clinical effectiveness of endovascular revascularization versus supervised hospital-based exercise training--randomized controlled trial. Radiology. 2009;250(2):586–95. doi:10.1148/radiol.2501080607.
6. McAllister RM, Terjung RL. Training-induced muscle adaptations: increased performance and oxygen consumption. J Appl Physiol (1985). 1991;70(4):1569–74.
7. Bebout DE, Hogan MC, Hempleman SC, Wagner PD. Effects of training and immobilization on VO2 and DO2 in dog gastrocnemius muscle in situ. J Appl Physiol (1985). 1993;74(4):1697–703.
8. Saltin B, Nazar K, Costill DL, Stein E, Jansson E, Essen B, et al. The nature of the training response; peripheral and central adaptations of one-legged exercise. Acta Physiol Scand. 1976;96(3):289–305. doi:10.1111/j.1748-1716.1976.tb10200.x.
9. Booth FW, Thomason DB. Molecular and cellular adaptation of muscle in response to exercise: perspectives of various models. Physiol Rev. 1991;71(2):541–85.
10. Krogh A. The number and distribution of capillaries in muscles with calculations of the oxygen pressure head necessary for supplying the tissue. J Physiol. 1919;52(6):409–15.
11. Krogh A. The supply of oxygen to the tissues and the regulation of the capillary circulation. J Physiol. 1919;52(6):457–74.
12. Hudlicka O, Brown M, Egginton S. Angiogenesis in skeletal and cardiac muscle. Physiol Rev. 1992;72(2):369–417.
13. Carrow RE, Brown RE, Van Huss WD. Fiber sizes and capillary to fiber ratios in skeletal muscle of exercised rats. Anat Rec. 1967;159(1):33–9. doi:10.1002/ar.1091590106.

14. Adolfsson J, Ljungqvist A, Tornling G, Unge G. Capillary increase in the skeletal muscle of trained young and adult rats. J Physiol. 1981;310:529–32.
15. Mai JV, Edgerton VR, Barnard RJ. Capillarity of red, white and intermediate muscle fibers in trained and untrained guinea pigs. Experientia. 1970;26(11):1222–3.
16. Andersen P. Capillary density in skeletal muscle of man. Acta Physiol Scand. 1975;95(2):203–5. doi:10.1111/j.1748-1716.1975.tb10043.x.
17. Andersen P, Henriksson J. Capillary supply of the quadriceps femoris muscle of man: adaptive response to exercise. J Physiol. 1977;270(3):677–90.
18. Ingjer F. Effects of endurance training on muscle fibre ATP-ase activity, capillary supply and mitochondrial content in man. J Physiol. 1979;294:419–32.
19. Frandsen U, Hoffner L, Betak A, Saltin B, Bangsbo J, Hellsten Y. Endurance training does not alter the level of neuronal nitric oxide synthase in human skeletal muscle. J Appl Physiol (1985). 2000;89(3):1033–8.
20. Hoier B, Nordsborg N, Andersen S, Jensen L, Nybo L, Bangsbo J, et al. Pro- and anti-angiogenic factors in human skeletal muscle in response to acute exercise and training. J Physiol. 2012;590(3):595–606. doi:10.1113/jphysiol.2011.216135.
21. Richardson RS, Wagner H, Mudaliar SR, Saucedo E, Henry R, Wagner PD. Exercise adaptation attenuates VEGF gene expression in human skeletal muscle. Am J Physiol Heart Circ Physiol. 2000;279(2):H772–8.
22. Klausen K, Andersen LB, Pelle I. Adaptive changes in work capacity, skeletal muscle capillarization and enzyme levels during training and detraining. Acta Physiol Scand. 1981;113(1):9–16. doi:10.1111/j.1748-1716.1981.tb06854.x.
23. Salmons S, Henriksson J. The adaptive response of skeletal muscle to increased use. Muscle Nerve. 1981;4(2):94–105. doi:10.1002/mus.880040204.
24. Tesch PA, Thorsson A, Kaiser P. Muscle capillary supply and fiber type characteristics in weight and power lifters. J Appl Physiol Respir Environ Exerc Physiol. 1984;56(1):35–8.
25. Luthi JM, Howald H, Claassen H, Rosler K, Vock P, Hoppeler H. Structural changes in skeletal muscle tissue with heavy-resistance exercise. Int J Sports Med. 1986;7(3):123–7. doi:10.1055/s-2008-1025748.
26. Gavin TP, Drew JL, Kubik CJ, Pofahl WE, Hickner RC. Acute resistance exercise increases skeletal muscle angiogenic growth factor expression. Acta Physiol (Oxf). 2007;191(2):139–46. doi:10.1111/j.1748-1716.2007.01723.x.
27. Jensen L, Bangsbo J, Hellsten Y. Effect of high intensity training on capillarization and presence of angiogenic factors in human skeletal muscle. J Physiol. 2004;557(Pt 2):571–82. doi:10.1113/jphysiol.2003.057711.
28. Hansen-Smith FM, Hudlicka O, Egginton S. In vivo angiogenesis in adult rat skeletal muscle: early changes in capillary network architecture and ultrastructure. Cell Tissue Res. 1996;286(1):123–36.
29. Zhou AL, Egginton S, Brown MD, Hudlicka O. Capillary growth in overloaded, hypertrophic adult rat skeletal muscle: an ultrastructural study. Anat Rec. 1998;252(1):49–63.
30. Egginton S, Zhou AL, Brown MD, Hudlicka O. Unorthodox angiogenesis in skeletal muscle. Cardiovasc Res. 2001;49(3):634–46.
31. Zhou A, Egginton S, Hudlicka O, Brown MD. Internal division of capillaries in rat skeletal muscle in response to chronic vasodilator treatment with alpha1-antagonist prazosin. Cell Tissue Res. 1998;293(2):293–303.
32. Djonov V, Baum O, Burri PH. Vascular remodeling by intussusceptive angiogenesis. Cell Tissue Res. 2003;314(1):107–17. doi:10.1007/s00441-003-0784-3.
33. Prior BM, Yang HT, Terjung RL. What makes vessels grow with exercise training? J Appl Physiol (1985). 2004;97(3):1119–28. doi:10.1152/japplphysiol.00035.2004.
34. Hudlicka O. Is physiological angiogenesis in skeletal muscle regulated by changes in microcirculation? Microcirculation. 1998;5(1):7–23.
35. Chien S. Mechanotransduction and endothelial cell homeostasis: the wisdom of the cell. Am J Physiol Heart Circ Physiol. 2007;292(3):H1209–24. doi:10.1152/ajpheart.01047.2006.

36. Rivilis I, Milkiewicz M, Boyd P, Goldstein J, Brown MD, Egginton S, et al. Differential involvement of MMP-2 and VEGF during muscle stretch- versus shear stress-induced angiogenesis. Am J Physiol Heart Circ Physiol. 2002;283(4):H1430–8. doi:10.1152/ajpheart.00082.2002.
37. Resnick N, Yahav H, Shay-Salit A, Shushy M, Schubert S, Zilberman LC, et al. Fluid shear stress and the vascular endothelium: for better and for worse. Prog Biophys Mol Biol. 2003;81(3):177–99.
38. Heil M, Schaper W. Influence of mechanical, cellular, and molecular factors on collateral artery growth (arteriogenesis). Circ Res. 2004;95(5):449–58. doi:10.1161/01.RES.0000141145.78900.44.
39. Silver AE, Vita JA. Shear-stress-mediated arterial remodeling in atherosclerosis: too much of a good thing? Circulation. 2006;113(24):2787–9. doi:10.1161/CIRCULATIONAHA.106.634378.
40. Tornling G, Adolfsson J, Unge G, Ljungqvist A. Capillary neoformation in skeletal muscle of dipyridamole-treated rats. Arzneimittelforschung. 1980;30(5):791–2.
41. Dawson JM, Hudlicka O. The effects of long term administration of prazosin on the microcirculation in skeletal muscles. Cardiovasc Res. 1989;23(11):913–20.
42. Roca J, Gavin TP, Jordan M, Siafakas N, Wagner H, Benoit H, et al. Angiogenic growth factor mRNA responses to passive and contraction-induced hyperperfusion in skeletal muscle. J Appl Physiol (1985). 1998;85(3):1142–9.
43. Williams JL, Cartland D, Hussain A, Egginton S. A differential role for nitric oxide in two forms of physiological angiogenesis in mouse. J Physiol. 2006;570(Pt 3):445–54. doi:10.1113/jphysiol.2005.095596.
44. Milkiewicz M, Kelland C, Colgan S, Haas TL. Nitric oxide and p38 MAP kinase mediate shear stress-dependent inhibition of MMP-2 production in microvascular endothelial cells. J Cell Physiol. 2006;208(1):229–37. doi:10.1002/jcp.20658.
45. Yun S, Dardik A, Haga M, Yamashita A, Yamaguchi S, Koh Y, et al. Transcription factor Sp1 phosphorylation induced by shear stress inhibits membrane type 1-matrix metalloproteinase expression in endothelium. J Biol Chem. 2002;277(38):34808–14. doi:10.1074/jbc.M205417200.
46. Williams JL, Weichert A, Zakrzewicz A, Da Silva-Azevedo L, Pries AR, Baum O, et al. Differential gene and protein expression in abluminal sprouting and intraluminal splitting forms of angiogenesis. Clin Sci (Lond). 2006;110(5):587–95. doi:10.1042/CS20050185.
47. Chang H, Wang BW, Kuan P, Shyu KG. Cyclical mechanical stretch enhances angiopoietin-2 and Tie2 receptor expression in cultured human umbilical vein endothelial cells. Clin Sci (Lond). 2003;104(4):421–8. doi:10.1042/.
48. Yamaguchi S, Yamaguchi M, Yatsuyanagi E, Yun SS, Nakajima N, Madri JA, et al. Cyclic strain stimulates early growth response gene product 1-mediated expression of membrane type 1 matrix metalloproteinase in endothelium. Lab Invest. 2002;82(7):949–56.
49. Nishimura K, Li W, Hoshino Y, Kadohama T, Asada H, Ohgi S, et al. Role of AKT in cyclic strain-induced endothelial cell proliferation and survival. Am J Physiol Cell Physiol. 2006;290(3):C812–21. doi:10.1152/ajpcell.00347.2005.
50. Egginton S, Hudlicka O, Brown MD, Walter H, Weiss JB, Bate A. Capillary growth in relation to blood flow and performance in overloaded rat skeletal muscle. J Appl Physiol (1985). 1998;85(6):2025–32.
51. Hellsten Y, Rufener N, Nielsen JJ, Hoier B, Krustrup P, Bangsbo J. Passive leg movement enhances interstitial VEGF protein, endothelial cell proliferation, and eNOS mRNA content in human skeletal muscle. Am J Physiol Regul Integr Comp Physiol. 2008;294(3):R975–82. doi:10.1152/ajpregu.00677.2007.
52. Hoier B, Rufener N, Bojsen-Moller J, Bangsbo J, Hellsten Y. The effect of passive movement training on angiogenic factors and capillary growth in human skeletal muscle. J Physiol. 2010;588(Pt 19):3833–45. doi:10.1113/jphysiol.2010.190439.
53. Adair TH, Gay WJ, Montani JP. Growth regulation of the vascular system: evidence for a metabolic hypothesis. Am J Physiol. 1990;259(3 Pt 2):R393–404.
54. Yang HT, Ogilvie RW, Terjung RL. Low-intensity training produces muscle adaptations in rats with femoral artery stenosis. J Appl Physiol (1985). 1991;71(5):1822–9.

55. Sundberg CJ. Exercise and training during graded leg ischaemia in healthy man with special reference to effects on skeletal muscle. Acta Physiol Scand Suppl. 1994;615:1–50.
56. Breen EC, Johnson EC, Wagner H, Tseng HM, Sung LA, Wagner PD. Angiogenic growth factor mRNA responses in muscle to a single bout of exercise. J Appl Physiol (1985). 1996;81(1):355–61.
57. Gustafsson T, Puntschart A, Kaijser L, Jansson E, Sundberg CJ. Exercise-induced expression of angiogenesis-related transcription and growth factors in human skeletal muscle. Am J Physiol. 1999;276(2 Pt 2):H679–85.
58. Richardson RS, Wagner H, Mudaliar SR, Henry R, Noyszewski EA, Wagner PD. Human VEGF gene expression in skeletal muscle: effect of acute normoxic and hypoxic exercise. Am J Physiol. 1999;277(6 Pt 2):H2247–52.
59. Semenza GL. Oxygen sensing, homeostasis, and disease. N Engl J Med. 2011;365(6):537–47. doi:10.1056/NEJMra1011165.
60. Richardson RS, Noyszewski EA, Kendrick KF, Leigh JS, Wagner PD. Myoglobin O2 desaturation during exercise. Evidence of limited O2 transport. J Clin Invest. 1995;96(4):1916–26. doi:10.1172/JCI118237.
61. Ameln H, Gustafsson T, Sundberg CJ, Okamoto K, Jansson E, Poellinger L, et al. Physiological activation of hypoxia inducible factor-1 in human skeletal muscle. FASEB J. 2005;19(8):1009–11. doi:10.1096/fj.04-2304fje.
62. Mason SD, Rundqvist H, Papandreou I, Duh R, McNulty WJ, Howlett RA, et al. HIF-1alpha in endurance training: suppression of oxidative metabolism. Am J Physiol Regul Integr Comp Physiol. 2007;293(5):R2059–69. doi:10.1152/ajpregu.00335.2007.
63. Green HJ, Sutton JR, Cymerman A, Young PM, Houston CS. Operation Everest II: adaptations in human skeletal muscle. J Appl Physiol (1985). 1989;66(5):2454–61.
64. Hoppeler H, Kleinert E, Schlegel C, Claassen H, Howald H, Kayar SR, et al. Morphological adaptations of human skeletal muscle to chronic hypoxia. Int J Sports Med. 1990;11(Suppl 1):S3–9. doi:10.1055/s-2007-1024846.
65. Lundby C, Pilegaard H, Andersen JL, van Hall G, Sander M, Calbet JA. Acclimatization to 4100 m does not change capillary density or mRNA expression of potential angiogenesis regulatory factors in human skeletal muscle. J Exp Biol. 2004;207(Pt 22):3865–71. doi:10.1242/jeb.01225.
66. Vigano A, Ripamonti M, De Palma S, Capitanio D, Vasso M, Wait R, et al. Proteins modulation in human skeletal muscle in the early phase of adaptation to hypobaric hypoxia. Proteomics. 2008;8(22):4668–79. doi:10.1002/pmic.200800232.
67. Handschin C, Spiegelman BM. Peroxisome proliferator-activated receptor gamma coactivator 1 coactivators, energy homeostasis, and metabolism. Endocr Rev. 2006;27(7):728–35. doi:10.1210/er.2006-0037.
68. Arany Z, Foo SY, Ma Y, Ruas JL, Bommi-Reddy A, Girnun G, et al. HIF-independent regulation of VEGF and angiogenesis by the transcriptional coactivator PGC-1alpha. Nature. 2008;451(7181):1008–12. doi:10.1038/nature06613.
69. Leick L, Hellsten Y, Fentz J, Lyngby SS, Wojtaszewski JF, Hidalgo J, et al. PGC-1alpha mediates exercise-induced skeletal muscle VEGF expression in mice. Am J Physiol Endocrinol Metab. 2009;297(1):E92–103. doi:10.1152/ajpendo.00076.2009.
70. Chinsomboon J, Ruas J, Gupta RK, Thom R, Shoag J, Rowe GC, et al. The transcriptional coactivator PGC-1alpha mediates exercise-induced angiogenesis in skeletal muscle. Proc Natl Acad Sci U S A. 2009;106(50):21401–6. doi:10.1073/pnas.0909131106.
71. Geng T, Li P, Okutsu M, Yin X, Kwek J, Zhang M, et al. PGC-1alpha plays a functional role in exercise-induced mitochondrial biogenesis and angiogenesis but not fiber-type transformation in mouse skeletal muscle. Am J Physiol Cell Physiol. 2010;298(3):C572–9. doi:10.1152/ajpcell.00481.2009.
72. Pogozelski AR, Geng T, Li P, Yin X, Lira VA, Zhang M, et al. p38gamma mitogen-activated protein kinase is a key regulator in skeletal muscle metabolic adaptation in mice. PLoS One. 2009;4(11):e7934. doi:10.1371/journal.pone.0007934.

73. Wright DC, Geiger PC, Han DH, Jones TE, Holloszy JO. Calcium induces increases in peroxisome proliferator-activated receptor gamma coactivator-1alpha and mitochondrial biogenesis by a pathway leading to p38 mitogen-activated protein kinase activation. J Biol Chem. 2007;282(26):18793–9. doi:10.1074/jbc.M611252200.
74. Hardie DG, Ross FA, Hawley SA. AMPK: a nutrient and energy sensor that maintains energy homeostasis. Nat Rev Mol Cell Biol. 2012;13(4):251–62. doi:10.1038/nrm3311.
75. Winder WW, Hardie DG. Inactivation of acetyl-CoA carboxylase and activation of AMP-activated protein kinase in muscle during exercise. Am J Physiol. 1996;270(2 Pt 1): E299–304.
76. Rasmussen BB, Winder WW. Effect of exercise intensity on skeletal muscle malonyl-CoA and acetyl-CoA carboxylase. J Appl Physiol (1985). 1997;83(4):1104–9.
77. Fujii N, Hayashi T, Hirshman MF, Smith JT, Habinowski SA, Kaijser L, et al. Exercise induces isoform-specific increase in 5′AMP-activated protein kinase activity in human skeletal muscle. Biochem Biophys Res Commun. 2000;273(3):1150–5. doi:10.1006/bbrc.2000.3073.
78. Jager S, Handschin C, St-Pierre J, Spiegelman BM. AMP-activated protein kinase (AMPK) action in skeletal muscle via direct phosphorylation of PGC-1alpha. Proc Natl Acad Sci U S A. 2007;104(29):12017–22. doi:10.1073/pnas.0705070104.
79. Ouchi N, Shibata R, Walsh K. AMP-activated protein kinase signaling stimulates VEGF expression and angiogenesis in skeletal muscle. Circ Res. 2005;96(8):838–46. doi:10.1161/01.RES.0000163633.10240.3b.
80. Miura S, Kawanaka K, Kai Y, Tamura M, Goto M, Shiuchi T, et al. An increase in murine skeletal muscle peroxisome proliferator-activated receptor-gamma coactivator-1alpha (PGC-1alpha) mRNA in response to exercise is mediated by beta-adrenergic receptor activation. Endocrinology. 2007;148(7):3441–8. doi:10.1210/en.2006-1646.
81. Norrbom J, Sallstedt EK, Fischer H, Sundberg CJ, Rundqvist H, Gustafsson T. Alternative splice variant PGC-1alpha-b is strongly induced by exercise in human skeletal muscle. Am J Physiol Endocrinol Metab. 2011;301(6):E1092–8. doi:10.1152/ajpendo.00119.2011.
82. Gidlund EK, Ydfors M, Appel S, Rundqvist H, Sundberg CJ, Norrbom J. Rapidly elevated levels of PGC-1alpha-b protein in human skeletal muscle after exercise: exploring regulatory factors in a randomized controlled trial. J Appl Physiol (1985). 2015;119(4):374–84. doi:10.1152/japplphysiol.01000.2014.
83. Idei N, Nishioka K, Soga J, Hidaka T, Hata T, Fujii Y, et al. Vascular function and circulating progenitor cells in thromboangitis obliterans (Buerger's disease) and atherosclerosis obliterans. Hypertension. 2011;57(1):70–8. doi:10.1161/HYPERTENSIONAHA.110.163683.
84. Maruhashi T, Nakashima A, Matsumoto T, Oda N, Iwamoto Y, Iwamoto A, et al. Relationship between nitroglycerine-induced vasodilation and clinical severity of peripheral artery disease. Atherosclerosis. 2014;235(1):65–70. doi:10.1016/j.atherosclerosis.2014.04.012.
85. Iwamoto A, Kajikawa M, Maruhashi T, Iwamoto Y, Oda N, Kishimoto S, et al. Vascular function and intima-media thickness of a leg artery in peripheral artery disease: a comparison of Buerger disease and atherosclerotic peripheral artery disease. J Atheroscler Thromb. 2016;23(11):1261–9. doi:10.5551/jat.35436.
86. Brendle DC, Joseph LJ, Corretti MC, Gardner AW, Katzel LI. Effects of exercise rehabilitation on endothelial reactivity in older patients with peripheral arterial disease. Am J Cardiol. 2001;87(3):324–9.
87. Olfert IM, Howlett RA, Tang K, Dalton ND, Gu Y, Peterson KL, et al. Muscle-specific VEGF deficiency greatly reduces exercise endurance in mice. J Physiol. 2009;587(Pt 8):1755–67. doi:10.1113/jphysiol.2008.164384.
88. Leung DW, Cachianes G, Kuang WJ, Goeddel DV, Ferrara N. Vascular endothelial growth factor is a secreted angiogenic mitogen. Science. 1989;246(4935):1306–9.
89. Wellmann S, Taube T, Paal K, Graf VEH, Geilen W, Seifert G, et al. Specific reverse transcription-PCR quantification of vascular endothelial growth factor (VEGF) splice variants by LightCycler technology. Clin Chem. 2001;47(4):654–60.
90. Findley CM, Mitchell RG, Duscha BD, Annex BH, Kontos CD. Plasma levels of soluble Tie2 and vascular endothelial growth factor distinguish critical limb ischemia from intermittent

claudication in patients with peripheral arterial disease. J Am Coll Cardiol. 2008;52(5):387–93. doi:10.1016/j.jacc.2008.02.045.
91. Makin AJ, Chung NA, Silverman SH, Lip GY. Vascular endothelial growth factor and tissue factor in patients with established peripheral artery disease: a link between angiogenesis and thrombogenesis? Clin Sci (Lond). 2003;104(4):397–404. doi:10.1042/CS20020182.
92. Bates DO, Cui TG, Doughty JM, Winkler M, Sugiono M, Shields JD, et al. VEGF165b, an inhibitory splice variant of vascular endothelial growth factor, is down-regulated in renal cell carcinoma. Cancer Res. 2002;62(14):4123–31.
93. Harper SJ, Bates DO. VEGF-A splicing: the key to anti-angiogenic therapeutics? Nat Rev Cancer. 2008;8(11):880–7. doi:10.1038/nrc2505.
94. Kikuchi R, Nakamura K, MacLauchlan S, Ngo DTM, Shimizu I, Fuster JJ, et al. An antiangiogenic isoform of VEGF-A contributes to impaired vascularization in peripheral artery disease. Nat Med. 2014;20(12):1464–71. doi:10.1038/nm.3703.
95. Norgren L, Hiatt WR, Dormandy JA, Nehler MR, Harris KA, Fowkes FG, et al. Inter-society consensus for the management of peripheral arterial disease (TASC II). J Vasc Surg. 2007;45(Suppl):S5–67. doi:10.1016/j.jvs.2006.12.037.
96. Jove M, Salla J, Planavila A, Cabrero A, Michalik L, Wahli W, et al. Impaired expression of NADH dehydrogenase subunit 1 and PPARgamma coactivator-1 in skeletal muscle of ZDF rats: restoration by troglitazone. J Lipid Res. 2004;45(1):113–23. doi:10.1194/jlr.M300208-JLR200.
97. Crunkhorn S, Dearie F, Mantzoros C, Gami H, da Silva WS, Espinoza D, et al. Peroxisome proliferator activator receptor gamma coactivator-1 expression is reduced in obesity: potential pathogenic role of saturated fatty acids and p38 mitogen-activated protein kinase activation. J Biol Chem. 2007;282(21):15439–50. doi:10.1074/jbc.M611214200.
98. Mensink M, Hesselink MK, Russell AP, Schaart G, Sels JP, Schrauwen P. Improved skeletal muscle oxidative enzyme activity and restoration of PGC-1 alpha and PPAR beta/delta gene expression upon rosiglitazone treatment in obese patients with type 2 diabetes mellitus. Int J Obes (Lond). 2007;31(8):1302–10. doi:10.1038/sj.ijo.0803567.
99. Hernandez-Alvarez MI, Thabit H, Burns N, Shah S, Brema I, Hatunic M, et al. Subjects with early-onset type 2 diabetes show defective activation of the skeletal muscle PGC-1{alpha}/Mitofusin-2 regulatory pathway in response to physical activity. Diabetes Care. 2010;33(3):645–51. doi:10.2337/dc09-1305.

Chapter 16
Nanoparticle-Mediated Endothelial Cell-Selective Drug Delivery System

Kaku Nakano, Jun-ichiro Koga, and Kensuke Egashira

Abstract Peripheral arterial disease (PAD) causes peripheral tissue ischemia, resulting in skin ulceration, gangrene, and limb amputation. The therapeutic goal is restoration of blood supply, and various treatments including exercise, medical therapy, and surgical revascularization are performed. The prognosis of critical limb ischemia, however, is still unsatisfactory, and there are unmet needs in developing novel therapeutics. We have been developing novel nanotechnology-based drug delivery system aiming efficient angiogenesis by delivering of HMG-CoA reductase inhibitor (statin) to endothelium. In this chapter, we summarized current developing status of nanotechnology-based drug delivery system including animal experiments and ongoing clinical trials (CLINICAL TRIAL REGISTRATION NUMBER: UMIN000008011).

Keywords Drug delivery system • Angiogenesis • Peripheral arterial disease • Statin

16.1 Peripheral Arterial Disease

Peripheral arterial disease (PAD) causes intermittent claudication, ulcer formation, and limb amputation by impairing blood supply to the peripheral organs/tissues. These symptoms have serious impacts on patients' quality of life (QOL). Atherosclerotic lesion formation results in organic stenosis/occlusion of the

K. Nakano, Ph.D. • J.-i. Koga, M.D., Ph.D. • K. Egashira, M.D., Ph.D. (✉)
Department of Cardiovascular Research, Development, and Translational Medicine,
Kyushu University, 3-1-1, Maidashi, Higashi-ku, Fukuoka 812-8582, Japan
e-mail: egashira@cardiol.med.kyushu-u.ac.jp

Y. Higashi, T. Murohara (eds.), *Therapeutic Angiogenesis*,
DOI 10.1007/978-981-10-2744-4_16

arteries and impairs peripheral blood supply. Various risk factors such as hypertension, diabetes mellitus, and dyslipidemia promote the pathobiology of atherosclerosis [1, 2]. PAD occurs as one of the phenotypes of systemic atherosclerosis. Actually, 75% PAD patients die of cerebrovascular diseases and coronary artery diseases [3–6]. Therefore, risk factor management is important to improve the prognosis. Therapeutic principles of PAD have been presented in various guidelines including “Management of Peripheral Arterial Disease” by TransAtlantic Inter-Society Consensus (TASC), TASC II, and ACC/AHA 2005 Guidelines [7]. Many epidemiologic studies reported that the prevalence of PAD is high in men and aged people. The prevalence of PAD has been reported to be 2.5% for people aged 40–59 years, 8.3% for 60–69 years, and 18.8% for 70 years or older [8, 9].

Critical limb ischemia (CLI) is a PAD with rest pain or ischemic skin lesions (ulcers, gangrene) classified as grade III or IV in the Fontaine classification. The primary goals of CLI treatment are improvement in QOL, i.e., relief of symptoms due to ischemia and prevention of limb amputation. Improvement of survival is also an important purpose of the treatment of PAD. TASC II reported that the 5- and 10-year survival of patients with CLI are 50% and 10%, respectively. Half of the CLI patients are attempted revascularization, and the others undergo primary amputation or medical treatment. Half of the patients, however, die or experience limb amputation within a year, and CLI is resolved in only 25% patients [3]. Therefore, there are unmet needs in the treatment of PAD, especially in CLI. Clinical trials have been already conducted that aims “therapeutic angiogenesis” in patients with CLI, but no clinically relevant drug therapy has been established and marketed so far.

16.2 Developmental Status of Novel Therapeutics and Problems

In recent years, there has been increasing attention on a new therapy, therapeutic angiogenesis, which increases collateral arteries by administrating growth factors or genes stimulating angiogenesis into the ischemic tissues. Table 16.1 shows the developmental status of new drugs or cell therapy for PAD in worldwide from the Thomson Reuters Pharma database. In addition, transplantation of autologous stem cells has been conducted in CLI patients. Autologous stem cell transplantation therapy is largely classified into two types according to the cell source—peripheral blood and bone marrow. Any type of autologous stem cell transplantation therapy, however, requires a cell processing center with dedicated instruments for cell preparation as well as a highly organized transplant team, making it difficult to establish the therapy as general/standard medicine.

Table 16.1 Developmental status of new drugs for peripheral arterial disease Data Source: Thomson Reuters Pharma (Cortellis Competitive Intelligence)

Drug code	Developer	Phase of development
DVC1-010	Dnavec Corp.	Phase 2 clinical
AMG-0001	AnGes MG Inc.	Phase 3 clinical
Ixmyelocel	Astrom Biosciences Inc.	Phase 2 clinical
Plasmin	Talecris Biotherapeutics	Phase 2 clinical
AX-200	SYGNIS Bioscience	Phase 2 clinical
Autologous fat-derived stem cells	RNL Bio Co, Ltd	Phase 2 clinical
Autologous CD133+ hematopoietic stem cells	University of Wisconsin-Madison	Phase 1 clinical
Autologous endothelial progenitor cell therapy	Kobe Institute of Biomedical Research and Innovation	Phase 2 clinical
ZFP-VEGF	Sangamo BioSciences Inc.	No development reported
Autologous CD34+ stem cell therapy	University of Debrecen	Clinical
AKB-9778	Procter & Gamble Pharmaceuticals Inc.	Discovery
VM-202	ViroMed Co Ltd	Phase 2 clinical

16.2.1 Statins

HMG-CoA reductase inhibitors, statins, are widely used cholesterol-lowering agents and reported to have pleiotropic effects independent of their cholesterol-lowering effect [10–12]. One of the pleiotropic effects of statins is their angiogenic action. For example, various animal experiments have shown that the systemic administration of a statin increases re-endothelialization of damaged blood vessels and promotes angiogenesis of ischemic tissue. However, daily administration of a high-dose statin is needed to obtain therapeutic effects, which may lead to serious adverse side effects in clinical settings [13–16]. Endothelial cell-targeting drug delivery of statins by nanotechnology-based drug delivery system (DDS) can optimize the therapeutic effects and avoid adverse effects of statin.

16.2.2 PLGA NPs for the Treatment of Cardiovascular Diseases

We have been developed drug delivery system based on lactide/glycolide copolymer (PLGA) nanoparticles [11, 17–19]. PLGA NPs were prepared by emulsion solvent diffusion method. The mean diameter of particles is approximately 200 nm

[20, 21]. In the bloodstream, PLGA NPs escape from clearances in the kidney and reticuloendothelial system, which prolong blood retention time and result in effective drug delivery to the site with increased vascular permeability [22]. PLGA NPs also have the following advantages for DDS: (1) PLGA has been already used in humans for over 40 years [23], and concerns about the safety are little; and (2) PLGA can encapsulate both hydrophobic and hydrophilic agents including chemicals and nucleotides.

16.2.3 HMG-CoA Reductase Inhibitory Action

To candidate for incorporated statins into the PLGA nanoparticle, the effects of statins were compared in vitro assays. Pitavastatin, atorvastatin, rosuvastatin, simvastatin, fluvastatin, and pravastatin were purchased, extracted from products, and purified.

First, we compared inhibitory activity of statins (pitavastatin, fluvastatin, rosuvastatin, atorvastatin, simvastatin, and pravastatin) on HMG-CoA reductase using rat liver microsomes. Statins (0.1–1000 nM) and [3-^{14}C]HMG-CoA (0.1 mM) were added to 0.1 mg of microsome protein. After 30 min, mevalonic acid was converted to mevalonolactone and then isolated by thin-layer chromatography to measure radioactivity. Pitavastatin, fluvastatin, rosuvastatin, atorvastatin, simvastatin, and pravastatin inhibited HMG-CoA reductase activity in a concentration-dependent manner, and the half maximal inhibitory concentrations (IC_{50}) were 1.88, 1.89, 3.09, 3.46, 5.12, and 40.8 nM, respectively.

16.2.4 Effects of Pitavastatin on Endothelial Cell Migration and Tube Formation

A scratch motility assay using human umbilical vein endothelial cells (HUVECs) was performed to assess the re-endothelialization response following endothelial denudation. HUVECs were seeded on collagen I-coated 24-well plates at a density of 2×10^4 cells per well and grown to be confluent with EGM2. The monolayers were scratched with a pipet tip and photographed before incubation with EBM containing 0.5% BSA containing VEGF165 (10 ng/mL, R & D) with or without statins. After 5 h, cells were photographed, and the number of cells migrating to the scratched area was counted. In the human endothelial cell scratch motility assay (re-endothelialization in vitro), only pitavastatin increased the re-endothelialization

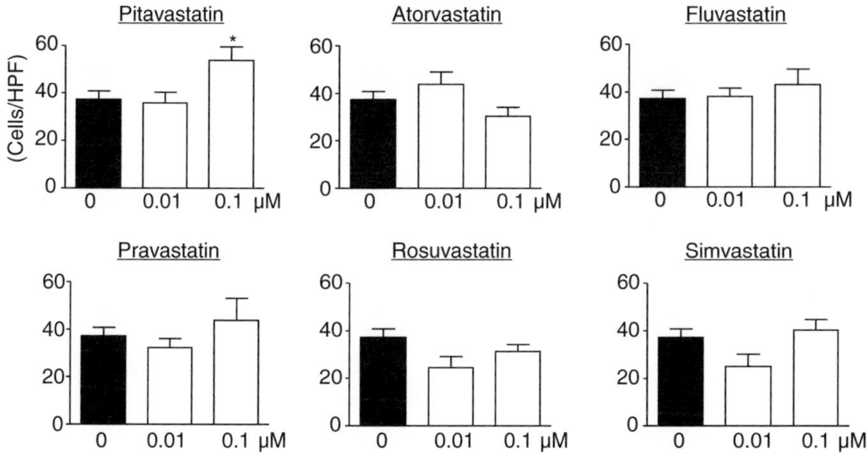

Fig. 16.1 In vitro human umbilical vein endothelial cell scratch motility assay. A vertical axis denotes the number of migrated cells into the scratched area. $*p < 0.05$ vs. no treatment by one-way ANOVA followed with Dunnett's multiple comparison test. $N = 6–8$ each

response following scratch injury; the other five statins showed no such effects (Fig. 16.1) [24].

16.2.5 Angiogenesis Activity of Human Endothelial Cells

Angiogenesis of human vascular endothelial cells was tested by two-dimensional Matrigel assay, as previously described [25]. HUVEC (2×10^4) were plated on an eight-well chamber slide pre-coated with 200 μL of Matrigel (BD Bioscience) in 500 μL of EBM-2 medium with supplements (Lonza) in the presence or absence of statins (1 or 10 nM). In another set of experiments, HUVECs were pretreated with NP incorporated with pitavastatin at 1 or 10 nM for 24 h, washed, and then suspended in Matrigel. Treatment with pitavastatin increased angiogenic activity in HUVECs, whereas other statins had no effect (Fig. 16.2) [26].

Based on these in vitro results, pitavastatin was selected as the nanoparticulation compound because this compound (1) elicited the strongest HMG-CoA reductase inhibitory effect (minimum IC_{50}) and (2) promotes endothelial cell migration as compared with other statins.

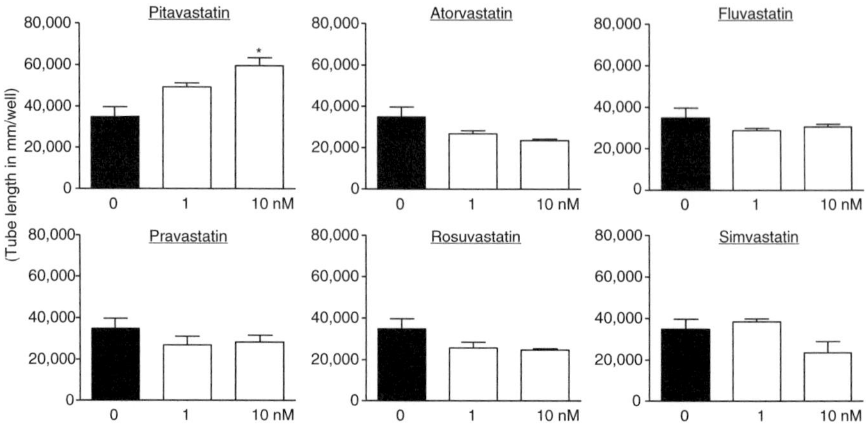

Fig. 16.2 Effects of statins on angiogenic capacity of human umbilical vein endothelial cells in vitro. Quantitative analysis of tube formation (tube length (mm) per well) of six independent experiments. $*p < 0.01$ vs. control by one-way ANOVA with Dunnett's multiple comparison test

16.3 Nanoparticle-Mediated Drug Delivery System

16.3.1 Therapeutic Neovascularization by Nanotechnology-Mediated Cell-Selective Delivery of Pitavastatin into the Vascular Endothelium in Mice

Restoration of tissue perfusion is a major goal in the treatment of patients with critical limb ischemia. Double-blinded placebo-controlled clinical trials designed to induce neovascularization by administering exogenous angiogenic growth factors failed to demonstrate a clinical benefit and, in contrast, showed some undesired side effects [27, 28]. These results were not consistent with the results observed in animal experiments and small-scale open-label clinical trials [29, 30]. These disappointing results may be resulted from, at least in part, less effective transfection efficacy of the genetic materials or the rapid washout of proteins. Because multiple endogenous angiogenic growth factors are induced for the development of functional collaterals, the strategy of intramuscular injection of single exogenous angiogenic growth factor may be insufficient. Excessive angiogenic growth factors in local tissues could increase the risks of edema, angioma-like capillary formation, atherosclerosis after vascular injury, and tumor angiogenesis. Hence, optimization of the distribution and pharmacokinetics of angiogenic growth factors by drug delivery system (DDS) could be a novel therapeutic strategy to improve the efficacy and decrease undesired effects.

The PLGA NP offers the advantages of safety, delivery of encapsulated drugs into the cellular cytoplasm, and slow cytoplasmic drug release. PLGA NP is effectively and rapidly taken up by vascular endothelial cells in vitro. To our knowledge, however, no prior studies have examined whether PLGA NPs are useful as an endothelial cell-selective DDS in vivo.

16.3.2 Drug Formulation According to GMP Regulation

PLGA with an average molecular weight of 20,000 and a lactide to glycolide copolymer ratio of 75:25 (PLGA7520; Wako Pure Chemical Industries, Osaka, Japan) was used as a substrate material for the NPs. The bioabsorption half-life of this product was 2 weeks in rat tissue. PLGA NPs incorporated with pitavastatin (Kowa Pharmaceutical Co. Ltd., Tokyo, Japan) were prepared by a previously reported emulsion solvent diffusion method in purified water, according to Good Manufacturing Practice [25]. Pitavastatin-loaded PLGA NPs (Investigational New Drug Code: NK-104-NP) contained 12% pitavastatin and were preserved as freeze-dried material. The mean particle size was analyzed by the light scattering method (Microtrack UPA150; Nikkiso, Tokyo, Japan). The mean diameters of pitavastatin-NP were 196 nm (Fig. 16.3a).

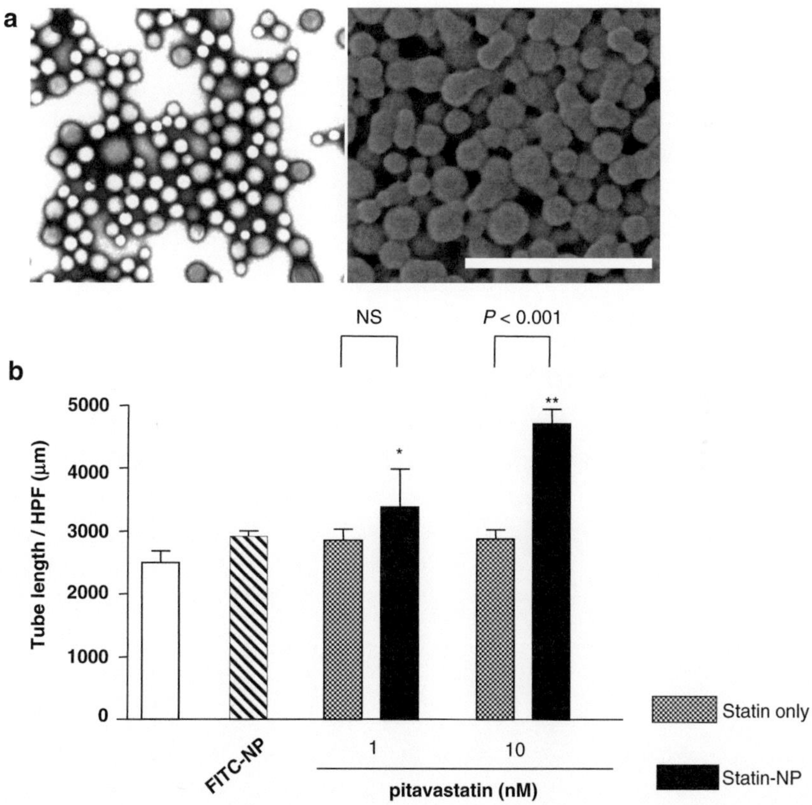

Fig. 16.3 PLGA NPs and the effects of statin-NP on angiogenic capacity of human umbilical vein endothelial cells in vitro. (**a**) PLGA NPs prepared by the emulsion solvent diffusion method. Images were taken by transmission electron microscope (*left*) and scanning electron microscope (*right*). Scale bar indicates 1 μm. (**b**) Quantitative analysis of tube formation (tube length) of four independent experiments. *$p < 0.01$, **$p < 0.001$ vs. control

16.3.3 Intracellular DDS

If the use of pitavastatin-loaded PLGA NPs drastically increases statin concentrations in cells, it may result in adverse drug reactions. Therefore, we compared the effect of pitavastatin-loaded PLGA NPs and pitavastatin alone on angiogenesis in HUVECs. Pitavastatin-loaded PLGA NPs induced angiogenesis at 10 nM, whereas equivalent concentration of pitavastatin alone does not show angiogenic effects. These in vitro results indicate that encapsulation of pitavastatin in PLGA NPs increases the efficacy of pitavastatin on angiogenesis (Fig. 16.3b). Our in vitro data suggest that nanoparticulation retards elimination of pitavastatin by gradually releasing incorporated pitavastatin to the cytosol (intracellular DDS) [25].

16.3.4 Endothelial Cell-Targeting Delivery of NPs In Vivo

Cellular distribution of FITC was examined 3, 7, and 14 days after intramuscular injection of FITC-NP. On day 3 postinjection, FITC signals were detected only in ischemic muscle of mice-injected FITC-NP, whereas no FITC signals were observed in control nonischemic muscle or in ischemic muscle injected with only FITC (Fig. 16.4). The FITC signals were localized predominantly in the capillaries and

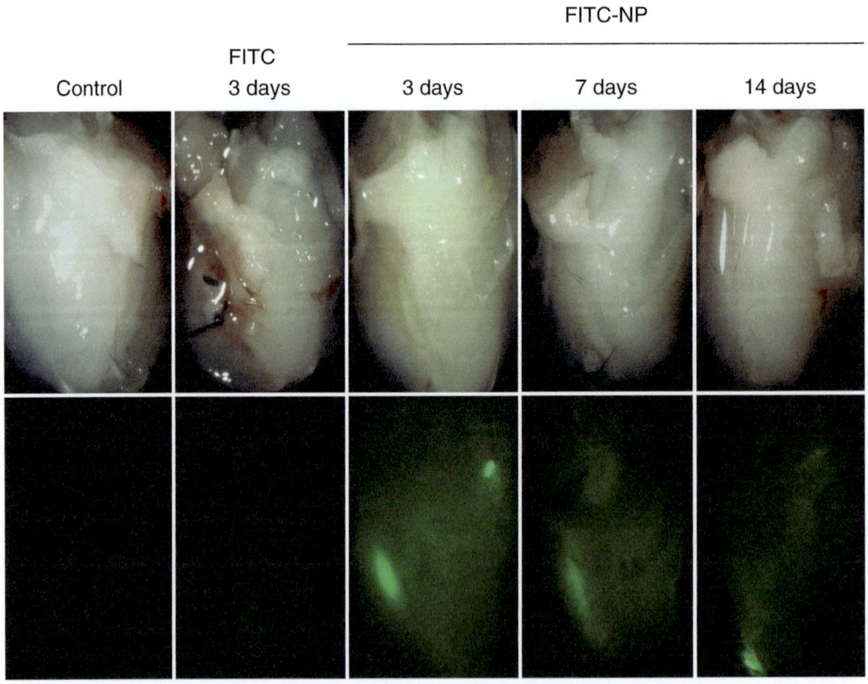

Fig. 16.4 Representative light and fluorescent stereomicrographs of gastrocnemius muscles from nonischemic hindlimb (control) and from ischemic hindlimb (FITC, FITC-NP)

arterioles. FITC signals were also detected in myocytes at this time point. These data suggest that NP solution might distribute to intra- and extracellular spaces of ischemic skeletal muscle tissues immediately after intramuscular injection of NP, and then the NP was uptaken by cells (endothelial cells, smooth muscle cells, myocytes, etc.) or retained in extracellular spaces at this early time point. On days 7 and 14, FITC signals were predominantly observed in capillaries and arterioles. Immunofluorescent staining revealed that FITC signals localized mainly in endothelial cells positive for CD31 in FITC-NP-injected ischemic muscle (Fig. 16.5). In contrast, no FITC signals were observed in myocytes. FITC signals were not detected in contralateral nonischemic hindlimb or in remote organs (liver, spleen, kidney, and heart) at any time point (data not shown).

A mouse model of lower limb ischemia was prepared and divided into four groups: an untreated group, an FITC-loaded PLGA nanoparticle administration group (0.18 mg/100 μL PLGA nanoparticle suspension), a pitavastatin-only administration group (0.01 mg/100 μL, 0.5 mg/kg of pitavastatin calcium), and a pitavastatin-loaded PLGA nanoparticle administration group (0.18 mg/100 μL pitavastatin-loaded PLGA nanoparticle suspension containing 0.01 mg of pitavastatin calcium or 0.5 mg/kg of pitavastatin calcium). Each of these groups except the untreated group was administered with the drug by single intramuscular injection at four sites of the lower limb immediately after femoral artery ligation, and blood flow was measured with a laser Doppler blood flowmeter after 7 and 14 days to evaluate recovery of blood flow (Fig. 16.6). The pitavastatin-loaded PLGA nanoparticle increased blood flow compared with the untreated group, the FITC-loaded

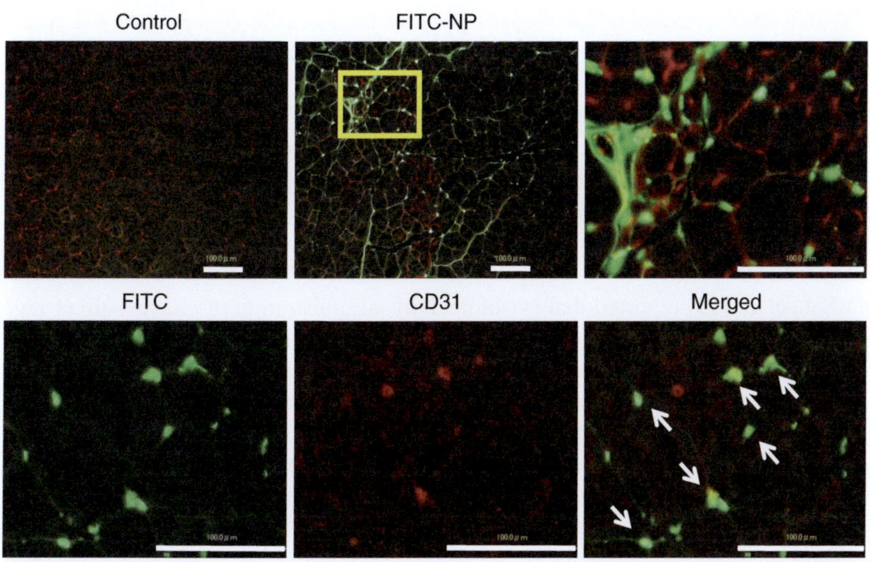

Fig. 16.5 Fluorescent micrographs of cross sections of the gastrocnemius muscle from nonischemic mice with no injection (control), ischemic muscles 14 days after the injection of FITC-NP (*middle*), and expanded view of *boxed area* of middle panel (*right*). Scale bars: 100 μm

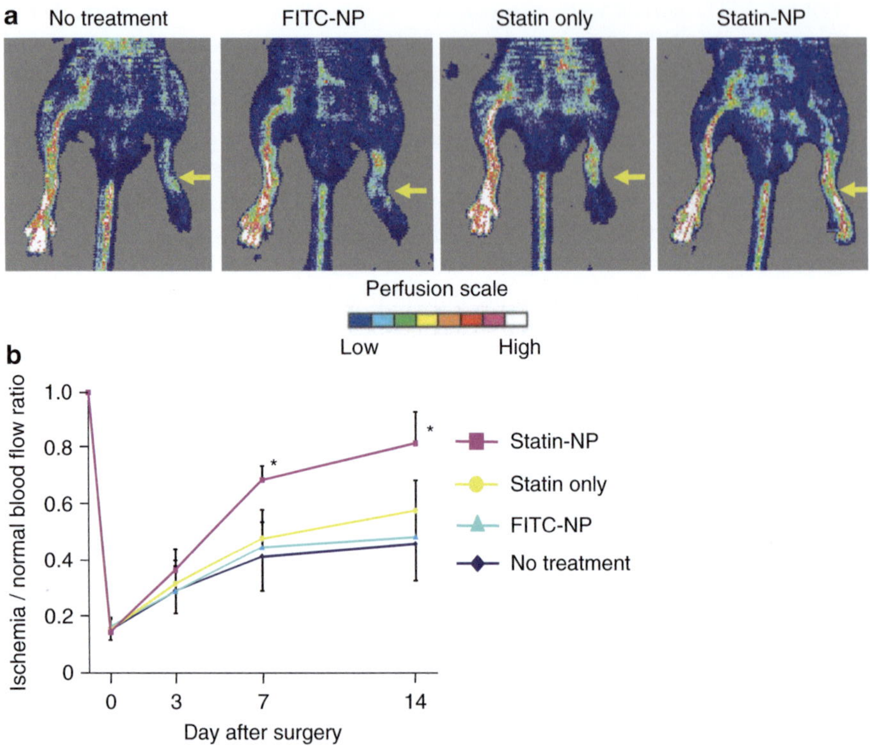

Fig. 16.6 (**a**) Representative laser Doppler perfusion imaging at 14 days postischemia. *Arrow* indicates ischemic limb. (**b**) Quantification of blood flow recovery

PLGA nanoparticle, and the pitavastatin alone. PLGA NPs induced angiogenesis and arteriogenesis without changing blood biomarker concentration (Fig. 16.7). In contrast, pitavastatin alone did not increase blood flow of the ischemic limb within the range of 4–20 mg/kg. An increase of vascular progenitor cells in peripheral blood induced by statins was not observed after injection of pitavastatin-loaded PLGA NPs.

Sata et al. [15] reported that systemic daily administration of pitavastatin (1 mg/kg per day × 49 days = 49 mg/kg) has significant therapeutic effects in mice with hindlimb ischemia. In the present study, we confirmed the study of Sata et al. [15] by showing that oral daily administration of pitavastatin for 14 days (1 and 10 mg/kg per day × 14 = 14 and 140 mg/kg, respectively) had significant therapeutic effects, as did statin-NP (0.4 mg/kg). Therefore, our NP-mediated delivery system seems to be as effective at an approximately 100 times lower dose than the cumulative systemic dose. Furthermore, measurement of the tissue and serum concentrations of pitavastatin confirmed the effective local retention of statin-NP in ischemic skeletal muscles in vivo. NP-mediated delivery of pitavastatin accelerated angiogenesis in human endothelial cells in vitro. Therefore, it is possible that after

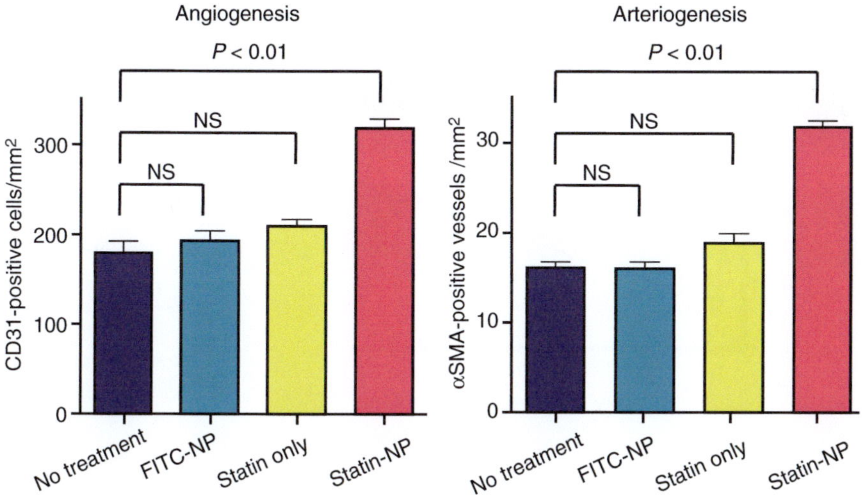

Fig. 16.7 Quantitative analysis of angiogenesis and arteriogenesis. $n = 4$ each. *$P < 0.01$ vs. no treatment group

NP-mediated endothelial delivery, pitavastatin was slowly released from the NP into the cytoplasm along with PLGA hydrolysis, resulting in significant therapeutic effects.

16.3.5 The Mechanisms of Pitavastatin Nanoparticle-Mediated Angiogenesis

Intramuscularly injected pitavastatin-NPs induced therapeutic angiogenesis without any effects on serum lipid levels, suggesting that pitavastatin-NPs demonstrated their efficacy by mechanisms independent with lipid-lowering effects. Furthermore, the effects of pitavastatin-NPs were abrogated by NO synthase inhibitor, L-NAME or genetic deletion of eNOS in wild-type mice. These findings support the hypothesis that eNOS pathway plays essential role in the mechanism of pitavastatin-NP-induced angiogenesis (Fig. 16.8). We have demonstrated that pitavastatin-NP activates the eNOS/PI3K/Akt pathway accompanied with induction of endogenous angiogenic factors, i.e., VEGF and FGF-2, as shown by other investigators, too [13, 15, 16, 31]. These therapeutic effects afforded by the NP-mediated delivery of pitavastatin were not associated with a further increase of circulating vascular progenitor cells. Intramuscular injection of high-dose pitavastatin solution (4 and 20 mg/kg) has no therapeutic effect, suggesting a specific advantage of endothelial cell-targeting delivery of pitavastatin by PLGA NPs. These findings suggest that pitavastatin-NP acted locally on ischemic vascular endothelium to induce therapeutic neovascularization and are consistent with the notion that NP-mediated

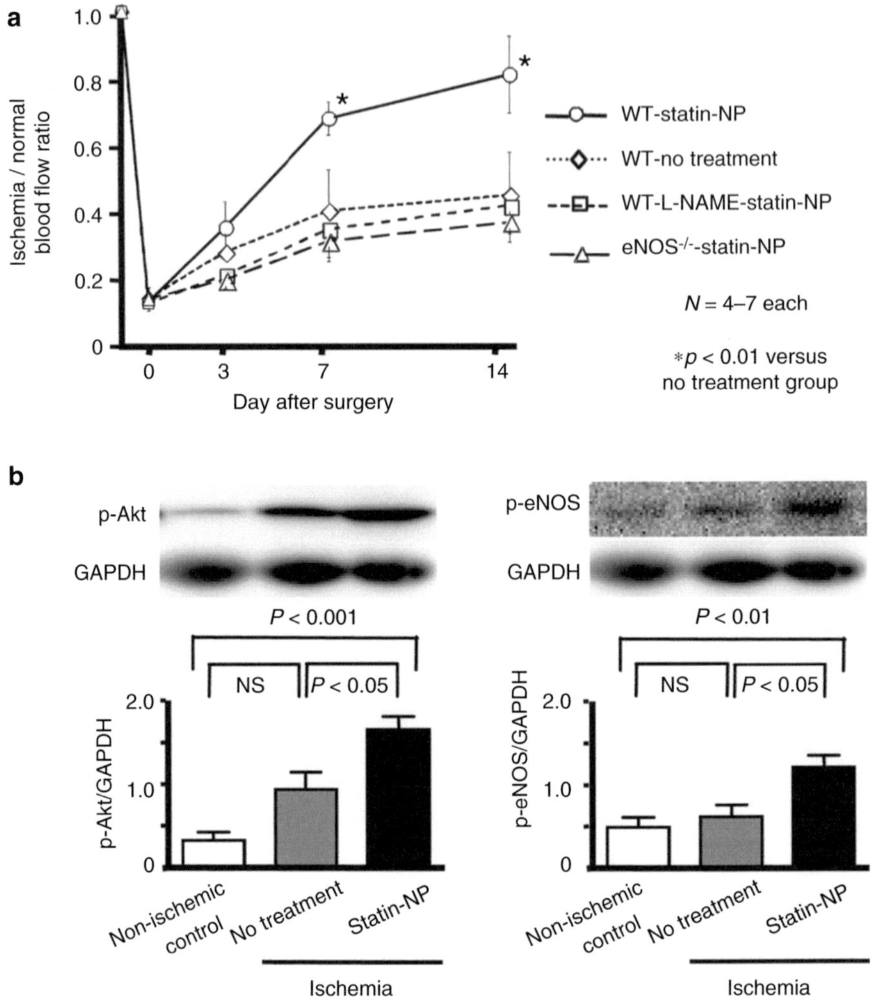

Fig. 16.8 (**a**) Quantification of blood flow recovery in wild-type (WT) mice with or without administration of L-NAME, a NOS inhibitor, and in eNOS$^{-/-}$ mice. (**b**) Western blot analysis of phosphorylated Akt and eNOS in ischemic and nonischemic muscles 7 days after ischemia. $n = 6$ each. *NS* not significant

endothelial cell-targeting delivery of statin produces a well-harmonized integrative angiogenesis to form functionally mature collaterals by controlled expression of endogenous multiple angiogenic growth factors and signals. Collectively, these results suggest that PLGA NP-mediated approach could be a more effective and clinically feasible approach for therapeutic neovascularization.

16.3.6 Nanoparticle-Mediated Endothelial Cell-Targeting Delivery of Pitavastatin Induces Functional Collateral Arteries (Therapeutic Arteriogenesis)

To translate our experimental findings in the murine model of acute hindlimb ischemia to clinically applicable approaches, it is desirable to determine whether NP-mediated statin delivery into vascular endothelial cells induces the development of collateral arteries (arteriogenesis) and thus restores tissue perfusion in a setting of chronic ischemia in larger animals. To address this point, male Japanese White rabbits were used. To induce chronic hindlimb ischemia, the left femoral artery was completely excised from its proximal origin at the branch point of the external iliac artery to the bifurcation of the saphenous and popliteal arteries. For intramuscular injections, drug-incorporating nanoparticles were suspended in 5 mL of phosphate-buffered saline (PBS) and injected into ten different sites of the left medial thigh muscles with a 27-gauge needle 7 days after femoral artery excision. At 7 days after ischemia, PBS, FITC-NPs, pitavastatin (0.5 mg/kg), and pitavastatin-NPs (0.05, 0.15, and 0.5 mg/kg as pitavastatin calcium) were injected into ischemic muscle. FITC-NPs were detected in vascular endothelial cells in the ischemic tissue for over 4 weeks. Induction of collateral arteries confirmed by angiography was demonstrated only in the pitavastatin-NP group (equivalent to 0.5 mg/kg of pitavastatin calcium) (Fig. 16.9) [26].

16.3.7 Preclinical POC (Proof of Concept) Study in Nonhuman Primate Critical Limb Ischemia

Although there is currently no clear consensus regarding which animal model is most appropriate for the evaluation of therapeutic angiogenesis, nonhuman primate models may have advantages over non-primate animal models, because the body size and pharmacokinetics in these animals are more similar than other small-sized primates. Therefore, male cynomolgous monkeys were subjected to critical lower limb ischemia. Pitavastatin-NPs (4 mg pitavastatin/body) injected in the left semimembranosus muscle 4 weeks after induction of limb ischemia with a 26-gauge injection needle (ten sites, 3-day and 6-day repeated doses). In addition, a single-dose experiment was also conducted at 4 mg of pitavastatin calcium. The 6-day protocol demonstrated development of collateral circulation at 4 weeks after induction of limb ischemia. The 3-day protocol demonstrated similar effects at 8 weeks after the ischemia with recovery of the mobility of lower limbs (Fig. 16.10).

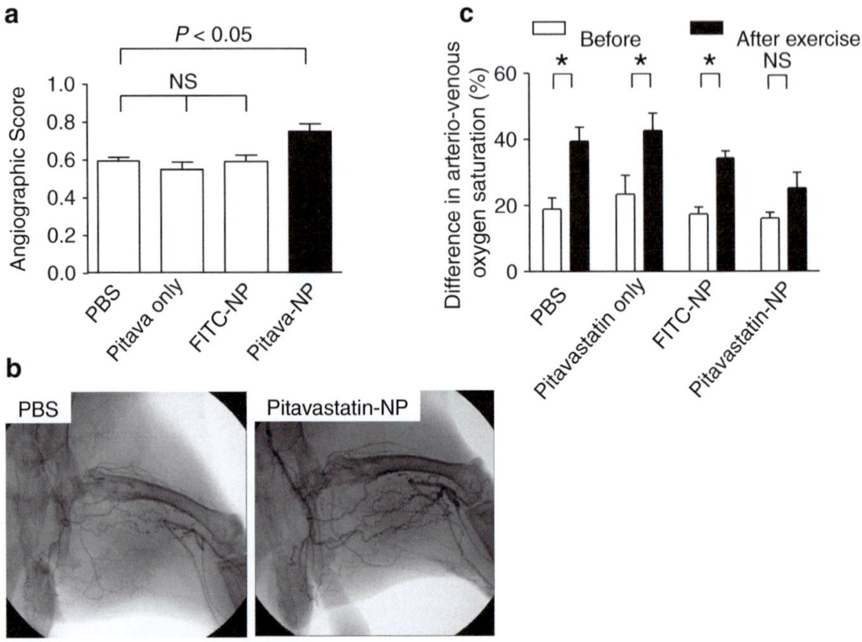

Fig. 16.9 Effects of pitavastatin nanoparticles (NPs) on angiographically visible collateral arterial development. (**a**) Quantitative data of the angiographic scores ($n = 6$ each). (**b**) Representative angiograms of ischemic rabbit lower extremities at 28 days after treatments. Corkscrew-like collateral arteries were observed only in the pitavastatin-NP group. (**c**) The difference of arterial and venous oxygen saturation after 30 min exercise is shown in the four groups ($n = 6$ each)

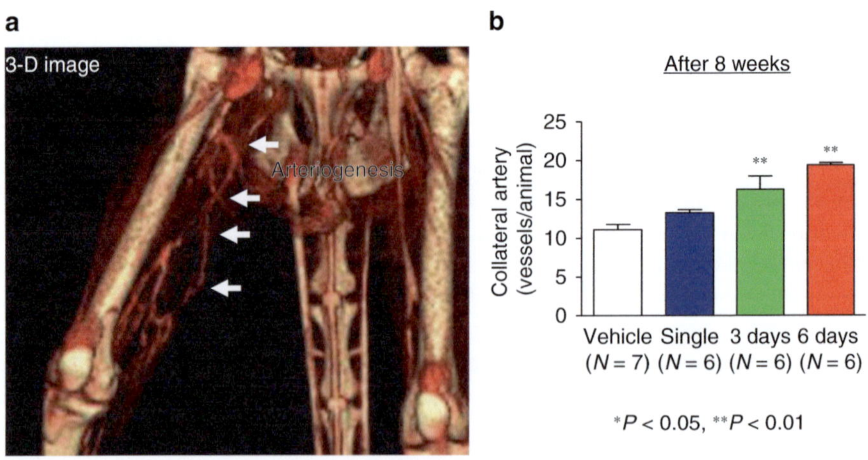

Fig. 16.10 Effects of pitavastatin nanoparticles (NPs) on angio-CT image-visible collateral arterial development are shown 28 days after treatment in the nonhuman primate model

16.3.8 Rationale for Dose and Mode of Administration

The highest dose set in this clinical trial, described later (4 mg of pitavastatin calcium per body), is the maximum dose of the oral pitavastatin calcium currently used in hospitals. Administrations were conducted at 20 sites in the lower limb for five consecutive days. The dose per site was approximately 1 mL, and markings were made on the lower limb so that injection sites will not overlap each other.

The results of three preclinical studies with animal models of lower limb ischemia suggested that intramuscular administration has a higher efficacy than oral administration, and in order to deliver a large amount of the drug to the ischemic area, intramuscular administration once daily at 20 injection sites for 5 days was considered as clinically feasible protocol.

16.4 Investigator-Initiated Clinical Trial for PAD

16.4.1 Study Design

This study was conducted as a phase I/IIa, multicenter, open-label, dose-escalation, and randomized trial investigating the safety and efficacy of NK-104-NP containing 0.5, 1, 2, and 4 mg of pitavastatin calcium. NK-104-NP was intramuscularly administered for five consecutive days in patients with CLI, and pharmacokinetics of the parent compound and the lactone bodies of pitavastatin in plasma/urine were evaluated. This clinical trial is a first-in-man study assessing the safety and efficacy of intramuscularly injected PLGA nanoparticles. This clinical trial was conducted in compliance with ethical principles of the Declaration of Helsinki, Guideline for Good Clinical Practice (GCP, Ordinance of the Ministry of Health, Labour and Welfare), and other related regulatory requirements. This trial was also conducted under the support of the Clinical and Translational Research Center of Kyushu University Hospital. The study is conducted in multicenter (Kyushu University Hospital and Nagoya University Hospital) after approval by the Pharmaceutical and Medical Device Agency (PMDA) and the ethics committee (Kyushu University Hospital and Nagoya University Hospital). The trial was registered to UMIN Clinical Trial Registration (UMIN000008011). Patients who satisfied all the criteria were enrolled with a written consent.

16.4.2 Study Population and Inclusion Criteria

The eligibility criteria of this study that the participant has symptoms of chronic limb ischemia (atherosclerosis obliterans (ASO)) include rest pain, nonhealing ischemic ulcers, and contraindicated to catheter interventions or surgical revascularizations. Patients with the following criteria were enrolled: (1) symptoms classified to

Fontaine grade III or IV, (2) contraindicated to revascularization of the femoral artery or below, (3) medical therapy (vasodilators, antiplatelet agents, or prostaglandins at least for 2 weeks)-resistant patients, and (4) 20 years or older at the time of consent.

16.4.3 Primary Endpoints

The primary endpoints of the study are the safety of the treatment, pharmacokinetics of pitavastatin and its metabolite in the plasma/urine, and efficacy of treatment (the changes of Fontaine classification and Rutherford classification).

16.4.4 Secondary Endpoints

The secondary endpoints include physiological tests (body weight, body temperature, blood pressure, and pulse), blood biochemistry, cardiac ultrasound imaging, and efficacy of treatment: ankle-brachial index (ABI), toe-brachial index (TBI), ankle pressure (AP), pulse volume recording (PVR), laser Doppler blood flow, angiography (IA-DSA), ulcer size, degree of pain (VAS), and transcutaneous oxygen pressure ($TcPO_2$). The number of cases that undergone either minor amputation or major amputation is also evaluated.

16.4.5 Adverse Effects Reporting

All the adverse events are documented in every patient. Based on these information, proper terms shall be selected from the Medical Dictionary for Regulatory Activities/Japanese version (MedDRA/JTM). For adverse events, the number of cases, number of patients, and incidence rates shall be summed for all subjects and for each administration condition. Summation shall also be performed for each system organ class (SOC) and for each preferred term (PT). For adverse drug reactions, the number of cases, number of subjects, and incidence rates are to be summed for all subjects. Summation shall also be performed for each SOC and for each PT.

16.4.6 Dose-Escalation Criteria

Dose escalation shall be decided by the person conducting the trial (principal investigator) on the basis of safety and pharmacokinetic results up to 2 weeks after the final administration to all four subjects in the group.

16.4.7 Observation Period

Follow-up shall be continued for 26 weeks after the final administration, as per the following clinical trial schedule.

16.4.8 Administration of NK-104-NP

NK-104-NP is supplied by Kowa Company Ltd. (Nagoya, Japan), as a sterile powder within vial. A specified amount of injectable saline was to be added to the vial, suspended, and administered intramuscularly. Administration is to be conducted at 20 sites on a lower limb once daily for 5 days. The 20 administration sites on a lower limb are to be determined in accordance with the separately prepared administration manual for the test product. The dose per site is to be 1 mL, and markings are to be made on the lower limb so that injection sites will not overlap. For the selection of the lower limb for administration, the lower limb in the worse condition is to be selected within the range of the inclusion criteria and the exclusion criteria.

16.4.9 Data Collection and Statistical Analysis in Clinical Trial

All data are collected at Data Center of the Center for Clinical and Translational Research (CCTR) in Kyushu University Hospital. Pharmacokinetic parameters are estimated in the Department of Clinical Pharmacokinetics (Drs. Ieiri and Fukae). Subjects on whom there is data and who were administered the test product are to be selected. Safety evaluation is to be performed with this analysis set.

Subjects on whom there is plasma pharmacokinetic data and who were administered the test product are to be selected. Pharmacokinetic evaluation is to be performed with this analysis set. For background information, the basic statistics of each group and the whole shall be calculated.

We are now conducting clinical trial testing the efficacy of pitavastatin-NPs nanoparticle-mediated delivery of pitavastatin in patients with critical limb ischemia.

16.5 Conclusion

In conclusion, this platform nanotechnology that aims vascular endothelial cell-targeting delivery of statin is a promising strategy to develop more effective and integrative nanomedicine in patients with severe organ ischemia. The nanotechnology platform may be developed further as an "integrative" approach for therapeutic neovascularization and extended to target other molecular signals specific to vascular endothelial cells. The results of the trial will set the stage for a larger trial to evaluate efficacy of NK-104-NP for CLI patients. Dynamic collaborative research beyond the fields of each researcher is necessary to promote rapid development and clinical application of these nanoparticle-based novel therapeutics.

Acknowledgments These studies are supported partly by Center for Clinical and Translational Research Kyushu University Hospital.

NK-104-NP investigators: principal investigator, T Matsumoto, MD, PhD.. Co-investigators: Y Maehara, MD, PhD.; T Furuyama, MD, PhD.; R Kyuragi, MD, PhD.; Y Aoyagi, MD; Y Matsubara, MD; K Inoue, MD; S Tanaka, MD; R Yoshiga, MD; K Yoshiya, MD (Department of Surgery and Science, Kyushu University Hospital); T Matoba, MD, PhD.; Y Mukai, MD, PhD. (Department of Cardiovascular Medicine, Kyushu University Hospital); J Koga, MD, PhD.; G Ikeda, MD, PhD. (Department of Cardiovascular Research, Development of Translational Medicine, Kyushu University); K Komori, MD, PhD.; A Kodama, MD, PhD.; A Koyama, MD, PhD. (Division of Vascular Surgery, Department of Surgery, Graduate School of Medicine, Nagoya University Hospital). Project Director: K Egashira, MD, PhD. (Department of Cardiovascular Research, Development and Translational Medicine, Kyushu University). Project Manager: K Nakano, PhD. (Department of Cardiovascular Research, Development and Translational Medicine, Kyushu University). Pharmacokinetic analysis: I Ieiri, PhD. (Chair) M Fukae, PhD. (Department of Clinical Pharmacokinetics, Graduate School of Pharmaceutical Sciences, Kyushu University). Pharmacovigilance: K Todaka, MD., PhD. (Chair); R Taketomi (Activities of Center for Clinical and Translational Research, Kyushu University). Clinical Research Coordinators: M Miura, RN; Y Matsuo, RN (EP-SOGO Co., Ltd.); C Akatsuka (Department of Cardiovascular Medicine, Kyushu University). Monitor: R Shibata, PhD. (EPI Japan Co., Ltd.). Data Managers: K Sakanashi, BE (Chair); S Ishida, LL.B.; K Sumino, RN (Department of Clinical Research Promotion, Kyushu University Hospital). Audit: K Hirano (EPI Japan Co., Ltd.). Statistical analysis: J Kishimoto, MA (Chair, Department of Research and Development of Next Generation Medicine, Kyushu University); H Watanabe, ME; G Eriguchi, MMS (Department of Clinical Research Promotion, Kyushu University Hospital).

References

1. Beckman JA, Creager MA, Libby P. Diabetes and atherosclerosis: epidemiology, pathophysiology, and management. JAMA. 2002;287(19):2570–81.
2. Libby P. Inflammation in atherosclerosis. Nature. 2002;420(6917):868–74. doi:10.1038/nature01323.
3. Norgren L, Hiatt WR, Dormandy JA, Nehler MR, Harris KA, Fowkes FG, et al. Inter-society consensus for the management of peripheral arterial disease (TASC II). J Vasc Surg. 2007;45(Suppl):S5–67. doi:10.1016/j.jvs.2006.12.037.
4. Steg PG, Bhatt DL, Wilson PW, D'Agostino Sr R, Ohman EM, Rother J, et al. One-year cardiovascular event rates in outpatients with atherothrombosis. JAMA. 2007;297(11):1197–206. doi:10.1001/jama.297.11.1197.
5. Ness J, Aronow WS. Prevalence of coexistence of coronary artery disease, ischemic stroke, and peripheral arterial disease in older persons, mean age 80 years, in an academic hospital-based geriatrics practice. J Am Geriatr Soc. 1999;47(10):1255–6.

6. Criqui MH, Langer RD, Fronek A, Feigelson HS, Klauber MR, McCann TJ, et al. Mortality over a period of 10 years in patients with peripheral arterial disease. N Engl J Med. 1992;326(6):381–6. doi:10.1056/NEJM199202063260605.
7. Hirsch AT, Haskal ZJ, Hertzer NR, Bakal CW, Creager MA, Halperin JL, et al. ACC/AHA 2005 Practice Guidelines for the management of patients with peripheral arterial disease (lower extremity, renal, mesenteric, and abdominal aortic): a collaborative report from the American Association for Vascular Surgery/Society for Vascular Surgery, Society for Cardiovascular Angiography and Interventions, Society for Vascular Medicine and Biology, Society of Interventional Radiology, and the ACC/AHA Task Force on Practice Guidelines (Writing Committee to Develop Guidelines for the Management of Patients With Peripheral Arterial Disease): endorsed by the American Association of Cardiovascular and Pulmonary Rehabilitation; National Heart, Lung, and Blood Institute; Society for Vascular Nursing; TransAtlantic Inter-Society Consensus; and Vascular Disease Foundation. Circulation. 2006;113(11):e463–654. doi:10.1161/CIRCULATIONAHA.106.174526.
8. Daskalopoulou M, George J, Walters K, Osborn DP, Batty GD, Stogiannis D, et al. Depression as a risk factor for the initial presentation of twelve cardiac, cerebrovascular, and peripheral arterial diseases: data linkage study of 1.9 million women and men. PLoS One. 2016;11(4):e0153838. doi:10.1371/journal.pone.0153838.
9. Gardner AW, Waldstein SR, Montgomery PS, Zhao YD. Effect of cognitive status on exercise performance and quality of life in patients with symptomatic peripheral artery disease. J Vasc Surg. 2016;63(1):98–104. doi:10.1016/j.jvs.2015.08.064.
10. Takemoto M, Liao JK. Pleiotropic effects of 3-hydroxy-3-methylglutaryl coenzyme a reductase inhibitors. Arterioscler Thromb Vasc Biol. 2001;21(11):1712–9.
11. Nakano K, Egashira K, Masuda S, Funakoshi K, Zhao G, Kimura S, et al. Formulation of nanoparticle-eluting stents by a cationic electrodeposition coating technology: efficient nanodrug delivery via bioabsorbable polymeric nanoparticle-eluting stents in porcine coronary arteries. JACC Cardiovasc Interv. 2009;2(4):277–83. doi:10.1016/j.jcin.2008.08.023.
12. Kitamoto S, Nakano K, Hirouchi Y, Kohjimoto Y, Kitajima S, Usui M, et al. Cholesterol-lowering independent regression and stabilization of atherosclerotic lesions by pravastatin and by antimonocyte chemoattractant protein-1 therapy in nonhuman primates. Arterioscler Thromb Vasc Biol. 2004;24(8):1522–8. doi:10.1161/01.ATV.0000134518.27241.da.
13. Llevadot J, Murasawa S, Kureishi Y, Uchida S, Masuda H, Kawamoto A, et al. HMG-CoA reductase inhibitor mobilizes bone marrow--derived endothelial progenitor cells. J Clin Invest. 2001;108(3):399–405. doi:10.1172/jci13131.
14. Altieri DC. Statins' benefits begin to sprout. J Clin Invest. 2001;108(3):365–6. doi:10.1172/JCI13556.
15. Sata M, Nishimatsu H, Suzuki E, Sugiura S, Yoshizumi M, Ouchi Y, et al. Endothelial nitric oxide synthase is essential for the HMG-CoA reductase inhibitor cerivastatin to promote collateral growth in response to ischemia. FASEB J. 2001;15(13):2530–2. doi:10.1096/fj.01-0415fje.
16. Kureishi Y, Luo Z, Shiojima I, Bialik A, Fulton D, Lefer DJ, et al. The HMG-CoA reductase inhibitor simvastatin activates the protein kinase Akt and promotes angiogenesis in normocholesterolemic animals. Nat Med. 2000;6(9):1004–10. doi:10.1038/79510.
17. Ikeda G, Matoba T, Nakano Y, Nagaoka K, Ishikita A, Nakano K, et al. Nanoparticle-mediated targeting of cyclosporine A enhances cardioprotection against ischemia-reperfusion injury through inhibition of mitochondrial permeability transition pore opening. Sci Rep. 2016;6:20467. doi:10.1038/srep20467.
18. Katsuki S, Matoba T, Nakashiro S, Sato K, Koga J, Nakano K, et al. Nanoparticle-mediated delivery of pitavastatin inhibits atherosclerotic plaque destabilization/rupture in mice by regulating the recruitment of inflammatory monocytes. Circulation. 2014;129(8):896–906. doi:10.1161/CIRCULATIONAHA.113.002870.
19. Kimura S, Egashira K, Chen L, Nakano K, Iwata E, Miyagawa M, et al. Nanoparticle-mediated delivery of nuclear factor kappaB decoy into lungs ameliorates monocrotaline-induced pulmonary arterial hypertension. Hypertension. 2009;53(5):877–83. doi:10.1161/HYPERTENSIONAHA.108.121418.

20. Kimura S, Egashira K, Nakano K, Iwata E, Miyagawa M, Tsujimoto H, et al. Local delivery of imatinib mesylate (STI571)-incorporated nanoparticle ex vivo suppresses vein graft neointima formation. Circulation. 2008;118(14 Suppl):S65–70. doi:10.1161/CIRCULATIONAHA.107.740613.
21. Koga JI, Matoba T, Egashira K. Anti-inflammatory nanoparticle for prevention of atherosclerotic vascular diseases. J Atheroscler Thromb. 2016;23(7):757–65. doi:10.5551/jat.35113.
22. Nagaoka K, Matoba T, Mao Y, Nakano Y, Ikeda G, Egusa S, et al. A new therapeutic modality for acute myocardial infarction: nanoparticle-mediated delivery of pitavastatin induces cardioprotection from ischemia-reperfusion injury via activation of PI3K/Akt pathway and anti-inflammation in a rat model. PLoS One. 2015;10(7):e0132451. doi:10.1371/journal.pone.0132451.
23. Lu JM, Wang X, Marin-Muller C, Wang H, Lin PH, Yao Q, et al. Current advances in research and clinical applications of PLGA-based nanotechnology. Expert Rev Mol Diagn. 2009;9(4):325–41. doi:10.1586/erm.09.15.
24. Tsukie N, Nakano K, Matoba T, Masuda S, Iwata E, Miyagawa M, et al. Pitavastatin-incorporated nanoparticle-eluting stents attenuate in-stent stenosis without delayed endothelial healing effects in a porcine coronary artery model. J Atheroscler Thromb. 2013;20(1):32–45.
25. Kubo M, Egashira K, Inoue T, Koga J, Oda S, Chen L, et al. Therapeutic neovascularization by nanotechnology-mediated cell-selective delivery of pitavastatin into the vascular endothelium. Arterioscler Thromb Vasc Biol. 2009;29(6):796–801. doi:10.1161/ATVBAHA.108.182584.
26. Oda S, Nagahama R, Nakano K, Matoba T, Kubo M, Sunagawa K, et al. Nanoparticle-mediated endothelial cell-selective delivery of pitavastatin induces functional collateral arteries (therapeutic arteriogenesis) in a rabbit model of chronic hind limb ischemia. J Vasc Surg. 2010;52(2):412–20. doi:10.1016/j.jvs.2010.03.020.
27. Losordo DW, Dimmeler S. Therapeutic angiogenesis and vasculogenesis for ischemic disease. Part I: angiogenic cytokines. Circulation. 2004;109(21):2487–91. doi:10.1161/01.CIR.0000128595.79378.FA.
28. Losordo DW, Dimmeler S. Therapeutic angiogenesis and vasculogenesis for ischemic disease: part II: cell-based therapies. Circulation. 2004;109(22):2692–7. doi:10.1161/01.CIR.0000128596.49339.05.
29. Baumgartner I, Isner JM. Stimulation of peripheral angiogenesis by vascular endothelial growth factor (VEGF). Vasa. 1998;27(4):201–6.
30. Marui A, Tabata Y, Kojima S, Yamamoto M, Tambara K, Nishina T, et al. A novel approach to therapeutic angiogenesis for patients with critical limb ischemia by sustained release of basic fibroblast growth factor using biodegradable gelatin hydrogel: an initial report of the phase I-IIa study. Circ J. 2007;71(8):1181–6.
31. Dimmeler S, Aicher A, Vasa M, Mildner-Rihm C, Adler K, Tiemann M, et al. HMG-CoA reductase inhibitors (statins) increase endothelial progenitor cells via the PI 3-kinase/Akt pathway. J Clin Invest. 2001;108(3):391–7. doi:10.1172/JCI13152.

MIX
Papier aus verantwortungsvollen Quellen
Paper from responsible sources
FSC® C105338

If you have any concerns about our products,
you can contact us on
ProductSafety@springernature.com

In case Publisher is established outside the EU,
the EU authorized representative is:
Springer Nature Customer Service Center GmbH
Europaplatz 3, 69115 Heidelberg, Germany

Printed by Libri Plureos GmbH
in Hamburg, Germany